An Introduction to UMTS Technology

An Introduction to UMTS Technology:
Testing, Specifications and Standard Bodies for Engineers and Managers

Dr. Faris Muhammad

BrownWalker Press
Boca Raton, Florida

An Introduction to UMTS Technology:
Testing, Specifications and Standard Bodies for Engineers and Managers

BrownWalker Press
Boca Raton, Florida • USA
2008

www.brownwalker.com

ISBN-10: 1-59942-458-4/ISBN-13: 978-1-59942-458-3 *(paperback)*
ISBN-10: 1-1599-446-0/ISBN-13: 978-159942-446-0 *(hardcover)*
ISBN-10: 1-1599-447-9/ISBN-13: 978-159942-447-7 *(ebook)*

Library of Congress Cataloging-in-Publication Data

Muhammad, Faris A., 1964-
  An introduction to UMTS technology : testing, specifications, and standard
bodies for engineers and managers / Dr Faris Muhammad.
      p. cm.
  Includes index.
  ISBN-13: 978-1-59942-446-0 (hbk. : alk. paper)
  ISBN-10: 1-59942-446-0 (hbk. : alk. paper)
  1. Universal Mobile Telecommunications System--Testing. 2. Universal
Mobile Telecommunications System--Standards. I. Title.

  TK5103.4883.M84 2008
  621.3845'6--dc22

                                        2008003584

Dedication

To *my late father and mother—*
    "My Lord! Have mercy on them both as they did care for me when I was little." May God bless their souls in heaven.

Table of Content

Chapter Two – UMTS Conformance Protocol Testing – Part One   **35**

Chapter Two – UMTS Conformance Protocol Testing – Part Two   109

Chapter Four – Testing Types and Stages   239

Chapter Five – Conformance Testing and TTCN   251

Chapter Six – Standardisation and Validation Bodies 275

Preface

The first Chapter of this book focuses on the basic concepts of W-CDMA technology. These principles are UMTS (CDMA) specific, and are unlikely used in other technologies; therefore some explanation may be necessary. Simplified introduction to these concepts will make reading this book easier.

Chapter 2 is divided into two parts, both covers protocol conformance testing. This type of testing is used throughout the initial stages of development to ensure the accuracy of a protocol implementation. It is also used in regression testing after initial product deployment to further verify any changes in the implementation/enhancement. Traditionally, conformance testing has been the domain of the telecommunications industry. The only way to ensure that standards are met is to test products in an effective way using the test specification. First part of chapter two mainly covers dle mode, dual RAT, RRC, RLC and MAC tests. Part two mainly concentrate on MBMS, MM, RB services, SMS, A-GPS, acoustic and IMS tests. This chapter lists all tests that are necessary to be conducted on different layers and modules of the UE under test. Further details of these tests can be found in the relevant standards.

Chapters 3 gives a unified and in-depth presentation of selections and RF conformance tests for UEs and NBs. In addition to the protocol conformance, the RF performance of the UE must also be verified. Many measurements of the transmitter and receiver performance are performed in a number of areas, e.g. out-of-band emissions. Measurements of the radio resource management (RRM) are performed to ensure that the control capability of the UE is operating according to the standards. The RRM is the component used to control the physical or RF layers in accordance with the requirements of the protocols from the upper layers. There are, for instance, very tight limits on the transmitter output power, as it is controlled to meet conditions such as variations in signal strength. This ensures that the handset only transmits sufficient power to maintain a reliable connection under the prevailing conditions.

Chapter 4 explains thouroghly all types of tests at different stages the system (whether UE or BS) need to go through throughout the lifecycle of the product. For example development testing is essential throughout the lifecycle of the product. Initial system validation of Layer 1 implementation is required at early stages of system development. Conformance testing has to cover L1/L2 protocol, Inter System Handover protocol, RRM protocol and RF performance and implementation of the system. Terminology may differ but the intended purposes of these test still widely used. Some of the tests may overlap in some cases for some products. This chapter provides good coverage of all tests at all stages.

Chapter 5 presents different languages and tools used in the conformance industry to carry out conformance testing. Standard bodies are using special languages like TTCN-2, TTCN-3 to specify the conformance test specifications but conventional languages like Basic/C/C++/Perl/Shell scripts still capture some of the market share for test suite implementation. The chapter gives more attention to TTCN-2 and TTCN-3 languages/tools.

Chapter 6 is the final chapter of this book. It covers many distinct processes apply to the provision of test cases. Test cases are written in prose, describing in detail how the test is carried out and the pass and fail criteria. To ensure that each test is a true representation of the original intent of the test, a validation and approval process has been set in place. Once the test case is verified by the supplier (3GPP), i.e. is satisfied that the test operates correctly, it is then given to an independent validation organisation to test for conformance with the original test specification and check for proper operation. When it has passed this test, it can be submitted to the relevant industry body for approval. After approval, it can be used in formal mobile terminal testing and certification. The standarisation and validation bodies and procedures are well detailed in this chapter.view of the procedure, as well as the equivalent but distinct procedures for the other systems is described in this chapter.

Acknowledgments

I would like to sincerely thank Mr Ian Poole and Dr Jafer H Hassan for their expert advice, guidance and encouragement throughout the preparation of this book.

I would like to express my deep gratitude to my wife Lehib for her support without which it would be impossible to complete the book.

I would also like to thank Phil Medd, Andy Summers, Kundan Sehmbey, Pradip Kar, Girish Kalra, my boss Simon Palmer, and close colleagues at Aeroflex for many interesting discussions and assistance.

I gratefully acknowledge 3GPP for granting me licence to use some of the 3GPP specifications.

Finally I would like to thank Dr Baha Hashimi for reviewing the manuscript.

Chapter One

Introduction

INTRODUCTION TO UMTS TECHNOLOGY

In this chapter basic concepts of W-CDMA are explained and discussed. These principles are UMTS (CDMA) specific, and are unlikely used in other technologies; therefore some explanation may be necessary. A simplified introduction to these concepts will make reading this book easier.

Principles of W-CDMA

Some mechanisms are required to share frequency resources in communication systems with multiple users. This mechanism is referred to as Multiple Access scheme. CDMA is a scheme where all users are transmitting on the same frequency at the same time but separated by codes.

Spread Spectrum

W-CDMA is a multiple access technique using a concept known as Spread Spectrum. In Spread Spectrum systems the information Bandwidth is spread across a wider transmission Bandwidth which is determined by a function that is independent of the transmitted information.

The original data sequence is binary multiplied with a spreading code. The bits in the spreading code are called chips and the data bits sequence are called symbols. Each user has its own spreading code. Application of the same spreading codes once again at the receiving end returns the transmitted signals to their original Bandwidths. The ratio between the transmission BW and the original BW is called the processing gain. The lower the SF the more data can by transported on the air interface. The relative strength of the desired signal and the rejection of other signals is proportionate to the number of chips over which the receiver has to integrate, which is the spreading factor. The larger the SF the larger the processing gain and hence the original signals do not need to be of high power to achieve a target quality level. The longer the symbol time the longer the integration process. This is referred to as processing gain and is directly proportional to the SF.

1

The spreading codes (referred to as channelization codes) are unique and have low cross-correlation with other spreading codes. This means that several wideband signals can co-exist on the same frequency without interference. When the combined signal is correlated with the particular spreading code, only the original signal with the corresponding spreading code is de-spread, while the remaining component of the signal remain spread.

The principle of correlation is used at the receiving end to recover the original signal out of the noise generated by all the other users' wideband signal. As illustrated in Figure 1.1 the original data is coded and the resulting signal is then transmitted. The received signal is multiplied by the code to recover the original data. If the receiver does not know the correct code the results will be a signal almost average to zero.

Table 1.1 illustrates the differences between the two types of codes that are used in W-CDMA and the way they are used in the downlink and the uplink.

OVSF (Orthogonal Variable Spreading Factor) are designed to allow the support of simultaneous variable data rate channels. The spreading factors are assigned so that they do not come from a parent or grandparent code on the same branch of the tree. This is to avoid code clashing.

Protocols and Channels
Protocol Architecture
The diagram in Figure 1.2 illustrates in details the protocol architecture that exist across the Uu interface in a FDD UMTS system. At the lower level there is the Physical Layer (L1). This is accessed by a number of SAPs by the MAC which in turn is accessed by the RLC Layer 2 protocols.

The interconnections between different layers of protocol are defined by means of the SAPs (Service Access Points). Figure 1.2 illustrates the basic SAPs in the control and the user plane for the radio interface . Two additional SAPs are provided for the two entities (BMC and PDCP). SAPs offer a range of well defined services that will be explored further in the following sections.

Signalling messages are divided into user plane and control plane at the higher layers. RRC is a L3 entity that exists in the control plane which provide some control services to higher layers.

The air interface is layered into three protocol layers:

Layer 1 (L1) or Physical Layer
It interfaces the MAC of L2 and the RRC of L3. It offers different transport channels to the MAC.

Layer 2 (L2) or Data Link Layer
It is split into MAC, RLC, PDCP and BMC. L3 and RLC are divided into User and Control planes while BMC and PDCP exist in the User plane only.

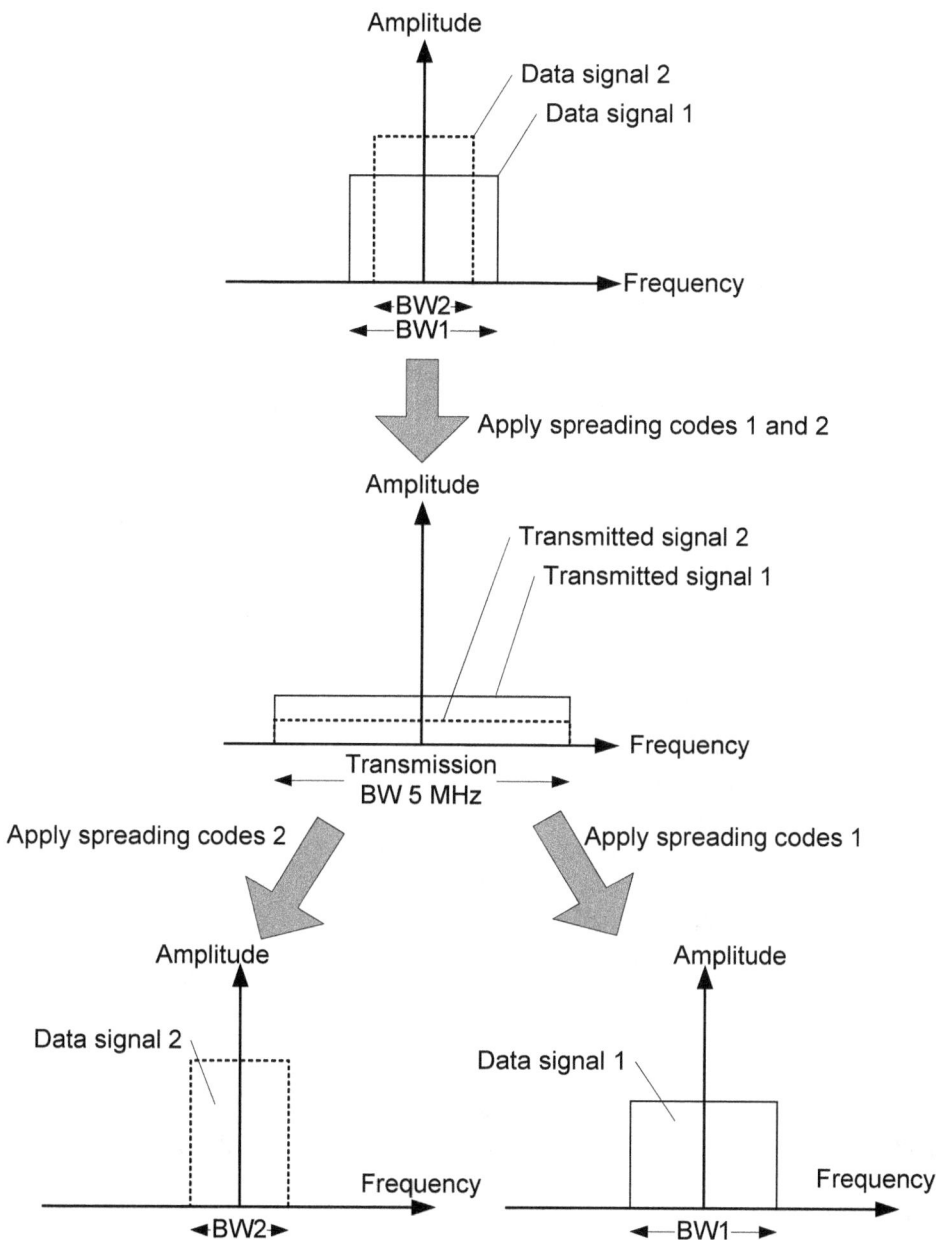

Figure 1.1   Spreading and data signal

Layer 3 (L3) or Network Layer

In the Control plane L3 is partitioned into RRC which interface with L2. The RLC provides ARQ functionality coupled with the radio transmission technique used.

Code type	UL	DL
Channelization (spreading)	4-256 Physical (user) data Control data Signalling data Form the same UE OVSF	4-512 (Release 4 onward SF range is (4 Ð 256) Control Channels Traffic Channels Different users on the same cell OVSF
Scrambling	To isolate users Large number of codes Use long codes (Gold) 38400 chips length in 10 ms frame Also use short codes (Kasami)	To prevent co-channel interference from adjacent cells Long codes Limited to 512 codes

Table 1.1 Channelization and scrambling codes

Transport channels

These are the SAPs that are at the output of the MAC and at the input of L1. They define the characteristics with which data is transported over the air interface. Each transport channel has associated with it a Transport Format set, which defines the coding, interleaving and mapping onto Physical Layer. Transport channels define the interface by which the MAC communicates with the physical layer. There exist two types exist Dedicated and Common Transport channels, DCH is the only dedicated channel. Dedicated means there is a point to point link between the UE and the network while Common means point to multipoint link. In general they map to a specific physical channel. Transport channels are:

Downlink Common and Dedicated

DCH is only one channel and transmitted over the entire cell. It is characterized by the possibility of fast rate change (every 10 ms), fast power control and inherent addressing of UE's. Used for bidirectional transfer of data and control.

  BCH is used for broadcasting system and cell specific information and is always transmitted over the entire cell with a low fixed bit rate.

  FACH is transmitted over the entire cell. FACH uses slow (open loop) power control only. Transfer of small amount of data

  PCH is transmitted over the entire cell. Broadcast of paging and notification messages while allowing for sleep mode.

Uplink Common and Dedicated

DCH is only one channel and transmitted over the entire cell. It is characterized by the possibility of fast rate change (every 10 ms), fast power control and inherent addressing of UE's. Used for bidirectional transfer of data and control.

Control Plane
Processes and protocols related to
Signalling and control of data transport

User Plane
Devoted to processes acting
on the actual data

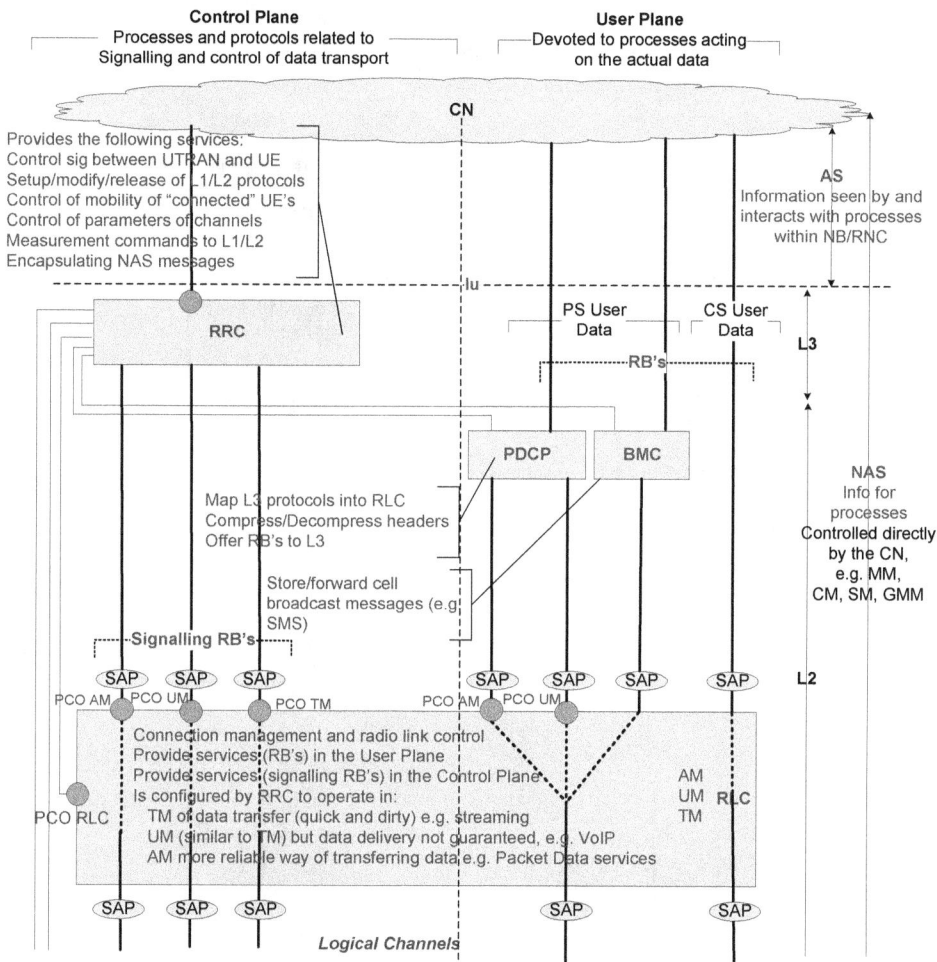

Figure 1.2a UMTS Protocols and layers

RACH is used for initial access or transfer of small amount of data. Open loop power control only.

Logical Channels

The MAC layer provides data transfer services on logical channels. Different logical channel types are defined for different kinds of data services as offered by the MAC. It is characterized by the type of information transferred. These are an information stream provided by the MAC dedicated to the transfer of a specific type of information over the radio interface. There are restrictions on the transport channel type that that can be used to carry a given logical channel. See channel mapping in Figure 1.3

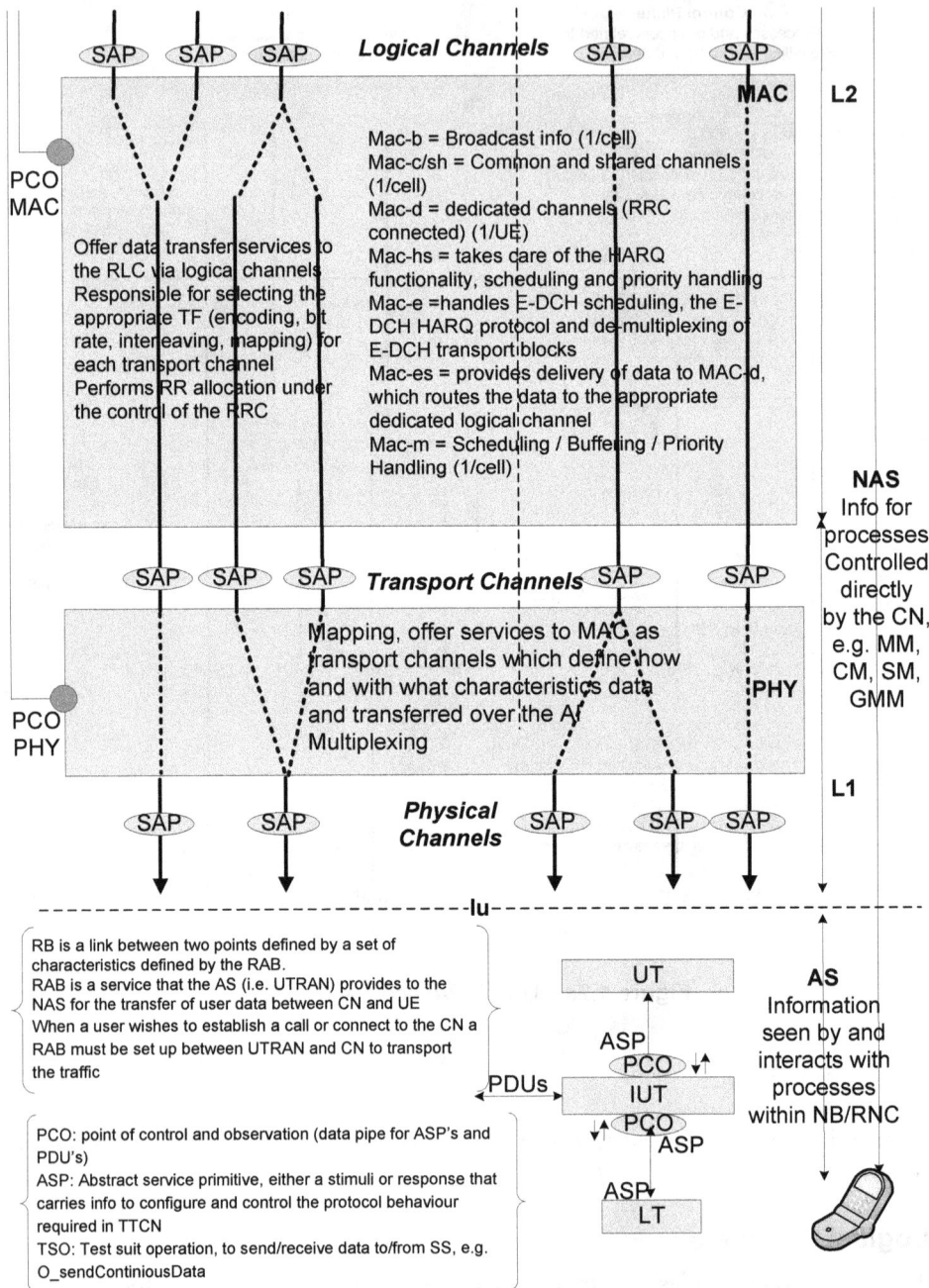

SAP   SAP   SAP   **Logical Channels**   SAP   SAP

MAC   L2

PCO
MAC

Offer data transfer services to
the RLC via logical channels
Responsible for selecting the
appropriate TF (encoding, bit
rate, interleaving, mapping) for
each transport channel
Performs RR allocation under
the control of the RRC

Mac-b = Broadcast info (1/cell)
Mac-c/sh = Common and shared channels
(1/cell)
Mac-d = dedicated channels (RRC
connected) (1/UE)
Mac-hs = takes care of the HARQ
functionality, scheduling and priority handling
Mac-e =handles E-DCH scheduling, the E-
DCH HARQ protocol and de-multiplexing of
E-DCH transport blocks
Mac-es = provides delivery of data to MAC-d,
which routes the data to the appropriate
dedicated logical channel
Mac-m = Scheduling / Buffering / Priority
Handling (1/cell)

NAS
Info for
processes
Controlled
directly
by the CN,
e.g. MM,
CM, SM,
GMM

SAP   SAP   SAP   **Transport Channels**   SAP   SAP

Mapping, offer services to MAC as
transport channels which define how
and with what characteristics data
and transferred over the Af
Multiplexing

PHY

PCO
PHY

L1

SAP   SAP   **Physical**   SAP   SAP   SAP
Channels

- Iu -

RB is a link between two points defined by a set of
characteristics defined by the RAB.
RAB is a service that the AS (i.e. UTRAN) provides to the
NAS for the transfer of user data between CN and UE
When a user wishes to establish a call or connect to the CN a
RAB must be set up between UTRAN and CN to transport
the traffic

UT

AS
Information
seen by and
interacts with
processes
within NB/RNC

ASP
PCO

PDUs

IUT
PCO
ASP

PCO: point of control and observation (data pipe for ASP's and
PDU's)
ASP: Abstract service primitive, either a stimuli or response that
carries info to configure and control the protocol behaviour
required in TTCN
TSO: Test suit operation, to send/receive data to/from SS, e.g.
O_sendContiniousData

ASP
LT

Figure 1.2b   UMTS Protocols and layers

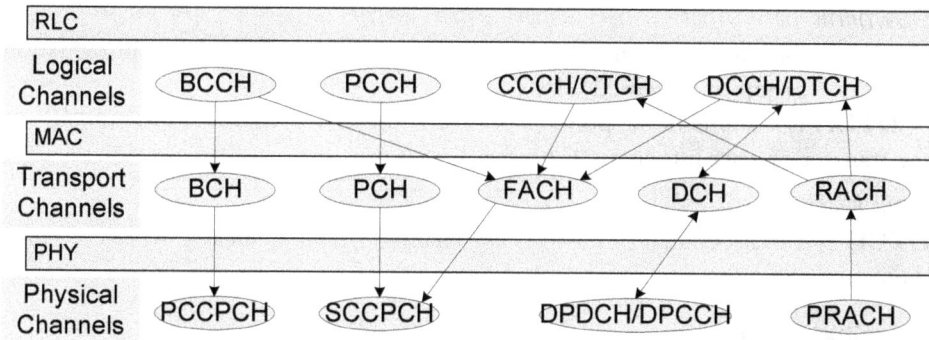

Figure 1.3   Channel mapping

Control Channels
These channels are used for control plane information

Downlink
BCCH is used to broadcast control and signalling information which is system and cell specific. Therefore it is only a downlink channel. It is carried via either the P-CCPCH or the S-CCPCH. This means that its messages can always be read by a UE, once the UE has detected a Node-B's unique scrambling code, which it does during its initial cell search.

PCCH is used to transfer paging information when the net does not know the location of the UE. Also is used when the UE is in the Cell Connected state. PCCH can be carried on PCH and it is a downlink channel only.

CCCH is a bidirectional point to multipoint channel that is used to transfer control information from the network to the UEs. It is used as part of the resource allocation procedure to carry the resource allocation procedure information. CCCH can be carried S-CCPCH.

DCCH is a point to multipoint channel that carries UE specific control information after an RRC connection is setup. DCCH can be carried on FACH or DCH (RNC to decide).

Uplink
CCCH on the uplink this is used by the UE having no RRC connection setup with the net. Also used by the UE's that are using common transport channels when accessing a new cell after cell reselection and has no resources allocated in the intended cell. CCCH can be carried on RACH.

DCCH is a control channel that can be used to transfer dedicated signalling messages from the UE to UTRAN. DCCH can be carried on RACH.

Traffic Channels
These channels are used for user plane information

Downlink

CTCH is point to multipoint channel used to transfer dedicated user information for all or a group of specified UEs.

DTCH is a bidirectional point to point channel that is dedicated to one UE for the transfer of user information on the downlink.

Uplink

DTCH is a bidirectional point to point channel that is dedicated to one UE for the transfer of user information on the uplink.

Physical Channels

Each physical channel has a specific characteristics and purpose and are defined by, carrier frequency, channelization code and relative phase i.e. I or Q (UL only).

UL Channels

PRACH is used to carry the random access transport channel RACH containing user specific information that required to contact UTRAN for starting a call setup, registration, cell update and location update. It uses open loop power control.

DPCCH This is used to carry necessary control information generated at L1 for the operation of DPDCH (i.e. known pilot bits to support channel estimation for coherent detection, TPC commands, FBI and optional TFCI). TFCI informs the receiver about the instantaneous parameters of the different transport channels multiplexed on the UL DPDCH, and corresponds to the data transmitted in the same frame. There is only one UL DPCCH for each L1 connection. This channel has a fixed data rate and SF.

DPDCH It is used to carry dedicated data and control information generated at L2 and above (i.e. DCH). There may be none, one or several of this UL channel on each L1 connection. Multiple data rates can exist within each DPDCH frame and the overall data rate per frame is variable. The SF varies according to the data rate.

DL Channels

DPCH contains two time-multiplexed elements DPDCH and DPCCH. On the DPCH the dedicated transport channel is transmitted time multiplexed with control information generated at L1 (known pilot bits, TPC, optional TFCI).

DPCCH This is used to carry necessary control information generated at L1 for the operation of DPDCH (i.e. known pilot bits to support channel estimation for coherent detection, TPC commands, FBI and optional TFCI). TFCI informs the receiver about the instantaneous parameters of the different transport channels multiplexed on the UL DPDCH, and corresponds to the data transmitted in the same frame. There is only one UL DPCCH for each L1 connection. This channel has a fixed data rate and SF.

DPDCH It is used to carry dedicated data and control information generated at L2 and above (i.e. DCH). There may be none, one or several of this UL channel on each L1 connection. Multiple data rates can exist within each DPDCH frame and the overall data rate per frame is variable. The SF is kept fixed and the rate variation is handled by rate matching or by DTX operations.

P-CPICH carries a predefined bit/symbol sequence, one/cell, broadcast over the entire cell, it is the phase reference for the SCH, P-CCPCH, AICH and PICH. It is also the default phase reference for al other DL Physical Channels. One channelization code is used and SF 256. It is used in measurements for HO and cell selection/reselection.

S-CPICH carries a predefined bit/symbol sequence, none, one or several per cell, may be transmitted over the entire or part of the cell. It is the reference phase for the S-CCPCH and the DL DPCH which carries the dedicated transport channel DCH in which the UE is informed by L3 signalling.

P-CCPCH used to carry the BCH, continuously transmitted over the entire cell for broadcasting cell specific information.

S-CCPCH used to carry FACH (only transmitted when there is a small amount of data available) and PCH (for paging and notification messages).

P-SCH, S-SCH The primary SCH carries an unmodulated code of length 256 chips that is transmitted every slot. This primary synchronisation code is the same for every Node-B in the system and used by the UE to determine the timing information for the S-SCH. Therefore they are used as part of the initial system acquisition process by the UE, by providing the code group of a cell and indicating the timeslot and code of the P-CCPCH.

AICH This is a physical channel used to carry the acquisition indicators which correspond to signatures on the PRACH or PCPCH, i.e. used by UTRAN to indicate back to the UE success/fail attempt to acquire random access by the UL RACH.

PICH is used to the page indicators, it is always associated with a S-CCPCH to which a PCH transport channel is mapped. It is used by UTRAN to indicate to the UE if there is a paging message. It has a fixed SF of 256. The UE start to decode the paging channels as soon as a paging indicator has been detected.

Radio Bearers

RB is defined as the service provided by L2 for transfer of user data between the UE and the network. The MAC entity at L2 accesses the services of the physical layer though entities known as Transport Channels.

HANDOVER

Soft Handover

In UMTS systems DL transmissions in adjacent cells can be on the same frequency (frequency re-use of 1) and therefore it is possible for the UE to connect to more than one Node-Bs at the same time. SHO allows the UE to communicate to the network through another cell to maximize the use of the signal. So rather than wait until signal strength in an adjacent cell exceeds that of the current cell, the

SHO state can be reached when certain threshold is exceeded. A different channel (pilot channel) is used to provide the signal strength measurements in handovers. SHO provides Macro Diversity and Softer HO provide Micro Diversity as it is a HO between cells belong to the same Node-B.

Hard Handover
Inter-Frequency HHO
HHO is more complicated procedure than SHO by which a measurements on the new proposed channel need to be performed before making HO. This is a HO between cells or Node-Bs operating at different frequencies. During this type of HO the used frequency by the UE changes. The UE ceases transmission on the used transmission before it switch to the new frequency and start transmission.

Intra-Frequency HHO
This is a HO between Node-Bs under the control of different RNCs where the Iur interface is unavailable.

Inter-System HHO
The introduction of the new 3G based services provides the user with many new multi-media based services, but this also presents the operators and equipment designers with many new challenges. One of these challenges is that until complete 3G coverage is achieved, users will need to utilise the existing 2G/2.5G networks to ensure complete coverage. To achieve this they will need to roam ubiquitously and seamlessly between the two different RATs.

A number of solutions have been developed to overcome the problems associated with inter-RAT hand over between UMTS and GSM/GPRS. These include the use of dual-mode UEs, CM channel measurements, cell re-selection, cell change order and inter-RAT. Also by embedding messages in one RAT as if it was of the other has made it possible to leave the GSM/GPRS network unchanged. Having a reduced handover message in length due to the GSM/GPRS bit rate limitations made it possible to maintain an acceptable Inter-RAT handover performance.

Inter-RAT, or sometimes referred to as Intersystem, handover is the process of maintaining a phone connection while moving from one cell to another of different RAT as illustrated in Figure 1.4. This section is looking at the GSM/GPRS-UMTS HO. This process is far from easy and there are many problems that need be to overcome.

In any handover process generally before the UE can start the HO process it must perform signal strength/quality measurements of the intended RAT and measure any other required HO parameter. Since the UE is occupied with the existing RAT and the measurement process has to be performed simultaneously a gap has to be made available to do the inter-frequency measurements. In GSM such idle gaps exist, as it is a Time Division Multiple Access (TDMA)-based

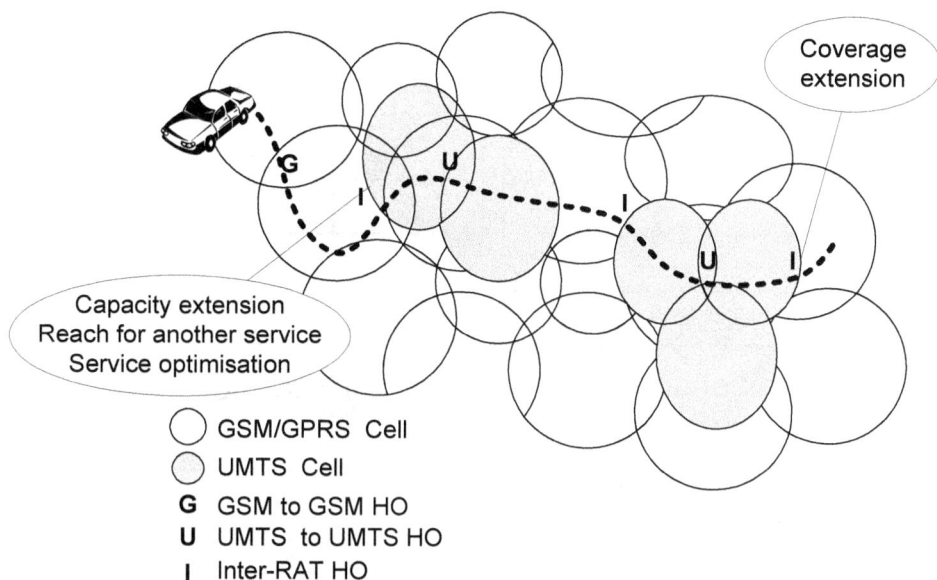

Figure 1.4   Inter system handover

technology. The UMTS Terrestrial Radio Access Network (UTRAN) system is Code Division Multiple Access (CDMA)- based system and measurement gaps need to be created via the CM method. An alternative less popular way is to design dual receiver UE's, one for the normal transmission and the other to perform the inter RAT measurements continuously or on demand. Normally inter-Rat HO depends on RF measurements of target cells. In cases of limited coverage, there might not be any margin for the peak power required by the other methods. In these cases, the UTRAN can be informed of suitable candidates (that is of surrounding cells) and HO without UE measurements (what is known as blind HO).

The UE also should be told the target cell frequency or spreading factor depending on the RAT type of the target cell, via the current cell, this is conveyed typically within the inter-system handover commands that are introduced specifically for this purpose.

Different RAT's adopt different data rates therefore the data rate may have to be downgraded due to HO to GSM/GPRS from UMTS. The HO procedure must handle this situation where the connection might end up with a different data rate.

Compressed mode
During the process of Inter-RAT HO the UE must be given time to make measurements on the other RAT by turning off transmission. When in CM the information that is normally transmitted in the 10ms frame must be squeezed in time to

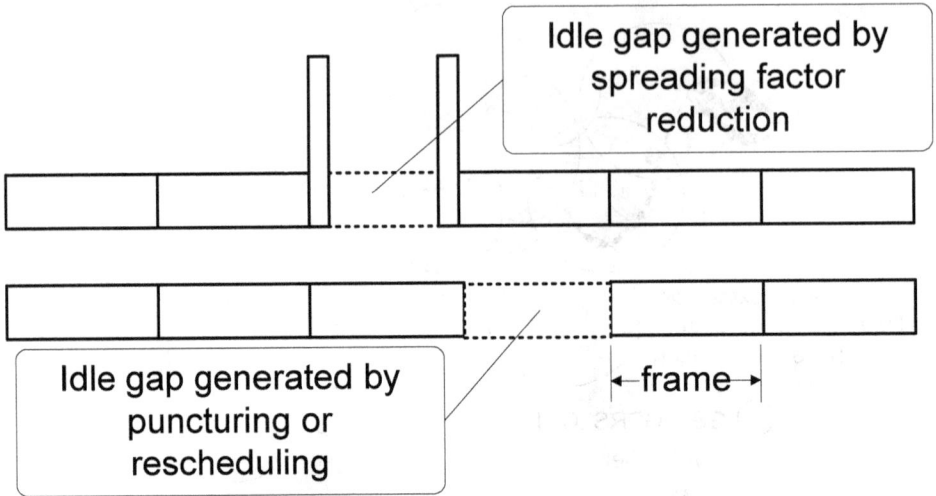

Figure 1.5 Compressed mode

make measurement gaps to a maximum of 7 slots per frame. This can be achieved in three ways;

The first method to compress data and produce some idle gap is to halve the spreading factor (and double the transmitting power, as illustrated in Figure 1.5) around the gap. During this period of reduced processing gain the power of the compressed frames is increased to maintain quality. **Reducing the spreading factor** by a factor of 2 will in turn double the data rate. This can only be done if the spreading factor is more than 4 and a further power increase is feasible.

Another way is the disposition of data by **higher layer scheduling** to allow fewer slots for user traffic. By setting restrictions only a subset of the allowed Transport Format Combinations (TFCs) are used. Therefore the amount of data for the physical layer is known and an idle gap can be created.

Removing various bits from the original data (transport channel **puncturing**) by rate matching (this method removed from R5). In UMTS rate matching is applied which repeats or punctures bits in the transport channel to fit in a physical channel. In the case of bit-repetition, e.g. 12.2kbps for speech in UTRA-FDD, puncturing these bits (see Figure 1.5) can easily reduce the amount of data. This type of CM is only used in the Down Link (DL).

There are two DL frame structures set by the higher layer, one to maximise the transmission gap and the other is optimised for power control. There is only one type of frame structure for the Uplink CM.

UTRAN transmits a "measurement control message" to the UE including the measurement ID and type of measurement to commence. Soon after reporting is complete, the UE sends a "measurement reporting message" to UTRAN including measurement ID and the results. The measurement control message is broadcast in idle mode within the System Information. When the UE monitors NBss using other RAT's, UTRAN must request the specific measurement needed to execute the requested handover.

UTRAN to GSM handover

When the UE is in a Circuit-Switched (CS) mode and the signal strength falls below a given threshold, the UTRAN orders the UE to perform GSM measurements. Typically, the UE is instructed to send a measurement report when the quality of a neighboring GSM cell exceeds a given threshold and the quality from UTRAN is unsatisfactory. When the UTRAN receives the measurement report message, it initiates the HO. The target Nobe-B Subsystem (BSS) sends a HO command message, which includes the details of the allocated resources, to the UE via the UTRAN radio interface. When the UE receives the HO command, it moves to the target GSM cell and establishes the radio connection according to the parameters contained in the HO command message. The UE then send a HO complete message to the BSS indicating a successful HO, after which the GSM network initiates the release of the UTRAN radio connection, as illustrated in Figure 1.6, 1.7a and 1.7b.

GSM to UTRAN handover

The UTRAN orders the UE to perform W-CDMA measurements by sending the measurement information message, which contains information on neighboring UTRAN cells. When the criteria for HO to UTRAN have been met, the BSS initiates the allocation of resources to the UTRAN cell. When the UE receives the HO to UTRAN command message from the BSS it tunes to the UTRAN frequency and begins radio synchronization. The UE then sends a message to the UTRAN indicating that the HO was successful and subsequently the GSM resources are released, see Figure 1.8a, 1.8b and 1.8c.

Inter RAT cell re-selection

The level of service that can be provided by the network to the UE in Idle mode or Connected Mode are limited service (emergency calls), normal service and operator service on a reserved cell. The UE performs cell re-selection in either idle mode, when the UE is in CS service, or connected mode, when common channels are used, for Packet-Switched (PS) service. The UE re-selects to a different RAT cell only when that cell is ranked higher than the current cells. The UE is only permitted to select a new RAT cell when the average received quality and average signal strength go beyond a minimum threshold. The quality and signal strength thresholds ensure that the UE can receive the information transmitted by the potential candidate cell. Unnecessary re-selections due to a ping-pong situation or fast moving UE's over small cells can be avoided by setting penalty time and temporary offset.

In **Idle mode** the Non Access Stratum (NAS) can control the RAT(s) in which the cell selection should be performed, i.e. by mapping RAT(s) to a selected Public Land Mobile Network (PLMN), and by maintaining a list of forbidden registration

Call setup procedure

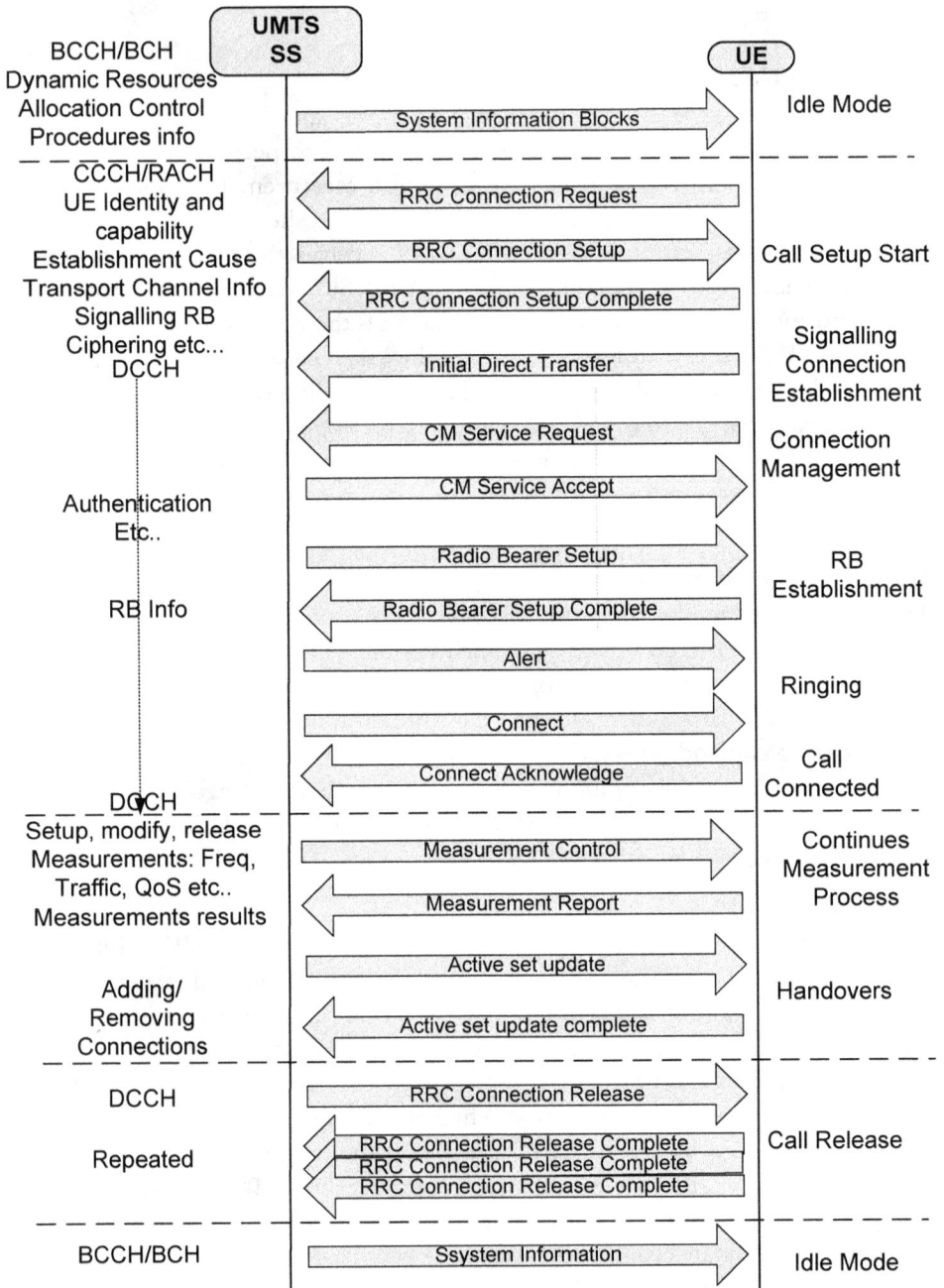

Figure 1.6 UMTS call set up procedure

UMTS to GPRS Handover

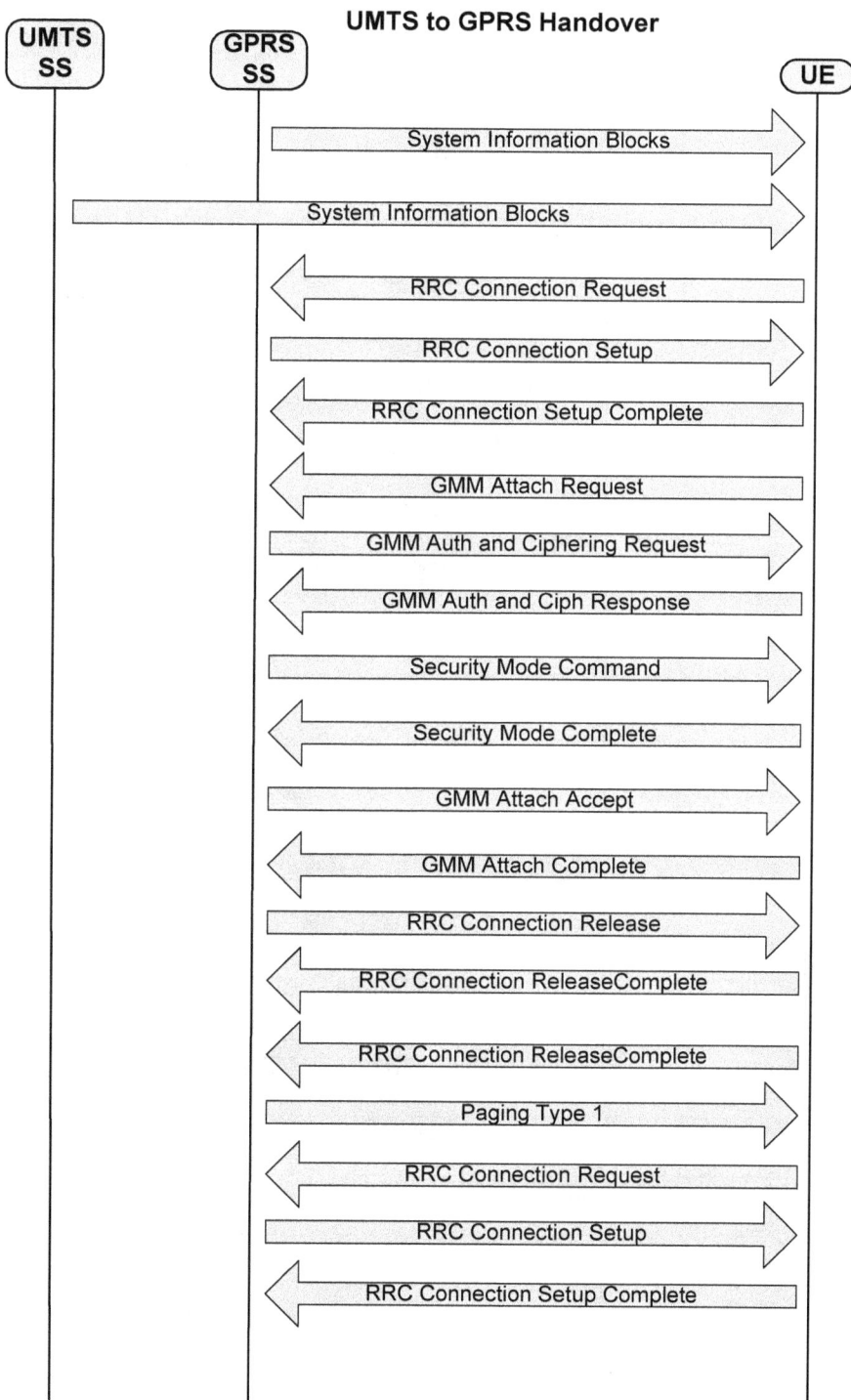

Figure 1.7a   UMTS to GPRS handover

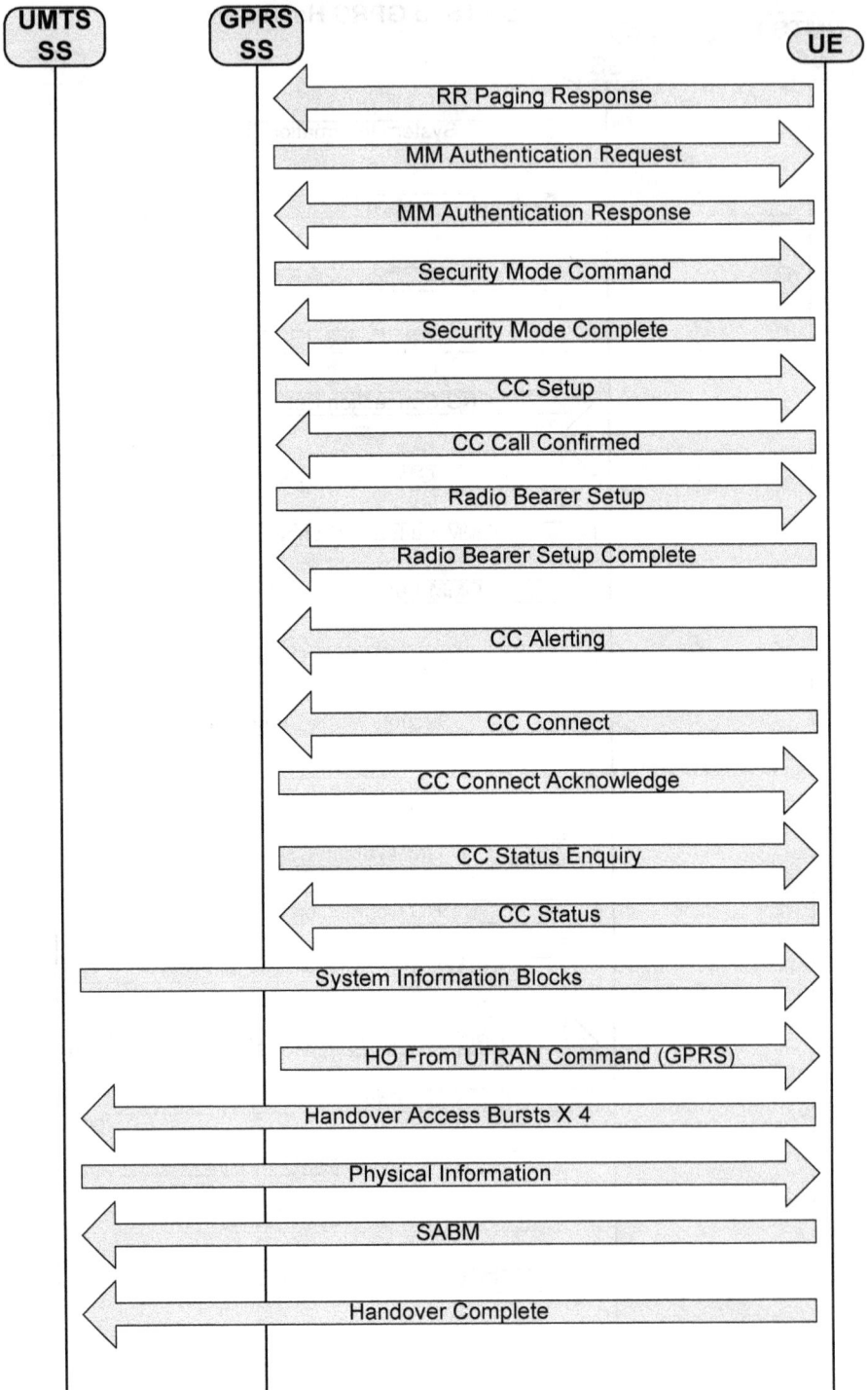

Figure 1.7b UMTS to GPRS handover

GPRS to UMTS Handover

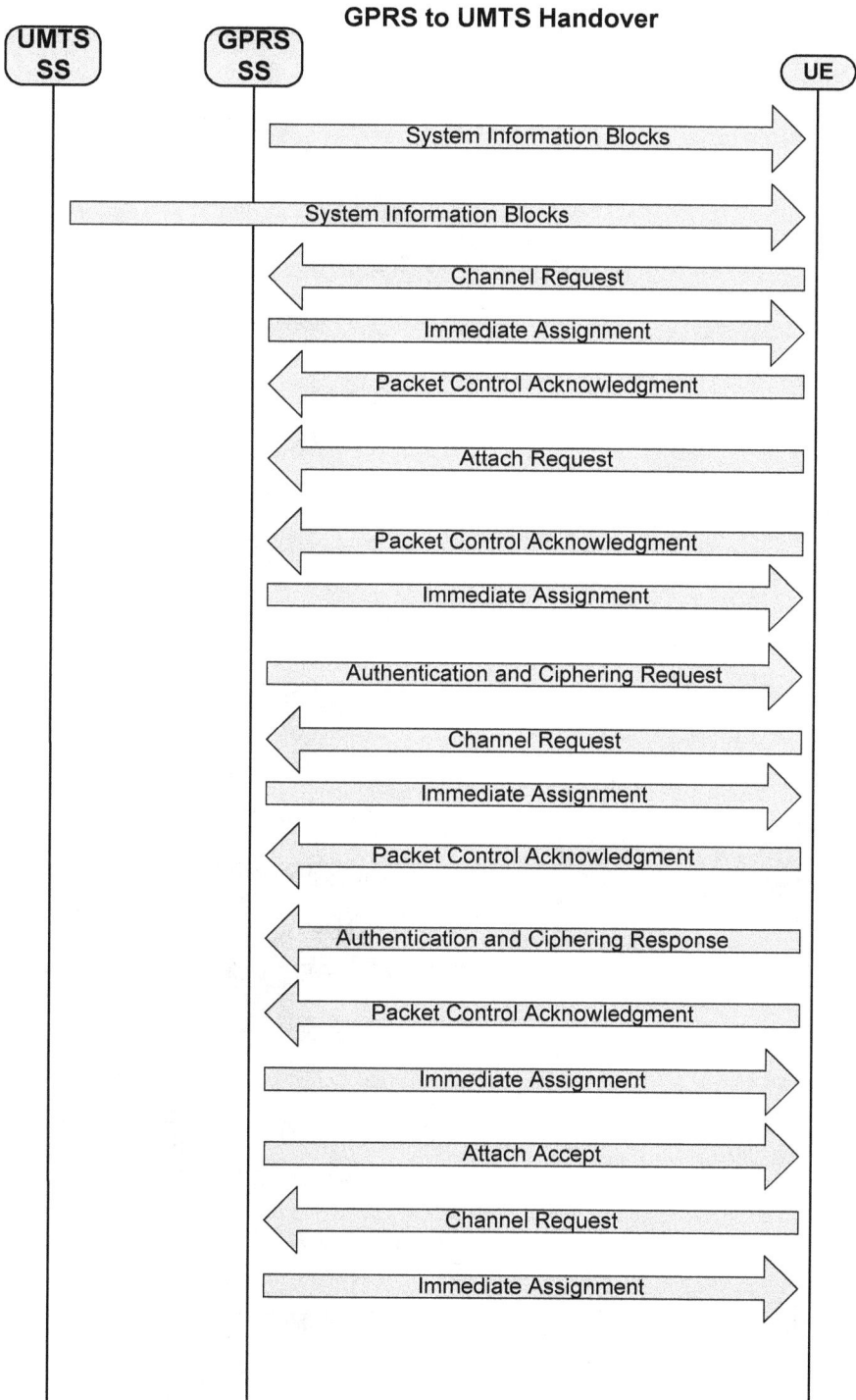

UMTS SS — GPRS SS — UE

System Information Blocks →

System Information Blocks →

← Channel Request

Immediate Assignment →

← Packet Control Acknowledgment

← Attach Request

← Packet Control Acknowledgment

Immediate Assignment →

Authentication and Ciphering Request →

← Channel Request

Immediate Assignment →

← Packet Control Acknowledgment

← Authentication and Ciphering Response

← Packet Control Acknowledgment

Immediate Assignment →

Attach Accept →

← Channel Request

Immediate Assignment →

Figure 1.8a GPRS to UMTS HO

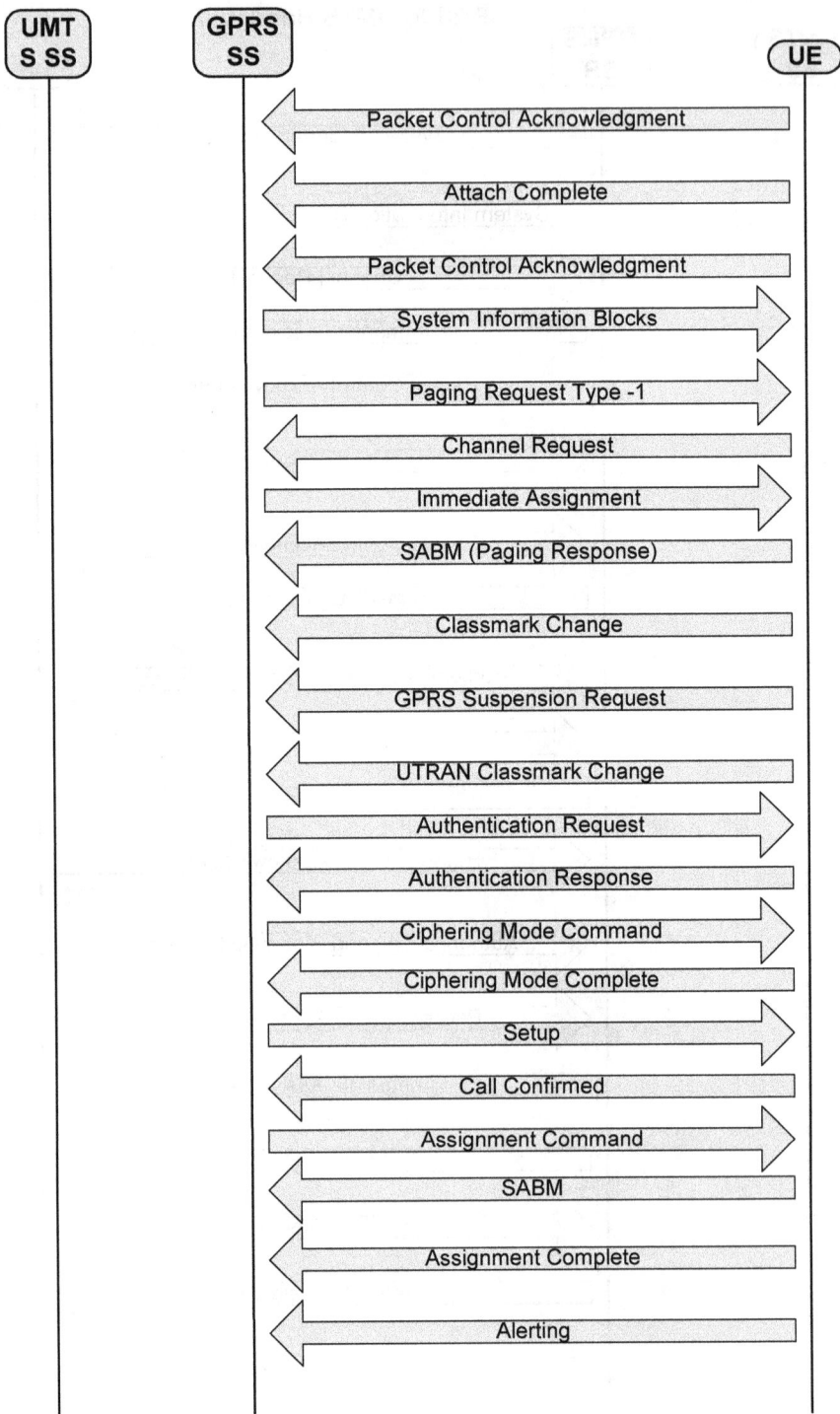

Figure 1.8b GPRS to UMTS HO

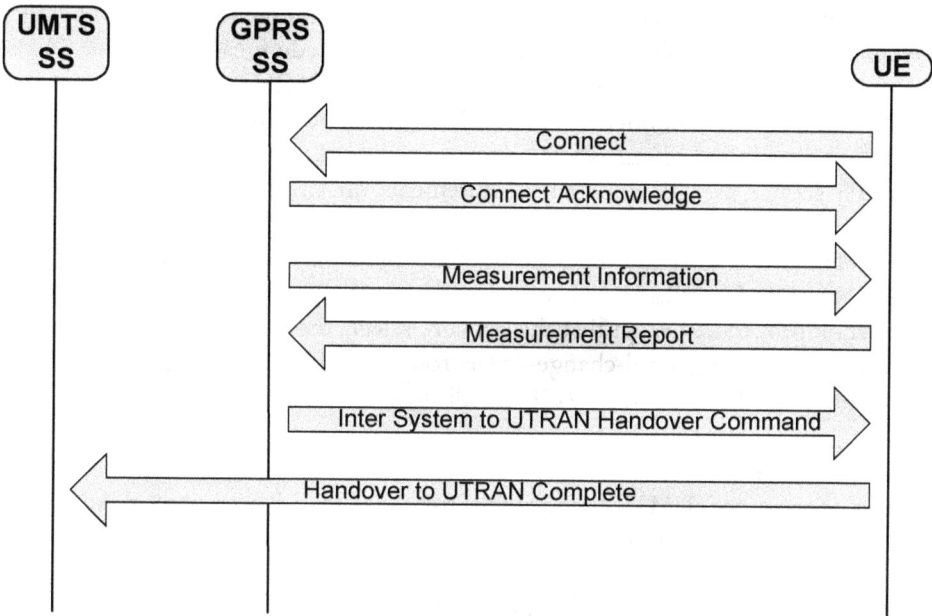

Figure 1.8c   GPRS to UMTS HO

area(s) and a list of equivalent PLMNs. The UE selects a suitable cell of a particular RAT based on idle mode measurements and cell selection criteria. The cell selection process can be expedited by saving information for several RATs on the UE. Once the UE is camped on a cell it regularly searches for a better cell according to the cell reselection criteria. The NAS is informed if the cell selection and reselection results in changes in the received system information (SI). For normal service, the UE has to camp on a suitable cell, tune to that cell's control channel(s). Once camped then the UE can receive SI from the PLMN (e.g. location area (LA), routing area (RA), Access Stratum and NAS information), and if registered it can receive paging and notification messages from the PLMN, and initiate call set-up for outgoing calls or other actions from the UE.

In **Connected mode** the UE is actively exchanging PS data with a UTRAN cell. If the UE re-selects a GSM/GPRS cell, it establishes the radio connection to the GSM/GPRS Base Station Subsystem (BSS) and then initiates the RA update procedure. During this procedure, the core network may retrieve information from UTRAN on the context of the UE including any awaiting data packets in the DL queue. When complete, the connection to UTRAN is released and the core network confirms the RA update.

Inter RAT cell change

The purpose of the Inter-RAT cell change order procedure is to transfer a PS connection between the UE and one RAT cell to another RAT cell. The transfer is controlled by the source RAT network, which specifies the identity of the target cell. UTRAN to GSM/GPRS cell change order procedure can be used in Dedicated Channel (DCH) or Forward Access Channel (FACH) state and only when there is a PS signaling connection established. The procedure can be used when no RABs are established or only PS RABs are established.

In PS service the UE conduct measurements on GSM/GPRS cells and sends the reports to the network, which orders the UE to switch to the candidate GSM/GPRS. The use of CM and the measurements procedure is similar to that used for UTRAN to GSM/GPRS HO procedure. The signaling is identical to that in the cell reselection procedure except that the network selects the target cell and initiates the procedure by sending a cell-change-order from the UTRAN message including e.g. information on the target GSM/GPRS cell.

Conclusions

Patchy coverage of UMTS is inevitable for sometime due to phased rollout of the network, therefore inter-RAT inter-working is required to enable 3G users to enjoy ubiquitous and seamless roaming from the early days of deployment. New procedures, methods and "workarounds" have been proposed and implemented to resolve inter-RAT inter-working and inter-operability challenges. These solutions include the use of dual-mode UEs, cell re-selection, cell change order and inter-RAT HO. CM channel measurements and embedding messages in one RAT as if it was of the other has made it possible to leave the GSM/GPRS network unchanged.

HSDPA

Many consider the HSDPA enhancements to W-CDMA as equivalent to EDGE to GSM. HSDPA is an enhanced technology that provides a smooth evolutionary path for UMTS (Universal Mobile Telecommunications System) networks to deliver higher capacity through improved spectral efficiency with improved higher data rates, shorter response time and better QoS (Quality of Service) to the user. The concept of sharing channels between users ensures that channel resources are utilised efficiently for packet data and hence should be cheaper to the user. HSDPA is an important enhancement to ensure high bit rate services are available at affordable prices.

In addition to other enhancements in 3GPP R5 HSDPA introduces a new DL transport channel that provides spectrally efficient means to support asymmetric and bursty packet data services, e.g. interactive, background and streaming services. This significantly increases peak data rates to about 10 Mb/s (theoretically 14 Mb/s) and improves DSCH throughput compared to that of using current R99 channels, see Table 1.4 for data rates corresponding to UE categories.

All of the above is supported by introducing fast and complex channel control mechanisms based on a short fixed packet 2ms TTI, AMC (Adaptive Modulation and Coding) and fast L1 H-ARQ (Layer 1 Hybrid Automatic Repeat reQuest). To facilitate this fast scheduling with a per-TTI resolution that corresponds to the instantaneous air interface load, the HSDPA related MAC (Medium Access Control) functionality is moved to the Node-B.

This section gives an overview, describes the basic principles, identifies the testing requirements and assesses the significance of this promising technology concept.

Technical motivation

The basic (R5) HSDPA introduces a shared MAC-hs (MAC-high speed) layer and a special HS-DSCH with the necessary control channels, which is similar to the (R99) DSCH but without fast power control. On the air interface, transmission time is allocated to the UE on a TTI (sub-frame) basis. Modulation, effective code rate, power and other transmission parameters can be adjusted dynamically by TTI based AMC (Adaptive Modulation and Coding). To increase peak data rates, cell throughput and to reduce transmission delay new techniques have been introduced in (R5) HSDPA. These include higher order modulation schemes and low redundancy coding combined with incremental redundancy. Further enhancements to reduce requirements on link performance include L1-based fast H-ARQ and Tx/Rx antenna diversity. Parallel channels are facilitated by multi-code in R5. Furthermore HSDPA in R6 will introduce antenna array processing technologies to enhance the peak data rate to ~30Mbit/s employing smart antennas using beam-forming for UEs with one antenna and MIMO (Multi Input Multi Output) for UEs with 2-4 antennas.

R99 uses several types of DL radio bearers to facilitate transportation of different service classes. Table 1.2 illustrates the different characteristics of these radio bearer channels and shows that HSDPA aims to provide a spectrally efficient means to support services such as Internet access and file download that require high data rate packet transport on the DL.

A simplified illustration of the general functionality of HSDPA is depicted in Figure 1.9. The Node-B determines the channel quality of each HSDPA-active UE based on power control, QoS, ACK/NACK (Acknowledge / Non-Acknowledge) ratio and UE specific quality feedback. Fast scheduling and link adaptation are then conducted promptly depending on the active scheduling algorithm and the user prioritization scheme employed by the Node-B.

High speed channels

A number of additional channels and a layer are introduced in order to implement HSDPA features.

HS-SCCH (High Speed Shared Control Channel) is the DL signalling channel that carries the key physical layer control information to enable demodulation of the data on HS-DSCH and to perform the possible physical layer combining of the data sent on HS-DSCH in the case of re-transmission or an erroneous packet.

Active HSDPA UE # n
  Active HSDPA UE # 2
   I Active HSDPA UE # 1

- - - - - - - Periodic CQI transmission by   Node-B
              HS-DPCCH Presenting
              DL channel quality   Scheduling DL
                                          packets based on the
                                          CQI info from all
                        HS-SCCH   UE's
              (control information for DL packets

              HS-PDSCH (carries the user DL packet data
              From the transport channel HS-DSCH)

Packet receive and
(N)/ACK decision

              (N)ACK transmission by HS-DPCCH

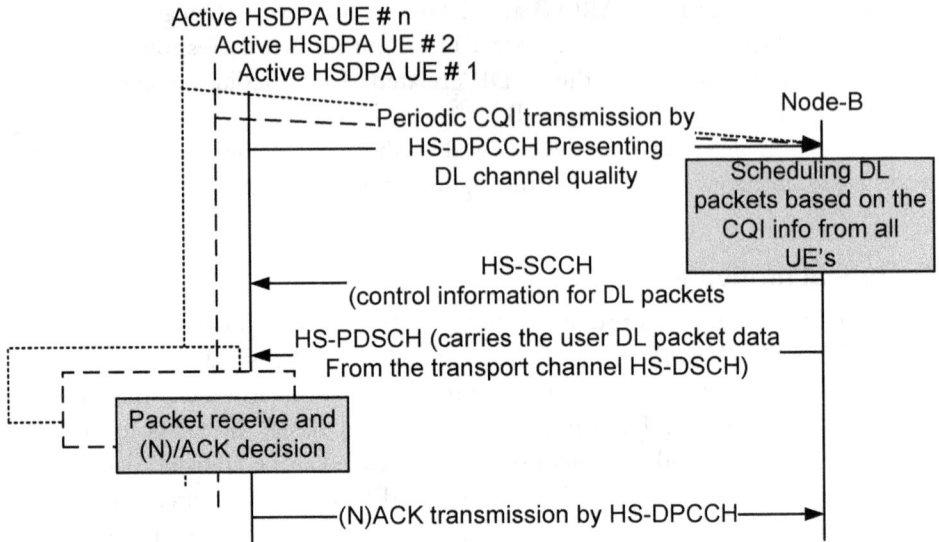

Figure 1.9   HSDPA functionality

The information includes channelization code set, modulation scheme, transport block size, HARQ process information, redundancy and constellation version, and new data indicator.

HS-DPCCH (High Speed Dedicated Physical Control Channel) is the UL signalling channel that carries the necessary control information in the UL, namely, ARQ acknowledgments and DL quality feedback information to be used in the Node-B scheduler to decide the UE and data rate to transmit to. It carries an ACK/NACK indication, to reflect the results of the CRC check after the packet decoding and combining. It also carries the DL CQI (Channel Quality Indicator) to indicate which estimated transport block size, modulation type and number of parallel codes could be received correctly with reasonable BLER (Block Error Rate) in the DL direction.

HS-PDSCH (High Speed Physical Downlink Shared Channel) is the physical channel that carries the actual user specific packet data in the DL from the transport channel **HS-DSCH** (High Speed Shared Downlink Channel). It is of fixed spreading factor 16 and of up to 15 codes multi-code transmission. The peak data rate is up to 10 Mbps range with 16 QAM (Quadrature Amplitude Modulation). The HS-DSCH interleaving period TTI is reduced to 2ms to achieve shorter round-trip time between the UE and Node-B for re-transmissions.

MAC-hs With the introduction of HS-DSCH, additional intelligence in the form of a new MAC-hs layer is required for the Node-B to be able to directly control re-transmission. This will expedite re-transmission and reduce delay with packet data operation. The prime functionality of the new MAC-hs is to take care of the HARQ functionality, scheduling and priority handling.

Adaptive modulation and coding schemes in HSDPA

Fast power control and VSF (Variable Spreading Factor) functionality of DSCH (R99) is replaced in HS-DSCH (R5) by variable modulation and coding schemes that cover a wide dynamic range in order to cope with the varying DL radio and channel quality conditions seen by the UE. The means of adaptation that are adopted by HSDPA, are the effective code rate, the modulation scheme, the number of codes used and power per code.

The scheme used by HS-DSCH is based on R99 rate 1/3 Turbo encoder but adds rate matching to obtain a finer resolution of the effective code rate. To achieve very high peak data rates, 16QAM is added besides the (R99) QPSK (Quadrature Phase Shift Keying) scheme. 16QAM modulation and fast link adaptation can be combined to optimise the instantaneous use of the varying radio channel. Since the TTI is only 2 ms (compared to 10 or 20 ms of R99) and scheduling and link adaptation are decided for each TTI, the variation in the channel can be tracked nearly instantaneously. 16QAM makes better use of the bandwidth but demands more receive power per bit therefore 16QAM is more suited for bandwidth-restricted cases (nearer to the Node-B) rather than for power-restricted cases.

Fast packet scheduling function

Fast packet scheduling functions are located in Node-B as part of the MAC-hs to manage the HS-DSCH resources, select the coding/modulation scheme and the Tx power for the HS-DSCH data packets. It simply decides the overall behaviour of the system. The UEs that should be scheduled within a particular TTI are decided using CQI reports coming from different UEs. Channel condition-dependant scheduling rather than sequential scheduling can increase the capacity significantly and better use air interface resources. Also fast scheduling allows guaranteed bit rate services using packet scheduling without the need for a dedicated channel. The UEs are prioritised by the scheduler according to the channel conditions, the amount of data pending in the buffer for each UE, the elapsed time for that UE since last served and pending re-transmissions for a UE. The scheduler criteria are merely vendor specific.

Fast hybrid-ARQ with soft combining or incremental redundancy

The H-ARQ scheme enables the UE to quickly request re-transmission to compensate for errors resulting from the link adaptation process. Also HARQ offers better error rate performance than conventional ARQ.

The UE tries to decode each received transport block and pass the result to the Node-B. The H-ARQ process allows Node-B to quickly re-transmit responding to the UE request. The UE then combines information from the previous transmission with the new re-transmission aiming for successful decoding of the transport block in a second attempt.

The HARQ process can employ one of two methods that is decided by Node-B depending on the available UE soft memory. The first is the Soft (Chase) combining method where the same packets (same set of coded bits as the first transmission, self-decodable) are re-sent to the UE. This method requires less UE buffer memory, and at highest data rate, only this method is used. The second method is the incremental redundancy where re-transmissions are not identical (different set of coded bits for the same information, not self-decodable). This method offers a slightly better performance but at the expense of the UE memory as re-transmissions require more processing than just a simple addition. This approach is more suited to lower data rates.

HSUPA

The improvements in efficiency and services provided by HSDPA encouraged 3GPP to apply similar concepts to the uplink. High-Speed Uplink Packet Access (HSUPA), sometimes termed Enhanced Uplink, represents the latest 3GPP Release 6 technology and aims to provide optimized packet data support in the uplink.

Some think it is the downlink that is critical with uplink usage being much less important. Consequently, is there any need to introduce this technology. While it is true that many broadband services currently are dominated by downlink data transfer, it is important to recognize the need for efficient packet data support on uplink.

Services such as multimedia messaging are already popular and likely to grow as high-quality cameras become standard in most handsets. Additionally, it is likely that the use of symmetric services such as voice and VoIP will start to increase, both offering more efficient use of bandwidth than their circuit-switched counterparts.

FDD Enhanced Uplink (also known as HSUPA) is a UMTS Release 6 feature which improves the performance of uplink dedicated transport channels, leading to increased capacity and throughput and reduced delay. This is achieved via a new uplink dedicated transport channel, E-DCH, which works in conjunction with techniques such as Node B controlled scheduling, HARQ and a shorter TTI. High uplink packet data rates of up to ~5.7Mbps are achievable.

New HSUPA Channels

In the HSUPA, the Enhanced Dedicated Channel (E-DCH) is introduced as a new transport channel for carrying user data on the uplink. On physical layer level, this translates into 2 new uplink channels and 3 downlink channels are introduced for control purposes:

DL L1 Channels

There are three DL physical channels:

E-RGCH is transmitted from all cells that are either in the serving radio link set (SRLS) or not. This channel is used (if the cell is in the SRLS) to incrementally (up, down, hold) adjust the serving grant of all UEs that controlled by the cell. It uses SF of 128 and 3 or 4 consecutive slots to cells in the SRLS.

If the cell is not in the SRLS, it sends a non-serving relative grant to enable it to adjust (hold, down) the transmission rate of UEs that are not under its control to avoid overload condition. It uses 15 consecutive slots to cells that are not in the SRLS.

E-AGCH is transmitted from a single serving cell in the SRLS and carry the absolute amount of uplink resources that can the UE uses for a specific HARQ process. Single or many UEs can use the absolute grant to select the TFC and hence the data rate for the next transmission for that HARQ process. This channel uses SF of 256 and transmitted over 1 sub-frame for 2ms TTI and over 1 frame for 10 ms TTI.

E-HICH This channel uses a SF of 128, 3 slots (2 ms TTI) and 12 slots (10 ms TTI). This channel is used to send the ACK/NACK commands for the HARQ process to the UE.

UL L1 Channels

E-DPDCH One or more of this channel carries the only E-DCH. E-DPDCH uses SF from 2-256.

E-DPCCH One of this channel carries the signalling information associated with E-DCH. E-DPCCH uses fixed SF of 256. This channel also carries the happy bit, the RSN and the TFCI information.

HSUPA Capability

The key features of the HSUPA functionality illustrated in Table 1.2 and Figure 1.10 are:

The HSUPA data channel is similar to HSDPA in that most of the parameters are fixed, e.g. CRC attachment is fixed at 24 bits and uses turbo channel coding (1/3 turbo coding).

The rate matching parameters including HARQ type to be used is defined by the L1 HARQ process. The HARQ process uses the redundancy version RV parameter to define whether puncturing or repetition are being applied. Puncturing and repetition may be applied equally to the symmetric data bits or the parity bits based on the RV value.

New uplink dedicated transport channel E-DCH is introduced at L1. The only one E-DCH is processed by one E-DCH CCTrCH. E-DCH is carried on one or more uplink E-DPDCHs. Uplink signalling information associated with E-DCH is carried by one uplink E-DPCCH.

E-DCH transfers one transport block per TTI (10ms or 2ms), depending on the UE capability Table 1.5. UE capability is divided into 6 FDD E-DCH physical layer categories, each category supporting particular maximum data rate.

E-DCH is intended for the transport of dedicated logical channels (DTCH, DCCH), so protocol layers above MAC are unchanged from R99/R4/R5. RLC UM and AM are used to transfer data on DTCH/DCCH mapped to E-DCH.

Non-serving EDCH (NB)   UE

   Non-serving EDCH (NB)

     Serving E-DCH (NB)

Generate scheduling feedback and happy bits

RSN, E-TFCI, Happy bit (E-DPCCH)

Relative grant E-RGCH

Relative grant E-RGCH

Absolute / relative grant (E-AGCH)

UL-Data E-DPDCH

Scheduling: Abs. / Rel. grants
HARQ: Soft combining, ACK/NACK

(N)ACK transmission by E-HICH

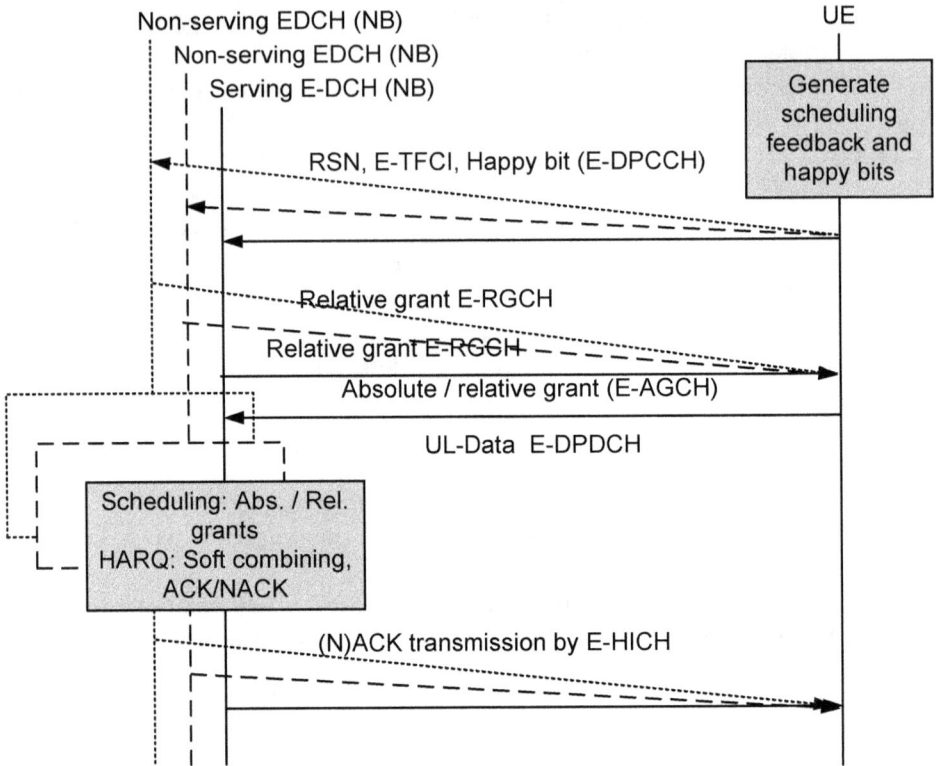

Figure 1.10   HSUPA functionality

The UE multiplexes dedicated logical channels on to MAC-d flows. A UE may be configured with multiple MAC-d flows with different QoS characteristics. Different MAC-d flows may be multiplexed on to E-DCH in the same TTI. The UTRAN performs the corresponding de-multiplexing.

MAC-e and MAC-es, are introduced into the MAC layer to support E-DCH. These help to provide the new functionalities of Node B controlled scheduling and HARQ. The MAC Control SAP is used to transfer control information to MAC-e and MAC-es. The MAC-e entity handles E-DCH scheduling, the E-DCH HARQ protocol and de-multiplexing of E-DCH transport blocks (MAC-e PDUs).

The HARQ entity supports multiple HARQ processes which acknowledge E-DCH transmissions from the UE via the E-HICH downlink physical channel.

E-DCH scheduling manages E-DCH cell resources for the UE by allocating Scheduling Grants via the E-AGCH and E-RGCH downlink physical channels.

De-multiplexing extracts the MAC-es PDUs from each MAC-e PDU and forwards them to the associated MAC-d flow. The MAC-es entity provides in-sequence delivery of data to MAC-d, which in turn routes the data to the appropriate dedicated logical channel.

Re-ordering queue distribution routes MAC-es PDUs to the correct re-ordering buffer. Re-ordering reorders MAC-es PDUs according to the received transmission sequence number TSN and timing information and forwards them for disassembly.

The disassembly function is responsible for extracting the MAC-d PDUs from each MAC-es PDU and delivers them to MAC-d.

Key Features of HSUPA

Uplink Scheduling

The scheduling mechanism is essential for the HSUPA, it is located in the Node B close to the air interface. Task of the uplink scheduler is to control the uplink resources the UEs in the cell are using. The scheduler therefore grants maximum allowed transmit power ratios to each UE. This effectively limits the transport block size the UE can select and thus the uplink data rate. The scheduling mechanism is based on absolute and relative grants. The absolute grants are used to initialize the scheduling process and provide absolute transmit power ratios to the UE, whereas the relative grants are used for incremental up- or downgrades of the allowed transmit power. Note that one UE has to evaluate scheduling commands possibly from different radio links. This is due to the fact that uplink macro diversity is used in HSUPA.

Hybrid Automatic Repeat Request (HARQ)

The HARQ concept is simply a retransmission protocol improving robustness against link adaptation errors. The Node B can request retransmissions of erroneously received data packets and will send for each packet either an acknowledgement (ACK) or a negative acknowledgement (NACK) to the UE. Furthermore, the Node B can do soft combining, i.e. combine the retransmissions with the original transmissions in the receiver. Due to uplink macro diversity, one UE has to evaluate ACK/NACK information for the same packet possibly from different radio links.

Reduction of Transmission Time Interval

To accelerate packet scheduling and reduce latency, HSUPA allows for a reduced transmission time interval (TTI) of 2 ms corresponding to 3 timeslots. A WCDMA radio frame of 10 ms therefore consists of 5 sub-frames. In HSUPA, however, the support of this 2 ms TTI in the UE is not mandatory, instead, it is a UE capability. It is configured at call setup whether 2 ms TTI or 10 ms TTI is to be used for HSUPA transmission.

Impact on Radio Access Network Architecture

Both the uplink scheduling and the HARQ protocol are located in the Node B, in order to move processing closer to air interface and be able to react faster on the radio link situation. This is illustrated in Figure 1.10.

Macro diversity is exploited for HSUPA, i.e. the uplink data packets can be received by more than 1 cell. There is one serving cell controlling the serving radio

	Release 99			HSDPA	HSUPA
Radio Bearer	FACH (Forward Link Access Channel)	DCH (DL) Dedicated Channel on the DL	DSCH	HS-DSCH	E-DCH
Summary characteristics and features	Low latency common channel for carrying small data volumes, non-real and bursty data, e.g. SMS Poor spectral efficiency	High parameter flexibility hence supports all traffic classes Slow channel re-configuration process hence limited to constant bit rate services , i.e. CS services that require a conversational QoS	Faster channel re-configuration and packet scheduling processes Better efficiency for bursty and higher data rate services than DCH 10 or 20 ms TTI MAC in RNC	Enhanced peak data rates, spectral efficiency, QoS control for bursty and asymmetric DL packet data AMC MAC in Node-B	Enhanced peak data rates and reduced latency 2 or 10 ms TTI
	Fixed spreading factor (4-256)	Fixed spreading factor (4-512)	VSF (4-256)	Fixed spreading factor (16)	Variable SF = 1-4
	No multi-coding	No multi-coding	Multi-code operation	Extended multi-code operation 2 ms TTI	Multi-code operation 1,2,4
	No ARQ	No ARQ	No ARQ	Fast L1 HARQ	Fast L1 HARQ
	No fast power control, does not support soft handover	Maintain constant data rate by fast power control and operates in soft handover	Maintain constant data rate by fast power control Does not support soft handover	No fast power control Does not support soft handover	SHO

Table 1.2 Methods for packet data transmission in R99 and R5

link assigned to the UE. The serving cell is having full control of the scheduling process and is providing the absolute grant to the UE.

The serving radio link set is a set of cells contains at least the serving cell and possibly additional radio links from the same Node B. The UE can receive and combine one relative grant from the serving radio link set. There can also be additional non-serving radio links at other Node Bs. The UE can have zero, one or several non-serving radio links and receive one relative grant from each of them.

MBMS

MBMS is a unidirectional point-to-multipoint service in which data is transmitted from a single source entity to a group of users in a specific area. There are two transmission modes to this: Point-to-Many and Point- to-Point.

R99/4	HSDPA/R5	HSUPA/R6
TTI = 10, 20, 40, 80 ms	TTI = 2 ms	TTI = 2 ms, 10ms, 80ms
Variable SF = 1-256	Fixed SF = 16	Variable SF = 1-4
More transport blocks per TTI	One transport block per TTI	One transport block per TTI
Convlolutional or turbo code	Turbo code only	Turbo code only, CRC of 24 bit
QPSK only	QPSK and 16QAM according to UE capability	QPSK and BPSK
Configurable CRC	CRC of 24 bits	
Scheduling in RNC	Scheduling in Node B	Node B based scheduling
Retransmissions in AM RLC	Physical layer retransmissions	Absolute Grants, Relative Grants, E-DPDCH/DPCCH Power
Power control	Adaptive modulation and coding	Node B based hybrid ARQ, synchronous, incremental redundancy
SHO	HHO	Supports SHO, active set is the same or subset of DPCH

Table 1.3 Comparison between R99/R5 and R6

HS-DSCH Category	Max number of HSDSCH codes (SF16) received	Minimum inter TTI interval	Modulation	Maximum peak rate
CAT 1	5	3	QPSK & 16-QAM	1.2Mbps
CAT 2				
CAT 3		2		1.8Mbps
CAT 4				
CAT 5		1		3.6Mbps
CAT 6				
CAT 7	10			7.3Mbps
CAT 8				
CAT 9	15			10.2Mbps
CAT 10				14.4Mbps
CAT 11	5	2	QPSK only	0.9Mbps
CAT 12		1		1.8Mbps

Table 1.4 UE categories for HSDPA

One important aspect of MBMS is flexibility MBMS can be set to use only a share of a cell carrier, leaving the rest for other services, e.g. voice and data. The MBMS share consists of a variable number of MBMS radio bearers, each radio bearer can have a different bit rate. Although MBMS supports user bit rates of up

E-DCH Category	Max number of transmitted E-DCH codes	Max SF	TTI	Max number of bits of E-DCH transport blocks transmitted within E-DCH TTI	Data rates Mbps
CAT 1	1	SF4	10 ms only	7110	0.71
CAT 2	2		10 ms	14484	1.45
			2 ms	2798	1.4
CAT 3			10 ms only	14484	1.45
CAT 4		SF2	10 ms	20000	2
			2 ms	5772	2.89
CAT 5			10 ms only	20000	2
CAT 6	4		10 ms	20000	
			2 ms	11484	5.74

Table 1.5 UE categories for HSUPA

to 256kbps, given current UE display sizes and resolutions, 64kbps is adequate e.g. for a news channel application and 128kbps for a sports channel application

The new M-Channels

MBMS mainly reuses existing logical and physical channels but new channels have been added, i.e.:

Logical channels

MCCH (MBMS point-to-multipoint control channel): contains details concerning current and upcoming MBMS sessions. MCCH can be sent in S-CCPCH carrying the DCCH of the UEs in CELL_FACH state, or in standalone S-CCPCH, or in same S-CCPCH with MTCH.

MSCH (MBMS point-to-multipoint scheduling channel): provides information on data scheduled on MTCH

MTCH (carries the actual MBMS application data). Used for a p-t-m DL transmission of user plane information between network and UEs in RRC Connected or Idle Mode. TCTF field is always used in MAC header. TTI value of 40 and 80ms have been introduced for MTCH.

Physical and indication channel

SCCPCH is used as a physical channel for FACH carrying MTCH or MCCH or MSCH

MICH (MBMS notification indicator channel): MBMS notification utilises a new MBMS specific PICH called the MBMS Notification Indicator Channel (MICH) in each cell. This channel is used by the network to inform UEs of available MBMS information on MCCH, i.e. MCCH, MSCH and MTCH reuse the forward access channel (FACH) transport and secondary common control physical channel (S-CCPCH).

MAC

The **RLC** and **MAC** layer reuse much of the existing protocol stacks.

MAC-m is added to the MAC-c/sh (one per cell) to do the following:

- Scheduling / Buffering / Priority Handling
- Channel mapping and TCTF multiplexing
- Addition of MBMS ID in the MAC for MBMS channels
- TFC selection

Transport channel

FACH is used as a transport channel for MTCH, MSCH and MCCH

Layer 1

Enhancements for the physical layer for MTCH are:

1. Soft-combining for FACH, SCCPCH physical channels from different radio links could be soft combined.
2. **Counting procedure** has been introduced in the standards to enable the network to keep track of the number of MBMS users in a cell, thereby helping the network to decide which bearer to use. With MBMS technology one 5MHz cell carrier can potentially support 16 point-to-multipoint MBMS channels at a user bit rate of 64kbps per channel.

MBMS Session Walkthrough

Information on a particular MBMS service is sent to a service provisioning server. This information is referred to as a service announcement. Service announcements provide information on the service and how UEs may access it. There are many ways on how to deliver MBMS service announcements to end-users, e.g.

- Save them on a web server from which they can be downloaded via HTTP or WAP.
- Push service announcements to UEs by means of existing mechanisms, such as SMS or MMS
- Delivered over a special MBMS service announcement channel.

If a user decides to use the broadcast service described in the service announcement then no further action required. If multicast is chosen a number of actions are required as below. Typical phases of an MBMS illustrated in Figure 1.11:

- The UE camps on the channel whose parameters are described in the service announcement.
- If a multicast service is chosen, a session join request must be sent to the network with parameters extracted from the service announcement.
- The UE becomes a member of the corresponding MBMS service group and receives all data delivered by the service.

Broadcast Mode	MBMS Connection Procedure		Multicast Mode
Operator may use different mechanisms to inform users about available MBMS services (service discovery) Announcements are available also for those who are not subscribed Location information may be used	Service announce-ment	Subscrip-tion	User subscribing to a specific MBMS service User may receive related MBMS service and operator may charge for it Subscription can be made for example on the Internet
Establishes the needed network resources for the MBMS data transfer Resources are reserved from both core network and radio access network	Session Start	Service announc-ement	Operator may use different mechanisms to inform users about available MBMS services (service discovery) Announcements are available also for those who are not subscribed Location information may be used
Informs users about on-going or forthcoming services available for all users with activated MBMS services	MBMS Notifica-tion	Joining	In the joining phase the user indicates to the network that he/she wants to specific multicast data stream Joining can be made at any time Used for charging so authentication is needed
The data is transferred to the user equipments (UEs) The UE is in connected mode No ciphering No retransmitting, so there may be some data loss	Data Transfer	Session Starts	Same as in broadcast mode with one exception: The network decides whether a common channel for all UEs in a cell is used, or separate channel for each UE
The reserved network resources are released when there?s no more data to be transferred	Session Stops	MBMS Notifica-tion	Informs users about on-going or forthcoming services available for all users with activated MBMS services
		Data Transfer	The data is transferred to the user equipments (UEs) The UE is in connected mode No ciphering No retransmitting, so there may be some data loss Data could be encrypted
		Session Stops	The reserved network resources are released when there?s no more data to be transferred
		Leaving	When user no longer wants to belong to a multicast group and receive data transfer Doesn?t unsubscribe the user from the service, just from the current data transfer Can be made any time

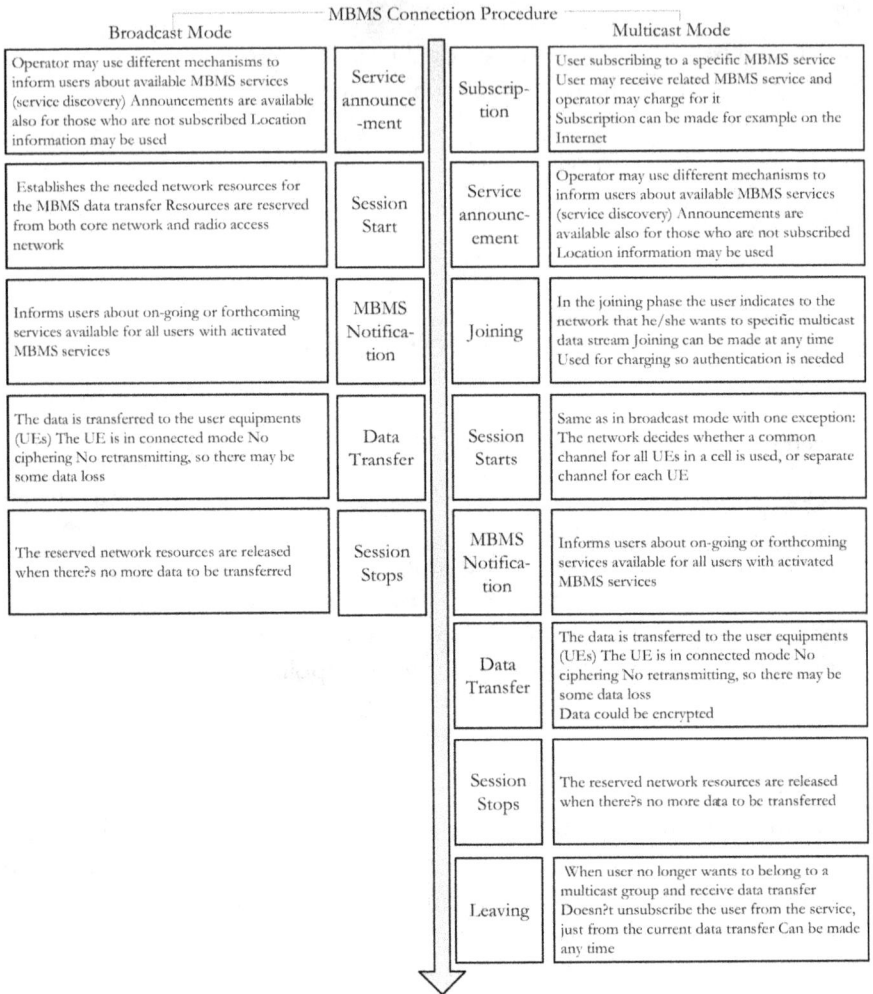

Figure 1.11   MBMS phases

- Prior to transmission the BM-SC (Broadcast/multicast service centre) must send a session start request to the GGSN in the core network.
- The GGSN allocates the required internal resources and forwards the request to the affected SGSNs.
- The SGSNs, in turn, request the allocation of radio resources necessary for providing the required quality of service (QoS).
- Finally, the UEs of the corresponding MBMS service group are notified that the service is about to deliver content.
- The server can then send multimedia data to the BM-SC, which forwards the data to the MBMS bearer. The data is transmitted to every UE taking part in the MBMS session.

- Then, the server sends a session stop notification to indicate that the data transmission phase has ended.
- End-users who want to leave an MBMS multicast service send a service leave request to the network, which removes the user from the related MBMS service group.

INTRODUCTION TO TESTING

Mobile subscribers expect that when they switch on their mobiles they will obtain high quality service. They expect to roam internationally and get the same level of service that they do at their home country. Expectation have been raised by the successful operation of these features and many more that has lead to the phenomenal success of GSM/GPRS that now has more than a billion subscribers worldwide. High quality of service is by no means easy to achieve, especially when the degree of complexity of mobiles and networks is considered.

Thousands pages of standards documentation is required because operations such as registration require many messages exchanged between the phone and the network. As a call proceeds, the network and phone have to communicate with one another to ensure that the call is maintained. Handovers and other scenarios must be specified very precisely and operate correctly. For UMTS, ISHO needs to be accommodated so mobiles can switch to the 2G or 2.5G network when no 3G network is available and vice versa.

Making sure that all the functionalities perform correctly before a mobile is introduced into the market is necessary. Utilizing a test methodology that enables phones to operate on any network in any country means that they have to be tested to ensure they conform to the agreed ETSI specifications. This conformance testing is at the heart of the success of the GSM/GPRS system.

A very similar approach has been adopted by the 3G UMTS or the WCDMA system on the hope that and expecting a similar degree of success will be met.

A number of areas comprise the overall testing of a new mobile. Safety, protocols, RF performance, USIM, audio performance, etc, is essential elements that need to be included in the overall selection of tests. The aforementioned areas will be explained further in the remaining chapters of the book.

Like any other electronic system, mobile phones and infrastructure equipment are no exception and must undergo rigorous safety tests to ensure that they will not cause harm or injury in any way. Also safety testing is undertaken to ensure that the level of RF absorption falls below the maximum permitted levels. The testing, known as SAR testing, involves the use of an anatomically correct model of the human head. Inside the model are sensors to measure the temperature rise. The test results obtained can be related to the levels of RF radiation that are being absorbed at different places on the head. The RF performance of the phone also must be checked. Many measurements of the transmitter and receiver performance are carried out in a variety of areas including the out-of-band emissions.

Measurements of the RRM are carried out to ensure that the control capability of the phone is operating correctly. The RRM is the entity used to control the physical or RF layers in accordance with the requirements of the protocols from the higher layers. There are, for instance, very tight limits on the transmitter output power as it is controlled to meet conditions such as variations in signal strength. This ensures that the phone only transmits sufficient power to maintain a reliable circuit under the prevailing conditions, and as a result, the overall level of noise in the phone bands is reduced to the minimum level. To achieve RF performance testing, a protocol tester often is used to control the phone and set up the relevant scenarios along with RF measurement and generation equipment. This equipment may include signal generators, power meters, analyzers, and noise generators. Additionally, to check operation of the phone with multipath and fading, special fading simulators are used.

In order to determine the overall operation of the mobile, protocol testing is essential. If the mobile protocol software operates incorrectly, consequently the phone will not operate properly on a life network. In view of the complexity of the protocols that are used, this testing can be very involved and requires the use of specialized network simulators. These testers emulate a variety of network entities, such as BTS, BSC, or in the case of UMTS, Node Bs and RNC. In this way, a host of scenarios from registering to terminating a call and all the different forms of handoff can be simulated. In fact, any situation that can be encountered needs to be fully tested.

Audio checks need to be completed. These ensure that the audio aspects of the phone meet the required standards both in terms of the microphone and earphone. A wide variety of checks is performed to ensure that audio characteristics are satisfactory under all conditions.

All testing performed on a mobile or BS must be repeatable and conform to the specifications regardless of the test equipment being used. To achieve this, formal test cases are written for each test. A variety of processes can be adopted for writing test cases. For GSM, test cases are written in prose, describing in detail the test setup, the stimuli that need to be applied, the way in which the test is carried out, and the pass and fail criteria. To ensure that each test is a true representation of the original intent of the test, a validation and certification process has been set in place.

Once the manufacturer is satisfied that the test operates correctly, it then is given to an independent validation organization to test for conformance with the original test specification and check for proper operation. Once it has passed this test, it can be submitted to the relevant industry body for certification. After certification, it can be used in formal handset testing. The 3GPP has overall control of the test cases for GSM and UMTS. Changes, however, are handled by the GERAN Working Group for GSM. For UMTS, changes are addressed by the RAN5 Group. The validation and approval of the implemented test cases then are handled by the GCF.

Chapter Two

UMTS Conformance Protocol Testing – Part One

PROTOCOL CONFORMANCE TESTING

The process of verifying that an implementation performs in accordance (conforms) to a particular standard/specification. In this case it is the 3GPP standards.

This type of testing is used throughout the initial stages of development to ensure the accuracy of a protocol implementation. It is also used in regression testing after initial product deployment to further verify any changes in the implementation/enhancement. Traditionally, conformance testing has been the domain of the telecommunications industry. The only way to ensure that standards are met is to test products in an effective way using the test specification.

In general, protocol conformance testing is appreciated in the telecommunications world. At ETSI, protocol conformance testing specifications have dominated the testing activities and will continue to do so in the future.

Conformance testing is able to determine whether the behaviour of an implementation conforms to the requirements laid out in its base specification including the full range of error and exception conditions which can only be induced or replicated by dedicated test equipment. It exercises most, if not all, of the possible ways of achieving each of a component's function. On the other hand, conformance testing does not prove end-to-end interoperability of functions between the two similar communicating systems, and does not exercise all system components and their interfaces together to determine whether the implementation works in a real-life environment. Also it does not prove the operation of proprietary features, functions, interfaces, and systems that are not in the public domain. However, these proprietary facilities may be exercised indirectly as part of the configuration or execution of the conformance tests.

The testers, which in a real testing environment may be distributed, execute test programs or scripts which are called Test Cases. The entire set of Test Cases is known as a Test Suite. ETSI develops Abstract Test Suites (ATS) written in the standardized testing language TTCN which can be compiled and run on a variety of real test systems.

In each case there may be different conformance test suites for the different protocols (components) that make up the products. At no time does an individual test suite test the product as a complete system.

While originally targeted at protocol testing, this methodology can be applied to other reactive systems such as services and APIs. For simplicity, we shall mainly consider conformance testing of protocols.

The Abstract Test Suite (ATS) is the entire collection of Test Cases. Each Test Case specifies the detailed coding of the Test Purposes, written usually in a test specification language such as the standardized TTCN2 or TTCN3.

The Executable Test Suite (ETS) can be quickly and easily implemented from the ATS using the TTCN compilers available on most modern test tool platforms (C++, Java etc.). Runtime support, such as message encoders/decoders, test control and adaptation layers is need to execute the tests in a real test system.

UE PROTOCOL TESTING

Variety of areas that need to be addressed when test program is to be arranged; protocols, RF performance, USIM, audio performance and RF. Protocol testing determines the overall operation of the UE. If the protocol software operates incorrectly, then the UE will not operate properly on a network. There even have been instances when the incorrect operation of UE has been the cause of problems with a network. Due the complexity of the protocols that are used, this testing can be very involved and requires the use of sophisticated network simulators. These system simulators emulate a variety of network entities, such as Node-Bs, RNCs. Scenarios from registering to terminating a call and all the different forms of handovers can be simulated, i.e. any situation that can be encountered in real operation can be fully tested.

The purpose of this chapter is to give a brief introduction to the conformance protocol test cases, and in way the intention is to fully explain these tests. For more details about these test refer to the relevant 3GPP specifications as listed at the end of the chapter.

UE in Idle mode (Single Radio Access Technology)
Public Land UE Network Selection
The following tests are an Idle Mode tests in single RAT FDD systems. Manual and automatic selection of the available PLMN is also covered. Test model is illustrated below in Figure 2.1.

Manual Selection of PLMN
PLMN selection of RPLMN, HPLMN, UPLMN and OPLMN. This is to make sure that the UE can provide the available and prioritised PLMNs list to the user when requested to do so in manual mode. Also the displayed PLMNs can be selected or reselected by the user. The RPLMN shall be selected at switch-on if

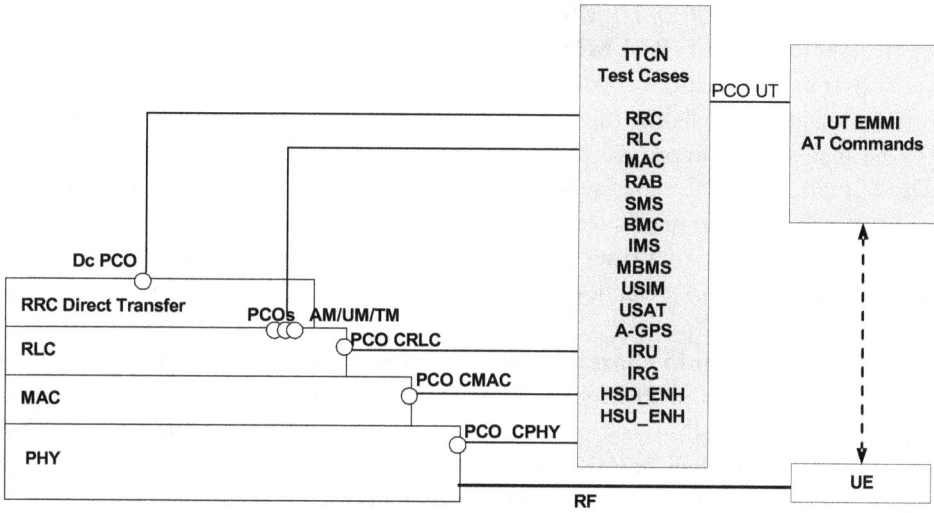

Figure 2.1   Layers test model (curtsey of 3GPP)

available, otherwise the displayed list shall include in priority order HPLMN, User-PLMN and Operator-PLMN. The last priority in the list is Other PLMN/RAT combinations

PLMN selection of Other PLMN / RAT combinations. UE should be able to present the available high quality signal PLMNs in random order to the user when requested to do so manually and that the displayed PLMNs can be selected or reselected by the user. Forbidden PLMNs shall also by displayed in the list.

PLMN selection; independence of RF level and preferred PLMN. The UE should be able to obtain normal service on a PLMN which is neither the better nor a preferred PLMN and that it tries to obtain service on a VPLMN if and only if the user selects it manually

Cell reselection of ePLMN. Test to verify that the UE shall be able to reselect to a cell of another PLMN declared as equivalent PLMN to the registered PLMN in the manual mode.

PLMN selection in shared network environment. Test to verify that the UE can present the available PLMNs in priority order to the user when asked to do so in manual mode and that the displayed PLMNs can be selected / reselected by the user. The test is performed in shared network environment. All PLMNs shall be displayed in the list, including forbidden PLMNs and PLMNs in the "Multiple PLMN list". If available, the RPLMN shall be selected at switch-on, otherwise the displayed list shall include in priority order HPLMN, User-PLMN, Operator-PLMN and "Other PLMN/access technology combinations. Only UTRAN cells and a UE equipped with a USIM with Radio Access Technology fields set to UTRAN are considered.

Automatic selection of PLMN

PLMN selection of RPLMN, HPLMN, UPLMN and OPLMN. Test to verify that in Automatic Network Selection Mode, the UE selects PLMNs in a prioritized order. If available, the RPLMN shall be selected at switch-on, otherwise the list shall include in priority order HPLMN, User-PLMN and Operator-PLMN. The last priority in the list is "Other PLMN/access technology combinations" which is not included in this test.

PLMN selection of Other PLMN / RAT combinations. Test to verify that in Automatic Network Selection Mode, the UE selects high quality signal PLMNs in a random order.

PLMN selection in shared network environment. Test to verify that when operating in automatic PLMN selection mode in shared network environment, the UE selects PLMNs in a prioritized order. Forbidden PLMNs shall not be selected. RPLMN shall be selected at switch-on, otherwise the list shall include in priority order HPLMN, User-PLMN and Operator-PLMN. The last priority in the list is "Other PLMN/access technology combinations" which is not included in this test.

Cell Selection and Re-Selection

Cell reselection. Test to verify that the UE performs the cell reselection correctly for intra/inter-frequency cells if the serving cell becomes barred or S<0 (where S is the Cell Selection value in dB).

Cell reselection for inter-band operation. Here the cell to be reselected is on a different frequency band.

Cell reselection using Qhyst, Qoffset and Treselection. Test to verify that the UE performs the cell reselection correctly if system information parameters Qoffset, Qhyst and Treselection are applied for non-hierarchical cell structures. TEMP_OFFSET and PENALTY_TIME are only applicable when HCS is applied.

HCS Cell reselection. Test to verify that the UE performs the cell reselection correctly for hierarchical cell structures. This shall be done according to the HCS priority, the received signal quality value Q and the quality level threshold criterion H.

HCS Cell reselection using reselection timing parameters for the H criterion. Test to verify that the UE performs the cell reselection correctly for hierarchical cell structures using TEMP_OFFSET and PENALTY_TIME applied to the H criterion.

HCS Cell reselection using reselection timing parameters for the R criterion. Test to verify that the UE performs the cell reselection correctly for hierarchical cell structures using TEMP_OFFSET and PENALTY_TIME applied to the R criterion.

Emergency calls. Test to verify that the UE shall be able to initiate emergency calls when no suitable cells of the selected PLMN are available, but at least one acceptable cell is available.

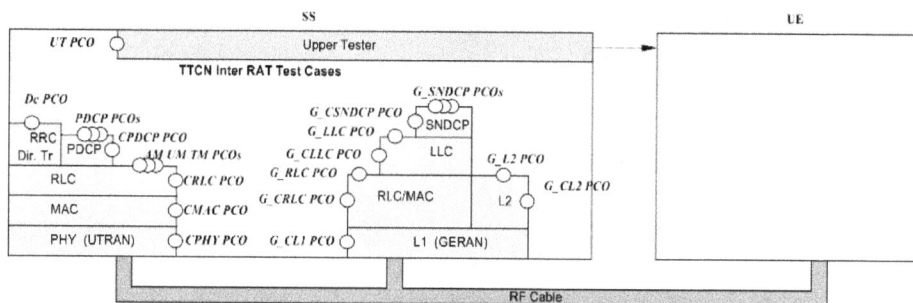

Figure 2.2 ISHO test model

Cell reselection: Equivalent PLMN. Test to verify that the UE performs the cell reselection correctly to a cell belonging to a PLMN Equivalent to the registered PLMN, if the serving cell of registered PLMN becomes barred or S<0.

Cell reselection using cell status and cell reservations – Type "A" and Type "B" USIM. Test to verify that the UE correctly interprets cell status and cell reservations when performing cell reselection.

HCS inter-frequency cell reselection. Test to verify that the UE performs inter-frequency cell reselection correctly for hierarchical cell structures in FDD. This shall be done according to the received signal quality value Q and the quality level thresholds Sintrasearch, Sintersearch and SsearchHCS.

Cell reselection in shared network environment. Test to verify that the UE performs cell reselection in a shared network environment (FDD). Correct handling of equivalent PLMNs is also verified.

Idle Mode UE and dual RAT

This section covers Inter RAT in idle mode situations, i.e. PLMN selection, RAT selection, cell selection/reselection. The test model used for TTCN test case implementation is depicted in Figure 2.2.

PLMN and RAT Selection

Selection of the correct PLMN and associated RAT. Test to verify that the UE selects the correct combination of PLMN and associated access technology according to the fields on the USIM.

Selection of RAT for HPLMN; Manual mode. Test to verify that the UE selects the HPLMN RAT according to the HPLMN RAT priority list on the USIM. If no RAT on the list is available, the UE shall try to obtain registration on the same PLMN using other UE-supported RATs.

Selection of RAT for UPLMN; Manual mode. Test to verify that the UE selects the UPLMN RAT according to the UPLMN RAT priority list on the USIM. If no PLMN/RAT on the UPLMN RAT priority list is available then the UE shall search for PLMNs in the OPLMN list.

Selection of RAT for OPLMN; Manual mode. Test to verify that the UE selects the OPLMN RAT according to the OPLMN RAT priority list on the USIM. If no PLMN/RAT on the OPLMN RAT priority list is available then the UE shall search for other PLMN/access technology combinations with received high quality signal in random order

Selection of "Other PLMN / RAT combinations"; Manual mode. Test to verify that if neither RPLMN, HPLMN, UPLMN nor OPLMN is available, the UE tries to obtain registration on "Other PLMN/access technology combinations with received high quality signal in random order". Forbidden PLMNs shall also by displayed in the list.

Selection of RAT for HPLMN; Automatic mode. Test to verify that the UE selects the HPLMN RAT according to the HPLMN RAT priority list on the USIM. If no RAT on the list is available, the UE shall try to obtain registration on the same PLMN using other UE-supported RATs.

Selection of RAT for UPLMN; Automatic mode. Test to verify that the UE selects the UPLMN RAT according to the UPLMN RAT priority list on the USIM. If no PLMN/RAT on the UPLMN RAT priority list is available then the UE shall search for PLMNs in the OPLMN list.

Selection of RAT for OPLMN; Automatic mode. Test to verify that the UE selects the OPLMN RAT according to the OPLMN RAT priority list on the USIM. If no PLMN/RAT on the OPLMN list is available then the UE shall search for other PLMN/access technology combinations with received high quality signal in random order.

Selection of "Other PLMN / access technology combinations"; Automatic mode. Test to verify that if neither RPLMN, HPLMN, UPLMN nor OPLMN is available, the UE tries to obtain registration on "Other PLMN/access technology combinations with received high quality signal in random order".

Selection of PLMN and RAT in shared network environment, Automatic mode. Test to verify that the UE selects the correct combination of PLMN and associated access technology according to the fields on the USIM in a shared network environment.

Selection of PLMN and RAT in shared network environment, Manual Mode. Test to verify that the UE can present the available PLMNs in priority order to the user when asked to do so in multi-mode Shared Network environment (Manual Mode). This includes PLMNs broadcasted in UTRAN cells (IE "PLMN Identity", UTRAN IE "Multiple PLMN list") and in GSM cells. The test also verifies that displayed PLMNs can be selected by the user and that the UE attempts to access the selected PLMN. After successful registration, the selected PLMN shall be displayed to the user.

Cell Selection and Reselection

Cell reselection if cell becomes barred or S<0; UTRAN to GSM. Test to verify that if both a GSM and UTRAN network is available, the UE performs cell reselection from UTRAN to GSM if the UTRAN cell becomes barred or S falls below zero.

Cell reselection if cell becomes barred or C1<0; GSM to UTRAN. Test to verify that if both a GSM and UTRAN network is available, the UE performs cell reselection from GSM to UTRAN if the GSM cell becomes barred or the path loss criterion C1 falls below zero for a period of 5 s.

Cell reselection timings; GSM to UTRAN. Test to verify that the UE meets the cell reselection timing requirements when both a GSM and UTRAN network is available.

Cell reselection in multi-mode shared network environment. Test to verify that the UE performs cell reselection in multi-mode shared network environment (FDD). It is verified that the UE reselects from a shared UTRAN cell in case of a higher ranked GSM cell or if the serving UTRAN cell becomes unsuitable. It is also verified that the UE is able to return to a shared UTRAN cell in case of loss of GSM coverage.

Cell reselection using SIB18; UTRAN to GSM. Test to verify that the UE performs cell reselection in multi-mode environment (FDD) when SIB18 present. It is verified that the UE reselects from a UTRAN cell to the GSM cell allowed by SIB18, even if other GSM cells are better ranked.

Packet Measurement order procedure / DL transfer / Normal case/ 3G cell reselection dedicated parameters. Test to confirm that the 3G search parameters and neighbour cell description are correctly used by the in order to reselect a 3G cell. Test to confirm that the individual parameters are used by the MS instead of broadcast 3G cell reselection parameters when the MS receives a PACKET MEASUREMENT ORDER message.

Intersystem Cell Reselection/Idle Mode/FDD_Qmin. Test to confirm that the MS uses the FDD_Qmin parameter at cell re-selection from GSM to UTRAN while in GSM Idle Mode.

Intersystem Cell Reselection/Idle Mode/FDD_Qoffset. Test to confirm that the MS uses the FDD_Qoffset parameter at cell re-selection from GSM/ GPRS to UTRAN while in Idle Mode.

Intersystem Cell Reselection/Idle Mode/Qsearch_I. Test to confirm that the MS uses the Qsearch_I parameter at cell re-selection from GSM to UTRAN while in Idle Mode.

Test of Measurement Report

Enhanced Measurement /all neighbours present. To test that, when the SS gives information about neighbouring cells, the MS reports appropriate results.

Test of Classmark

Classmark interrogation / UTRAN Classmark Change. Test to confirm that if the network requests the MS to supply all its classmark information, including the UTRA Classmark information, then this information is communicated on the DCCH to the network.

Early UTRAN Classmark Sending. This procedure allows the network to request the MS to supply all its classmark information to the network. In addition the network may request a MS supporting UTRAN to send the UTRAN classmark information. Networks may systematically use this procedure (e.g. during location updating) and, it if is incorrectly implemented in the MS, the basic connection establishment procedure may systematically fail. Test to confirm that if the network requests the MS to supply all its classmark information, including the UTRAN Classmark information, then this information is communicated on the DCCH to the network. The request of the classmark information is indicated in SYSTEM INFORMATION TYPE 3.

DTM

Uplink TBF establishment with no reallocation of CS resources / Abnormal cases / Inter System to UTRAN Handover Command. Verifying that the MS aborts the Packet Access procedure and proceed with the handover to UTRAN, upon reception of an INTER SYSTEM TO UTRAN HANDOVER COMMAND message.

Measurement reports and Cell change order procedures

Inter-RAT Cell Change Order (Known Cell) – UL Data Transfer. Test to confirm the when NC2 is commanded, the MS sends PACKET ENHANCED MEASUREMENT REPORT messages, in which both the serving and non-serving cells are reported. Test to confirm that when the cell change order procedure is started by sending a PACKET CELL CHANGE ORDER message with the IMMEDIATE_REL value set to 1, the MS shall abort any TBF in progress and stop transmitting. Test to confirm that the MS switches to the commanded UTRAN cell.

Inter-RAT Cell Change Order (Unknown Cell) – UL Data Transfer. Test to confirm that when the cell change order procedure is started by sending a PACKET CELL CHANGE ORDER message with the IMMEDIATE_REL value set to 1, the MS shall abort any TBF in progress and stop transmitting. Test to confirm that the MS switches to the commanded UTRAN cell.

Inter-RAT Cell Change Order (Known Cell) – DL Data Transfer. Test to confirm that when the cell change order procedure is started when the MS receives PACKET CELL CHANGE ORDER message with the IMMEDIATE_REL value set to 1. Test to confirm that the MS switches to the commanded UTRAN cell.

Inter-RAT Cell Change Order (Known Cell) – Simultaneous UL and DL transfer. Test to confirm that when the cell change order procedure is started by sending a PACKET CELL CHANGE ORDER message with the IMMEDIATE_ REL value set to 1, the MS shall abort any TBF in progress and stop transmitting. Test to confirm that the MS switches to the commanded UTRAN cell.

Inter-RAT (GPRS to UTRAN) Cell Change Order (Known cell) / Failure / UL transfer / T3174 expiry. Test to confirm that an MS in uplink packet transfer mode when commanded by a PACKET CELL CHANGE ORDER, reports a

PACKET CELL CHANGE FAILURE on the old GPRS cell due to the expiry of T3174 before receiving response to RRC CONNECTION REQUEST on target UTRAN cell.

Inter-RAT (GPRS to UTRAN) Cell Change Order (Known cell) / Failure/ DL transfer / REJECT from target UTRAN cell with Inter-RAT info set to GSM. Test to confirm that an MS in downlink packet transfer mode when commanded by a PACKET CELL CHANGE ORDER, reports PACKET CELL CHANGE FAILURE on the old GPRS cell if an RRC CONNECTION REJECT with Inter-RAT info set to GSM is received on the target UTRAN cell.

Handover to UTRAN while in DTM

Handover to UTRAN while in DTM / DL TBF. Verifying that the MS aborts Packet resources while in DTM and proceeds with the handover to UTRAN, upon reception of an INTER SYSTEM TO UTRAN HANDOVER COMMAND message.

Handover to UTRAN while in DTM / UL TBF. Verifying that the MS aborts Packet resources while in DTM and proceeds with the handover to UTRAN, upon reception of an INTER SYSTEM TO UTRAN HANDOVER COMMAND message.

Inter-System Hard Handover from GSM to UTRAN

ISHO to UTRAN/From GSM/Speech/Success. To test that MS supporting both GSM and UTRAN hands over to the indicated channel in the UTRAN target cell when it is in the speech call active state in the GSM serving cell and receives a INTERSYSTEM TO UTRAN HANDOVER COMMAND. It is also verified that the MS performs measurements on the target UTRAN cell before the handover.

ISHO to UTRAN/From GSM/Data/Same data rate/Success. To test that the MS hands over to the indicated UTRAN target cell and the data rate of the target channel is the same as the old channel when it is in the data call active state in the GSM serving cell and receives a INTER SYSTEM TO UTRAN HANDOVER COMMAND. It is also verified that the MS performs measurements on the target UTRAN cell before the handover.

ISHO to UTRAN/From GSM/Data/Same data rate/Extended Rates/ Success. To test that the MS hands over to the indicated UTRAN target cell and the data rate of the target channel is the same as the old channel when it is in the data call active state in the GSM serving cell and receives a INTER SYSTEM TO UTRAN HANDOVER COMMAND. It is also verified that the MS performs measurements on the target UTRAN cell before the handover.

ISHO to UTRAN/From GSM/Data/Data rate upgrading/Success. To test that the MS being in the data call active state hands over from the GSM serving cell to the indicated channel of a higher data rate in the UTRAN target cell after it receives a INTER SYSTEM TO UTRAN HANDOVER COMMAND.

It is also verified that the MS performs measurements on the target UTRAN cell before the handover.

ISHO to UTRAN/From GSM/Data/Data rate upgrading/Extended Rates/Success. To test that the MS being in the data call active state hands over from the GSM serving cell to the indicated channel of a higher data rate in the UTRAN target cell after it receives a INTER SYSTEM TO UTRAN HANDOVER COMMAND. It is also verified that the MS performs measurements on the target UTRAN cell before the handover.

ISHO to UTRAN/From GSM/SDCCH/CC Establishment/Success. To test that the MS supporting both GSM and UTRAN handovers from the GSM serving cell to the indicated channel in UTRAN target cell when the MS is on SDCCH during call establishment phase and receives an INTER SYSTEM TO UTRAN HANDOVER COMMAND. It is also verified that the MS performs measurements on the target UTRAN cell before the handover.

ISHO to UTRAN/From GSM/Speech/Blind HO/Success. To test that the MS handovers from the GSM serving cell to the indicated channel of UTRAN target cell when it is in the speech call active state without any knowledge of the target system (blind handover) and receives an INTER SYSTEM TO UTRAN HANDOVER COMMAND.

ISHO to UTRAN/From GSM/Speech/Failure. To test that the MS reactivates the old channel and transmits HANDOVER FAILURE message to the network on the old channel in the GSM cell when it received INTER SYSTEM TO UTRAN HANDOVER COMMAND towards a non-existing UTRAN cell.

ISHO to UTRAN/From GSM/Failure/Cause: Frequency not implemented. To test that the MS reactivates the old channel and transmits HANDOVER FAILURE message to the network on the old channel in the GSM cell when it received HANDOVER TO UTRAN COMMAND and the handover to UTRAN failed because of frequency not implemented.

ISHO to UTRAN/From GSM/Failure/Cause: UTRAN configuration unknown. To test that the MS reactivates the old channel and transmits HANDOVER FAILURE message to the network on the old channel in the GSM cell when it receives HANDOVER TO UTRAN COMMAND and the handover to UTRAN failed because of UTRAN configuration unknown.

ISHO to UTRAN/From GSM/Failure/Cause: Protocol Error. To test that the MS reactivates the old channel and transmits RR Status message to the network on the old channel in the GSM cell when it receives HANDOVER TO UTRAN COMMAND and the handover to UTRAN failed because of protocol error.

ISHO to UTRAN/From GSM/Integrity Protection Activation. To test that MS supporting both GSM and UTRAN applies the correct CS Security START value after a successful handover from GSM to UTRAN when Integrity protection is activated by the NW.

Layer 2 – Medium Access Control (MAC)

This is section concentrates on the testing aspects of L2 and its sub-layers and channels. Figure 2.3 is provided to remind you of the position of the layers in the overall architecture and its inter-relationship.

Testing of Logical – Transport Channels Mapping

CCCH mapped to RACH/FACH / Invalid TCTF. This tests that the MAC applies the correct header to the MAC PDU according to the type of logical channel carried on the RACH/FACH transport channel. Incorrect application of MAC headers would result in wrong operation of the UE.

DTCH or DCCH mapped to RACH/FACH / Invalid TCTF. This tests that the MAC applies the correct header to the MAC PDU according to the type

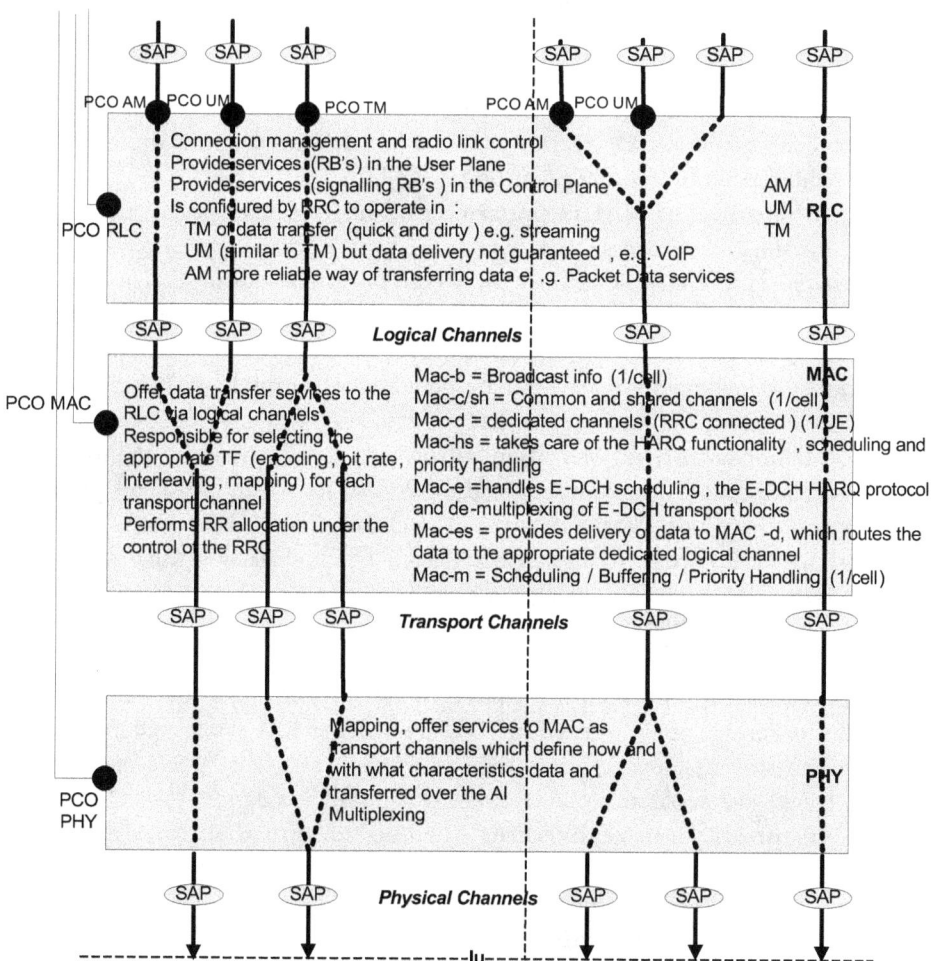

Figure 2.3   Diagram illustrates RLC/MAC and Physical layers

of logical channel carried on the RACH/FACH transport channel. Incorrect application of MAC headers would result in wrong operation of the UE.

DTCH or DCCH mapped to RACH/FACH / Invalid C/T Field. This tests that the MAC applies the correct header to the MAC PDU according to the type of logical channel carried on the RACH/FACH transport channel. Incorrect application of MAC headers would result in wrong operation of the UE.

DTCH or DCCH mapped to RACH/FACH / Invalid UE ID Type Field. This tests that the MAC applies the correct header to the MAC PDU according to the type of logical channel carried on the RACH/FACH transport channel. Incorrect application of MAC headers would result in wrong operation of the UE.

DTCH or DCCH mapped to RACH/FACH / Incorrect UE ID. This tests that the MAC applies the correct header to the MAC PDU according to the type of logical channel carried on the RACH/FACH transport channel. Incorrect application of MAC headers would result in wrong operation of the UE.

DTCH or DCCH mapped to DCH / Invalid C/T Field. This tests that the MAC applies the correct header to the MAC PDU according to the type of logical channel carried on the DCH transport channel. Incorrect application of MAC headers would result in wrong operation of the UE.

Correct Selection of RACH parameters. This is to test the UE determines the ASC for the given Access Class (AC)

Access Service Class selection for RACH transmission. Test to confirm that MAC selects ASC correctly.

Priority handling between data flows of one UE. Test to confirm that the UE prioritise signalling compared to data on a lower priority logical channel.

TFC Selection. Test to confirm that the UE supports a TFCS that does not allow simultaneous transmission of max data rate on all transport channels and that the UE selects a TFC according to the rule that no other TFC shall allow the transmission of highest priority data than the chosen TFC. Also to verify that the UE selects a TFC according to the rule that no other TFC shall allow the transmission of more data from the next lower priority logical channels.

High Speed - DSCH Layer 2 / MAC-hs

MAC-hs reordering and stall avoidance. Test to confirm that the UE performs MAC-hs reordering and delivers RLC PDUs in order to RLC. Also to confirm that the UE performs stall avoidance in case of missing MAC-hs PDUs based on a) window based stall avoidance and b) timer based stall avoidance.

MAC-hs priority queue handling. Test to confirm that the UE handles several priority queues, where different radio bearers are mapped to different queues.

MAC-hs PDU header handling. Test to confirm that the UE discards PDUs with reserved values of the fields in the MAC header. Test to confirm that the UE discards PDUs with values in the MAC header that are inconsistent with the RRC configuration. Test to confirm that the UE correctly reads the MAC header and disassembles the MAC-hs PDU into MAC-d PDUs and delivers the MAC-d PDUs to the RLC layer.

MAC-hs retransmissions. Test to confirm that the UE correctly transmit positive and negative acknowledgements when receiving MAC-hs PDUs

MAC-hs reset. Test to confirm that the UE flushes the reordering buffer and delivers all MAC-d PDUs in the buffer to higher layers upon reset. Test to confirm that the UE initializes the TSN and next_expected_TSN to their initial values. Test to confirm that the UE sends an RLC status report after the reset.

MAC-hs transport block size selection. Test to confirm that the UE selects the correct transport block size based on the TFRI value signalled on the HS-SCCH.

Enhanced-DCH Layer 2 / MAC-es/e

MAC-hs reordering and stall avoidance. Test to confirm that the UE performs MAC-hs reordering and delivers RLC PDUs in order to RLC. Test to confirm that the UE performs stall avoidance in case of missing MAC-hs PDUs based on a) window based stall avoidance and b) timer based stall avoidance.

MAC-es/e multiplexing without RRC restrictions. The purpose of this test case is to verify that the UE multiplexes data from different logical channels in the same TTI when no restriction on the multiplexing is configured by RRC.

MAC-es/e multiplexing with RRC restrictions. The purpose of this test case is to verify that the UE does not multiplex data from different logical channels in the same TTI when the multiplexing has been restricted by RRC.

Correct settings of MAC-es/e header fields. The purpose of this test case is to verify that the UE sets the MAC-es/e header fields in a correct way.

Correct settings of MAC-es/e scheduling information. The purpose of this test case is to verify that the UE sends the E-DCH scheduling information with correct content and with correct triggers

Happy bit setting. The purpose of this test case is to verify that the UE sets the happy bit correctly.

MAC-es/e non-scheduled transmissions. Test to confirm that the UE when RRC is configured for non-scheduled transmissions sends data without scheduling grant. Test to confirm that no SI is sent for non-scheduled transmissions.

MAC-es/e correct handling of scheduled transmissions when absolute grant varies. Test to confirm that the UE transmits different amount of data when the absolute grant varies.

MAC-es/e de-activation and re-activation of HARQ processes. The purpose of this test case is to verify the selective de-activation and re-activation of a HARQ process

MAC-es/e correct handling of relative grants. The purpose of this test case is: Test to verify that the UE acts on serving and non-serving relative grants when the UE is using the Primary E-RNTI. Test to confirm that the UE only acts on non-serving relative grants when the UE is using the Secondary E-RNTI. Test to confirm that the UE does not use a Serving_Grant value greater than Maximum_Serving_Grant when the Non_Serving_RG_Timer has not expired.

MAC-es/e correct handling of absolute grants on Primary and Secondary E-RNTI. The purpose of this test case is to verify that the UE acting on absolute grants given on the Primary and Secondary E-RNTI and switching between the two cases.

MAC-es/e combined non-scheduled and scheduled transmissions. Test to confirm that the UE is able to handle combined non-scheduled and scheduled transmissions.

MAC-es/e Correct handling of HARQ profile power offsets. The purpose of this test case is to verify that the UE applies different HARQ profiles from different MAC-d flows to E-DCH transmissions accordingly and in case data from two MAC-d flows is transmitted in the same E-DCH transmission, the UE selects the correct power offset.

MAC-es/e Correct handling of minimum set of E-TFCI. The purpose of this test case is to verify that the UE considers the minimum set of E-TFCI as always supported.

MAC-es/e E-TFC priority. Test to confirm that the UE transmits data in order of priority

MAC-es/e transport block size selection. Test to confirm that the UE transmits all possible transport block sizes within the UE capability.

MAC-es/e process handling. Test to confirm that the UE performs transmissions and retransmissions in the correct MAC-es process. Test to confirm that the UE uses only the allowed HARQ processes for scheduled and non-scheduled transmissions.

MAC-es/e maximum number of retransmissions. Test to confirm that the UE, when 2 MAC d flows are multiplexed, follows the maximum number of retransmissions according to the HARQ profiles (different values configured for maximum number of retransmissions for each HARQ profile).

MAC-es/e Correct handling of MAC-es/e reset. The purpose of this test case is to verify that the UE correctly handles a MAC-es/e reset procedure.

Radio Link Control (RLC)

Segmentation and reassembly / 7-bit "Length Indicators" / Padding. The RLC segments and concatenates SDUs into UMD PDUs according to the PDU size requested by MAC. "Length Indicators" are added to allow correct reconstruction of SDUs. Incorrect operation of segmentation, concatenation, or coding of "Length Indicators" will result in failure of the UE to communicate. To test that the UE correctly segments a large SDU, includes a "Length Indicator" indicating padding in the RLC PDU carrying the last SDU segment, and adds padding at the end. To test that the UE correctly deals with a 7-bit padding "Length Indicator" when present in a received PDU.

Segmentation and Reassembly / 7-bit "Length Indicators" / LI = 0. The RLC segments and concatenates SDUs into UMD PDUs according to the PDU size requested by MAC. A pre-defined "Length Indicator" value is used to indicate when a SDU ends coincident with the end of the previous PDU. Incorrect operation of segmentation, concatenation, or coding of "Length Indicators" will result in failure of the UE to communicate. To test that where a SDU exactly fills a PDU, a "Length Indicator" of all 0's is placed by the transmitter as the first "Length Indicator" in the next PDU. To test that where a SDU exactly fills a PDU, the receiver accepts a "Length Indicator" of all 0's, placed as the first "Length Indicator" in the next PDU.

Reassembly / 7-bit "Length Indicators" / Invalid LI value. The RLC segments and concatenates SDUs into UMD PDU according to the PDU size requested by MAC. "Length Indicators" are added to allow correct reconstruction of SDUs. The behaviour of the RLC on reception of an invalid "Length Indicator" value has been specified. Incorrect operation of segmentation, concatenation, or coding of "Length Indicators" will result in failure of the UE to communicate. To test that PDUs with invalid "Length indicator" '111 1110' are discarded by the receiving RLC.

Reassembly / 7-bit "Length Indicators" / LI value > PDU size. The RLC segments and concatenates SDUs into UMD PDUs according to the PDU size requested by MAC. "Length Indicators" are added to allow correct reconstruction of SDUs. The behaviour of the RLC on reception of an invalid "Length Indicator" value has been specified. Incorrect operation of segmentation, concatenation, or coding of "Length Indicators" will result in failure of the UE to communicate. To test that PDUs with "Length Indicators" that point beyond the end of the PDU are ignored by the receiving RLC entity.

Reassembly / 7-bit "Length Indicators" / First data octet LI. The RLC segments and concatenates SDUs into UMD PDUs according to the PDU size requested by MAC. "Length indicators" are added to allow correct reconstruction of SDUs. A special "Length Indicator" is defined to indicate that the start of an SDU is coincident with the start of the PDU. The special "Length Indicator" is needed to avoid discarding of an SDU when the first received PDU has a sequence number different from zero. Incorrect operation of segmentation, concatenation, or coding of "Length Indicators" will result in failure of the UE to communicate. To test that a UE in unacknowledged mode correctly handles a received RLC PDU with a 7-bit "Length Indicator" having its value equal to the special "Length Indicator" value 1111100 when the sequence number of the first received PDU is different from zero.

Segmentation and Reassembly / 7-bit "Length Indicators" / LI = 0. The RLC segments and concatenates SDUs into AMD PDUs according to the PDU size configured by RRC. A pre-defined "Length Indicator" value is used to indicate when an SDUs ends coincident with the end of the previous PDU. Incorrect operation of segmentation, concatenation, or coding of "Length Indicators" will result in failure of the UE to communicate. To test that where an SDU exactly fills a PDU, an "Length Indicator" of all 0's is placed by the transmitter as the first "Length Indicator" in the next PDU. To test that where an SDU exactly fills a PDU, and an "Length Indicator" of all 0's is the first "Length Indicator" in the next PDU, the receiver correctly reassembles the SDU.

Reassembly / 7-bit "Length Indicators" / Reserved LI value. The RLC segments and concatenates SDUs into AMD PDUs according to the PDU size configured by RRC. "Length Indicators" are added to allow correct reconstruction of SDUs. The behaviour of the RLC on reception of a reserved "Length Indicator" value is specified in the conformance requirement below. Incorrect operation of segmentation, concatenation, or coding of "Length Indicators" will result in failure of the UE to communicate. To test that PDUs with reserved "Length Indicators" are discarded by the receiving RLC.

Reassembly / 7-bit "Length Indicators" / LI value > PDU size. The RLC segments and concatenates SDUs into AMD PDUs according to the PDU size configured by RRC. "Length Indicators" are added to allow correct reconstruction of SDUs. The behaviour of the RLC on reception of an invalid "Length Indicator" value where the value is too large is specified in the conformance requirement below. Incorrect operation of segmentation, concatenation, or coding of "Length Indicator" will result in failure of the UE to communicate. To test that PDUs with "Length Indicators" that point beyond the end of the PDU are discarded by the receiving RLC.

Correct use of Sequence Numbering. Peer RLC entities use sequence numbering to detect missing PDUs, and for flow control purposes. Incorrect operation of sequence numbering will result in failure of the UE to communicate. 1. Test to confirm that the UE transmits the first PDU with the Sequence Number field equal to 0. 2. Test to confirm that the UE increments the Sequence Number field according to the number of PDUs transmitted. 3. Test to confirm that the UE wraps the Sequence Number after transmitting the 212-1th PDU. 4. Test to verify that the UE receiver accepts PDUs with SNs that wrap around every 212-1th PDU.

Control of Transmit Window. This test is to check that the UE is able to correctly control its RLC transmission window. Correct operation of RLC windowing is critical for acknowledged mode operation. Test to confirm that the UE does not transmit PDUs with sequence numbers outside of the transmit window, except the PDU with SN=VT(S)-1, even when the transmit window size is changed by the receiver.

Control of Receive Window. This test is to check that the UE is able to correctly control its RLC receive window. Correct operation of RLC windowing is critical for acknowledged mode operation. Test to confirm that the UE discards PDUs with sequence numbers outside the upper boundary of the receive window.

Polling for status / Last PDU in transmission queue. This case tests that the UE will poll for a status request on the last PDU in its transmission queue when that mode is enabled. Incorrect operation of polling will cause degradation of service, or at worst service failure. Test to confirm that a poll is performed when only one PDU is available for transmission, and the poll prohibit timer function is not used. Test to confirm that a poll is performed when only one PDU is available for transmission, and the poll prohibit timer function is used, but inactive.

Polling for status / Last PDU in retransmission queue. This case tests that the UE will poll for a status request on the last PDU in its retransmission queue when that mode is enabled. Incorrect operation of polling will cause degradation of service, or at worst service failure. Test to confirm that a poll is performed when only one PDU is available for retransmission, and the poll prohibit timer function is not used. Test to confirm that a poll is performed when only one PDU is available for retransmission, and the poll prohibit timer function is used, but inactive.

Polling for status / Poll every Poll_PDU PDUs. This case tests that the UE will poll for a status request every Poll_PDU PDUs when that mode is enabled. Incorrect operation of polling will cause degradation of service, or at worst service failure. Test to confirm that a poll is performed when the state variable VT(PDU) reaches Poll_PDU. Test to confirm VT(PDU) is incremented for both new and retransmitted PDUs.

Polling for status / Poll every Poll_SDU SDUs. This case tests that the UE will poll for a status request every Poll_SDU SDUs when that mode is enabled. Incorrect operation of polling will cause degradation of service, or at worst service failure. Test to confirm that a poll is performed when VT(SDU) reaches Poll_ SDU. Test to confirm that the poll is sent in the last PDU of the SDU.

Polling for status / Timer triggered polling (Timer_Poll_Periodic). This case tests that the UE will poll for a status request every Timer_Poll_Periodic ms when that mode is enabled. Incorrect operation of polling will cause degradation of service, or at worst service failure. Test to confirm that the UE polls the SS in the next PDU to be transmitted or retransmitted each time the Timer_Poll_Periodic timer expires. Test to confirm that if there is no PDU to be transmitted or retransmitted, the timer is restarted, but no poll is sent.

Polling for status / Polling on Poll_Window% of transmission window. This case tests that the UE will poll for a status request when it has reached Poll_ Window% of the transmission window, when that mode is enabled. Incorrect operation of polling will cause degradation of service, or at worst service failure. Test to confirm that the UE polls the SS when the window based polling condition J >= Poll_Window is fulfilled.

Polling for status / Operation of Timer_Poll timer / Timer expiry. This case tests that the UE will retransmit a poll for status if it does not receive a STATUS PDU within Timer_Poll ms after a poll for status is transmitted. Incorrect operation of polling will cause degradation of service, or possible service failure. Test to confirm that if the timer expires and no STATUS PDU containing an acknowledgement or negative acknowledgement of the AMD PDUs up to that which triggered the timer has been received, the receiver is polled once more.

Polling for status / Operation of Timer_Poll timer / Stopping Timer_ Poll timer. This case tests that the UE will stop the Timer_Poll timer if it receives a STATUS PDU within Timer_Poll ms after a poll for status is transmitted. Incorrect operation of polling will cause degradation of service, or possible service failure. Test to confirm that the Timer_Poll timer is stopped when receiving a STATUS PDU that acknowledges all AMD PDUs with SN up to and including VT(S)-1 at the time the poll was transmitted.

Polling for status / Operation of Timer_Poll timer / Restart of the Timer_Poll timer. This case tests that the UE will restart the Timer_Poll timer if another poll request is transmitted whilst the timer is running. Incorrect operation of polling will cause degradation of service, or possible service failure. Test to confirm that if a new poll is sent when the timer is running it is restarted.

Polling for status / Operation of timer Timer_Poll_Prohibit. This case tests that the UE will not send a poll request within Timer_Poll_Prohibit ms of a previous poll request when this mode of operation is enabled. Incorrect operation of polling will cause degradation of service, or possible service failure. Test to confirm that no poll is transmitted if one or several polls are triggered when the Timer_Poll_Prohibit timer is active and has not expired. Test to confirm that the UE polls only once after Timer_Poll_Prohibit expires even though triggered several times during the prohibit time.

Receiver Status Triggers / Detection of missing PDUs. This case tests that the UE transmits a status report whenever it detects that a PDU is missing, if this mode of operation is enabled. Incorrect operation of status reporting will cause degradation of service, or possible service failure. Test to confirm that a status report is transmitted if there are one or more missing PDUs.

Receiver Status Triggers / Operation of timer Timer_Status_Periodic. This case tests that the UE transmits a status report every Timer_Status_Periodic ms when this mode of operation is enabled. Incorrect operation of status reporting will cause degradation of service, or possible service failure. Test to confirm that a status report is transmitted each time the Timer_Status_Periodic timer expires.

Receiver Status Triggers / Operation of timer Timer_Status_Prohibit. This case tests that the UE does not transmit a status report more often than every Timer_Status_Prohibit ms when this mode of operation is enabled. Incorrect operation of status reporting will cause degradation of service, or possible service failure. Test to confirm that a status report is not transmitted while the Timer_Status_Prohibit timer is active. Test to confirm that only one status report is sent on the expiry of the Timer_Status_Prohibit timer if several triggers occur while it is active.

Status reporting / Abnormal conditions / Reception of LIST SUFI with Length set to zero. Peer RLCs use STATUS PDUs to manage flow control and retransmission. On a STATUS report PDU with an invalid LIST SUFI the RLC must behave as specified. Incorrect behaviour may result in degradation of QoS, or failure of the UE to communicate. Test to confirm that if a STATUS PDU is received with a LIST SUFI and the LENGTH field is set to "0000" that the list is discarded.

SDU discard after MaxDAT-1 number of transmissions. This case tests that if a PDU is unsuccessfully transmitted MaxDAT-1 times, the SDU it carries, and therefore all other associated PDUs, are discarded by the transmitter and receiver. This mode of SDU discard is used to minimize data loss, and incorrect operation will effect the quality of service. Test to verify that if $VT(DAT) = MaxDAT$ for any PDU the sender initiates the SDU discard with explicit signalling procedure

Operation of the RLC Reset procedure / UE Originated. This case tests that when the maximum number of retransmissions is exceeded, the UE initiates and performs the RLC Reset procedure. Incorrect operation of this procedure may cause loss of service. Test to confirm that the Reset procedure is initiated when the maximum number of retransmissions has been exceeded (Reset trigger condition 1). Test to confirm that the sender resets state variables to their initial value and resets configurable parameters to their configured value. Test to confirm that RSN is updated correctly. Test to confirm operation of Timer_RST.

Operation of the RLC Reset procedure / UE Terminated. This case tests that the UE responds correctly to the RLC Reset procedure initiated by the network. Incorrect operation of this procedure may cause loss of service. Test to confirm that upon reception of a RESET PDU the receiver responds with a RESET ACK PDU. Test to confirm that the receiver resets its state variables to their initial value and resets configurable parameters to their configured value.

Reconfiguration of RLC parameters by upper layers. This case verifies the UE behaviour after a reconfiguration of RLC parameters on an established RLC AM entity. Test to confirm that the UE starts to use the new set of RLC parameters when an already established AM RLC radio bearer is reconfigured.

MBMS

MCCH RLC Re-establishment. Test to confirm MCCH RLC re-establishment.

MTCH duplicate avoidance and reordering. Test to confirm the MTCH duplicate avoidance and reordering procedure

MCCH Out Of Sequence Delivery handling. Test to confirm MCCH Out Of Sequence Delivery handling.

Packet Data Convergence Protocol (PDCP)

IP Header Compression and PID assignment. The first test procedure verifies, that the "PDCP Data" PDU is used for uncompressed IP header packets, if no IP header compression is configured by higher layers. The second test procedure verifies, that the "PDCP No header" PDU is used for uncompressed IP header packets, if no IP header compression is configured by higher layers. Test to confirm, that the UE transmits and receives in acknowledged mode (RLC AM) TCP/IP and UDP/IP data packets without IP header compression as configured by higher layers. Test to confirm, that PID assignment rules are correctly applied, if usage of "PDCP Data" PDU are negotiated, i.e. the UE shall recognize PID value = 0 for a received TCP/IP and UDP/IP data packet and it shall use PID=0 to transmit IP data packets, if no IP header compression is negotiated. If usage of "PDCP No Header" PDU is negotiated, no PID assignment is used for transmitting and receiving TCP/IP and UDP/IP data packets.

Transmission of compressed Header when the UE in RLC AM. The UE shall be capable to deal with compressed TCP/IP and UDP/IP data packets and furthermore to establish a PDCP entity which applies IP header compression protocol RFC 2507. Test to confirm, that the UE transmits and receives in acknowledged mode (RLC AM) TCP/IP and UDP/IP data packets by using IP header compression protocol as described in RFC2507 as configured by higher layers. Test to confirm, that the PID assignment rules are correctly applied by the UE. The UE as shall use the correct PID value for the applied optimisation method for transmitting and receiving TCP/IP and UDP/IP data packets.

Transmission of uncompressed Header when the UE in RLC UM. The first test procedure verifies, that the "PDCP Data" PDU is used for uncompressed IP header packets, if no IP header compression is configured by higher layers. The second test procedure verifies, that the "PDCP No header" PDU is used for uncompressed IP header packets, if no IP header compression is configured by higher layers. Test to confirm, that the UE transmits and receives in unacknowledged mode (RLC UM) TCP/IP and UDP/IP data packets without IP header compression as configured by higher layers. Test to confirm, that PID assignment rules are correctly applied, if usage of "PDCP Data" PDU are negotiated, i.e. the UE shall recognize PID value = 0 for a received TCP/IP and UDP/IP data packet and it shall use PID=0 to transmit IP data packets, if no IP header compression is negotiated. If usage of "PDCP No Header" PDU is negotiated, no PID assignment is used for transmitting and receiving TCP/IP and UDP/IP data packets.

Transmission of compressed Header when the UE in RLC UM. The UE shall be capable to deal with compressed TCP/IP and UDP/IP data packets and furthermore to establish a PDCP entity which applies IP header compression protocol RFC 2507. Test to confirm, that the UE transmits and receives in unacknowledged mode (RLC UM) TCP/IP and UDP/IP data packets by using IP header compression protocol as described in RFC2507 as configured by higher layers. Test to confirm, that the PID assignment rules are correctly applied by the UE. The UE as shall use the correct PID value for the applied optimisation method for transmitting and receiving TCP/IP and UDP/IP data packets.

Extension of used compression methods when the UE in RLC UM. Test to confirm, that the UE is able to handle an extended PID value allocation table by header compression protocol IETF RFC 2507 after PDCP reconfiguration as configured by RRC.

Compression type used for different entities when the UE in RLC UM. Test to confirm, that a configured IP header compression protocol are applied to compress and decompress TCP/IP data packets by several PDCP entities in parallel, if more than one entities are established, i.e. the UE uses the same PID to transmit two TCP/IP data packets with the same content in parallel using two Radio Bearer configurations.

Reception of not defined PID values. Test to confirm, that a UE considers a received PDCP PDU message with not defined PID value as invalid, i.e. such an invalid PDCP PDU is not forwarded to the Radio Bearer entity on UE side. Therefore the UE using test loop mode 1 does not return such data packet to the SS.

Data transmission if lossless SRNS Relocation is supported. Test to confirm, that a UE supporting lossless SRNS relocation is able to receive and to send IP data packets by using PDCP Sequence Numbering as configured by higher layers.

Synchronisation of PDCP sequence numbers. Test to confirm, that the UE supporting lossless SRNS relocation as configured by higher layers is able to handle the "PDCP SeqNum" PDU to synchronize the used PDCP Sequence Number after reconfiguration of the Radio Bearer.

UTRAN MOBILITY INFORMATION: Lossless SRNS relocation in CELL_FACH (without pending of ciphering). Test to confirm that the UE that support lossless SRNS relocation, sends the correct expected downlink PDCP sequence number to SS after a successful SRNS relocation. Test to confirm that the UE sends calculated START values for each CN domain to SS after a successful SRNS relocation. In the case that ciphering is applied by the network, to confirm that the UE applies the new ciphering algorithm following a successful SRNS relocation.

Cell Update: Lossless SRNS relocation in CELL_FACH (without pending of ciphering). Test to confirm that the UE executes a cell update procedure after the successful reselection of another UTRA cell. Test to confirm that the UE that support lossless SRNS relocation, sends the correct expected downlink PDCP sequence number to SS after a successful SRNS relocation. Test to confirm that the UE sends calculated START values for each CN domain to SS after a successful SRNS relocation. In the case that ciphering is applied by the network, to confirm that the UE applies the new ciphering algorithm following a successful SRNS relocation.

URA Update: Lossless SRNS relocation in CELL_FACH (without pending of ciphering). Test to confirm that the UE executes a URA update procedure after the successful reselection of another UTRA cell. Test to confirm that the UE that support lossless SRNS relocation, sends the correct expected downlink PDCP sequence number to SS after a successful SRNS relocation. Test to confirm that the UE sends calculated START values for each CN domain to SS after a successful SRNS relocation.

Radio Bearer Establishment for transition from CELL_DCH to CELL_DCH: Success (Lossless SRNS relocation) (without pending of ciphering).

Test to confirm that the UE performs a combined hard handover and SRNS relocation and then transmit a RADIO BEARER SETUP COMPLETE message in the new cell. In the case that ciphering is applied by the network, to confirm that the UE applies the new ciphering algorithm following a successful SRNS relocation.

Radio Bearer Reconfiguration for transition from CELL_DCH to CELL_DCH: Success (Lossless SRNS relocation) (without pending of ciphering). Test to confirm that the UE performs a combined hard handover and SRNS relocation and then transmit a RADIO BEARER RECONFIGURATION COMPLETE message in the new cell. In the case that ciphering is applied by the network, Test to confirm that the UE applies the new ciphering algorithm following a successful SRNS relocation.

Radio Bearer Release for transition from CELL_DCH to CELL_DCH: Success (Lossless SRNS relocation) (without pending of ciphering). Test to confirm that the UE performs a combined hard handover and SRNS relocation and then transmit a RADIO BEARER RELEASE COMPLETE message in the new cell.

Transport Channel Reconfiguration for transition from CELL_DCH to CELL_DCH: Success (Lossless SRNS relocation) (without pending of ciphering). Test to confirm that the UE performs a combined hard handover and SRNS relocation and then transmit a TRANSPORT CHANNEL RECONFIGU-RATION COMPLETE message in the new cell.

PDCP Robust Header Compression (RoHC)

Base test of ROHC RTP O-mode compressor. The purpose of the base test is to verify that the compressor implements an active and efficient compression for a regular IP/UDP/RTP packet stream, i.e. that it makes use of the most efficient compressed packet formats provided by ROHC RTP for O-mode. Test to confirm that the ROHC compressor successfully transfers to O-mode operation and makes use of efficient compressed packet formats available to ROHC RTP.

Base test of ROHC RTP R-mode compressor. The purpose of the base test case is to verify that the compressor properly implements compression for a well-behaved IP/UDP/RTP packet flow, i.e. that it makes use of efficient compressed packet formats available to ROHC RTP when operating in R-mode.

Compressor response to single lost packets in O-mode. The purpose of this test is to verify that the compressor does not panic just because there is a single missing packet, i.e. the compressed packet size should not increase due to such events.

Broadcast Multicast Control (BMC)

BMC message reception when the UE in RRC Idle mode. Test to confirm, that a BMC configuration for a UE is able to receive activated CBS messages when in RRC Idle mode.

BMC message reception when the UE in RRC Connected mode, state CELL_PCH. Test to confirm, that a BMC configuration for a UE is able to receive activated CBS messages when in RRC Connected mode, state CELL_PCH.

BMC message reception when the in RRC Connected mode, state URA_PCH. Test to confirm, that a BMC configuration for a UE is able to receive activated CBS messages when in RRC Connected mode, state URA_PCH.

BMC message reception when the UE in RRC Idle mode (ANSI-41 CB data). Test to confirm, that a BMC configuration supporting ANSI-41 CB Data is able to receive activated CBS41 messages when in RRC Idle mode.

BMC message reception when the UE in RRC Connected mode, state CELL_PCH (ANSI-41 CB data). Test to confirm, that a BMC configuration supporting ANSI-41 CB Data is able to receive activated CBS41 messages when in RRC Connected mode, state CELL_PCH.

BMC message reception when the UE in RRC Connected mode, state URA_PCH (ANSI-41 CB data). Test to confirm, that a BMC configuration supporting ANSI-41 CB Data is able to receive activated CBS41 messages when in RRC Connected mode, state URA_PCH.

Reception of certain CBS message types. 1. Test to verify, that a UE supporting CBS ignores a deactivated CBS message type which has been broadcasted by SS. Test to confirm, that a UE only stores Serial Numbers of a newly transmitted CBS messages. This shall be verified by indication of a received CBS message with changed Serial Number as indication for the storage of Serial Numbers.

Radio Resource Control RRC

RRC Connection Management Procedure

Paging for Connection in idle mode. For the CS domain Test to confirm that the UE establishes an RRC connection after it receives a PAGING TYPE 1 message which includes IE "UE identity"(in IE "Paging Record") set to the IMSI of the UE, and responds with a correct INITIAL DIRECT TRANSFER message. For the PS domain. Test to confirm that the UE establishes an RRC connection after it receives a PAGING TYPE 1 message which includes IE "UE identity"(in IE "Paging Record") set to the P-TMSI allocated by SS at initial attach and responds with a correct INITIAL DIRECT TRANSFER message.

Paging for Connection in connected mode (CELL_PCH). Test to confirm that the UE enters the CELL_FACH state after it receives a PAGING TYPE 1 message which indicates that the paging has originated from UTRAN. Test to confirm that the UE performs cell update procedure after entering the CELL_FACH state.

Paging for Connection in connected mode (URA_PCH). Test to confirm that the UE enters the CELL_FACH state after it receives a PAGING TYPE 1 message in which the IE "Used paging identity" is set to "UTRAN identity", and the UE takes the U-RNTI value assigned to it in the IE "U-RNTI".

Paging for notification of BCCH modification in idle mode. Paging for notification of BCCH modification in idle mode. Test to confirm that the UE checks the new value tag of the master information block and reads the updated SYSTEM INFORMATION BLOCK messages after it receives a PAGING TYPE 1 message which includes the IE "BCCH Modification Information".

Paging for notification of BCCH modification in connected mode (CELL_PCH). Test to confirm that the UE, in addition to any actions caused by the IE "Paging record" occurrences in the PAGING TYPE 1 message, checks the new value tag of the master information block, and read the SYSTEM INFORMATION messages after it receives a PAGING TYPE 1 message which includes the IE "BCCH Modification Information".

Paging for notification of BCCH modification in connected mode (URA_PCH). Test to confirm that the UE checks the included new value tag of the master information block and reads the relevant SYSTEM INFORMATION block(s) after it receives a PAGING TYPE 1 message which includes the IE "BCCH Modification Information".

Paging for Connection in connected mode (CELL_DCH). Test to confirm that the UE responds to a PAGING TYPE 2 message which includes the IE "Paging Cause" and the IE "Paging Record Type Identifier". Test to confirm that the UE responds with a RRC STATUS message after it has received an invalid PAGING TYPE 2 message. To Page with the Paging Record Type Identifier set to "IMSI", in order to test the UEs behaviour to this situation which may occur when details of the temporary identity have been lost in the core network.

Paging for Connection in connected mode (CELL_FACH). Test to confirm that the UE responds to a PAGING TYPE 2 message, which includes a matching value for IE "Paging Record Type Identifier".

Paging for Connection in idle mode (multiple paging records). For the CS domain   Test to confirm that the UE establishes an RRC connection after it receives a PAGING TYPE 1 message which contains multiple paging records and includes IE "UE identity"(in IE "Paging Record") set to the IMSI of the UE, and responds with a correct INITIAL DIRECT TRANSFER message. For the PS domain Test to confirm that the UE establishes an RRC connection after it receives a PAGING TYPE 1 message which contains multiple paging records and includes IE "UE identity"(in IE "Paging Record") set to the P-TMSI allocated by SS at initial attach and responds with a correct INITIAL DIRECT TRANSFER message.

Paging for Connection in connected mode (URA_PCH, multiple paging records). Test to confirm that the UE enters the CELL_FACH state after it receives a PAGING TYPE 1 message in which the IE "Used paging identity" is set to "UTRAN identity", and the UE takes the U-RNTI value assigned to it in the IE "U-RNTI".

Paging for Connection in idle mode (Shared Network Environment). For the CS domain Test to confirm that the UE establishes an RRC connection after it receives a PAGING TYPE 1 message which includes IE "UE identity"(in IE "Paging Record") set to the IMSI of the UE, and responds with a correct

INITIAL DIRECT TRANSFER message. For the PS domain Test to confirm that the UE establishes an RRC connection after it receives a PAGING TYPE 1 message which includes IE "UE identity"(in IE "Paging Record") set to the P-TMSI allocated by SS at initial attach and responds with a correct INITIAL DIRECT TRANSFER message. For both CS and PS domain, it is verified that the UE is able to read and interpret the Multiple PLMN list broadcasted on the BCCH. It is also verified that the correct information is inserted into the IE "PLMN Identity" in the INITIAL DIRECT TRANSFER message.

RRC Connection Establishment in CELL_DCH state: Success. Test to confirm that the UE leaves the Idle Mode and correctly establishes signalling radio bearers on the DCCH. Test to confirm that the UE indicates the requested UE radio access capabilities and UE system specific capabilities (may be used by UTRAN e.g. to configure inter RAT- measurements). Test to confirm that the UE does not include the IE "UE Specific Behaviour Information 1 idle" in the RRC CONNECTION REQUEST message.

RRC Connection Establishment: Success after T300 timeout. Test to confirm that the UE retries to establish the RRC connection until V300 is greater than N300 after the expiry of timer T300 when the SS transmits no response for an RRC CONNECTION REQUEST message.

RRC Connection Establishment: Failure (V300 is greater than N300). Test to confirm that the UE stops retrying to establish the RRC connection if V300 is greater than N300 and goes back to idle mode.

RRC Connection Establishment: Reject ("wait time" is not equal to 0). Test to confirm that the UE retries to establish the RRC connection after the "wait time" elapses, if the UE receives an RRC CONNECTION REJECT message which includes the IE "wait time" not set to 0. Test to confirm that the UE performs a cell reselection when receiving an RRC CONNECTION REJECT message, containing relevant frequency information of the target cell to be re-selected.

RRC Connection Establishment: Reject ("wait time" is not equal to 0 and V300 is greater than N300). Test to confirm that the UE retries to establish the RRC connection after the "wait time" elapses if the UE receives an RRC CONNECTION REJECT message which specifies a non-zero IE "wait time". Test to confirm that the UE stops retrying to establish the RRC connection if V300 is greater than N300 and goes back to idle mode.

RRC Connection Establishment: Reject ("wait time" is set to 0). Test to confirm that the UE goes back to idle mode, if the SS transmits an RRC CONNECTION REJECT message which includes IE "wait time" set to 0. Test to confirm that the UE ignores an RRC CONNECT REJECT message not addressed to it. Test to confirm that the UE is capable of handling an erroneous RRC CONNECTION REJECT message correctly.

RRC Connection Establishment in CELL_FACH state: Success. Test to confirm that the UE is able to enter CELL_FACH state and setup signalling radio bearers using common physical channels. Test to confirm that the UE indicates the requested UE radio access capabilities (used by UTRAN to decide which RAB to establish) and UE system specific capabilities (may be used by UTRAN to configure inter RAT- measurements).

RRC Connection Establishment: Success after Physical channel failure and Failure after Invalid configuration. Test to confirm that the UE retries to establish the RRC connection until V300 is greater than N300 when a physical channel failure occurs because SS does not configure the physical channel that is specified in the transmitted RRC CONNECTION SETUP message. Test to confirm that the UE retries to establish the RRC connection until V300 is greater than N300 when the transmitted RRC CONNECTION SETUP message causes invalid configuration in the UE.

RRC connection establishment in CELL_DCH on another frequency. Test to confirm that the UE manages to establish an RRC CONNECTION on another frequency when so required by SS in the RRC CONNECTION SETUP message.

RRC connection establishment in CELL_DCH on another frequency. Test to confirm that the UE manages to establish an RRC CONNECTION on another frequency when so required by SS in the RRC CONNECTION SETUP message. Same test as above except that the target cell is on a different frequency band.

RRC Connection Establishment in FACH state (Frequency band modification): Success. Test to confirm that the UE enters to CELL_FACH state and correctly establishes signalling radio bearers using common physical channels of a cell within the frequency band specified by SS in RRC CONNECTION SETUP message.

RRC Connection Establishment: Reject with interRATInfo is set to GSM. Test to confirm that the UE shall select the GSM cell when RRC Connection Reject with Inter-RAT info set to GSM is received in response to RRC connection request.

RRC Connection Establishment: Reject with InterRATInfo is set to GSM and selection to the designated system fails. Test to confirm that the UE upon receiving RRC Connection Reject with Inter-RAT info set to GSM and failing to select the designated GSM system, shall reselect UTRAN cell only after the wait time specified in RRC Connection Reject. The UE shall then continue with the RRC CONNECTION establishment procedure.

RRC Connection Establishment using the default configuration for 3.4 kbps signalling bearers. Test to confirm that the UE establishes the radio bearer and transport channel configuration for 3.4kbps signalling radio bearers in accordance with the stored default parameters as identified by the IE "Default configuration identity" specified in the RRC Connection Setup Message

RRC Connection Establishment using the default configuration for 13.6 kbps signalling bearers. Test to confirm that the UE establishes the radio bearer and transport channel configuration for 13.6kbps signalling radio bearers in accordance with the stored default parameters as identified by the IE "Default configuration identity" specified in the RRC Connection Setup Message

RRC Connection Establishment / Domain Specific Access Control: Success. Test to confirm that the UE establishes a RRC Connection on the non-barred domain under the Domain Specific Access Restriction.

RRC Connection Release in CELL_DCH state: Success. Test to confirm: that the UE when receiving an RRC CONNECTION RELEASE message transmits N308+1 RRC CONNECTION RELEASE COMPLETE messages before release of radio resources and entering into idle mode that the time between UE transmissions of the RRC CONNECTION RELEASE COMPLETE message is equal to the value of the T308 timer.

RRC Connection Release using on DCCH in CELL_FACH state: Success. Test to confirm that the UE releases the L2 signalling radio bearer and resources and goes back to the idle state after it receives an RRC CONNECTION RELEASE message on downlink DCCH from the SS. It shall transmit an RRC CONNECTON RELEASE COMPLETE message using acknowledged mode on uplink DCCH to the SS.

RRC Connection Release using on CCCH in CELL_FACH state: Success. Test to confirm that the UE releases all its radio resources upon the reception of a RRC CONNECTION RELEASE message on the downlink CCCH, without transmitting RRC CONNECTION RELEASE COMPLETE message on the uplink.

RRC Connection Release in CELL_FACH state: Failure. Test to confirm that the UE releases all its radio resources and enters idle mode when the UE does not succeed in transmitting the RRC CONNECTION RELEASE COMPLETE message using acknowledged mode to the SS (i.e. the UE-RLC does not receive an acknowledgement for the transmission of the RRC CONNECTION RELEASE COMPLETE message from SS).

RRC Connection Release in CELL_FACH state: Invalid message. When the UE receives an invalid RRC CONNECTION RELEASE message on the downlink DCCH, it shall transmit an RRC CONNECTION RELEASE COMPLETE message that includes the appropriate error cause on the uplink DCCH.

RRC Connection Release in CELL_DCH state (Frequency band modification): Success. Test to confirm that when the UE receives an RRC CONNECTION RELEASE message the UE transmits N308+1 RRC CONNECTION RELEASE COMPLETE messages using UM on DCCH. Test to confirm that the UE enters into idle mode with performing cell-selection and selecting new cell configured by SS.

RRC Connection Release in CELL_FACH state (Frequency band modification): Success. Test to confirm that when the UE receives an RRC CONNECTION RELEASE message, the UE releases signalling radio bearer and its radio resources and goes back to the idle. Test to confirm that the UE enters into idle mode by performing cell-selection and selecting other cell than the UE selecting cell in connected mode.

RRC Connection Release in CELL_DCH state (Network Authentication Failure): Success. Test to confirm that when the upper layers request the release of the RRC connection, the UE releases signalling radio bearer and its radio resources and goes back to idle mode. Test to confirm that the UE enters idle mode, bars the cell for a period Tbarred and hence performs cell-selection to another (non-barred) cell.

UE Capability in CELL_DCH state: Success. Test to confirm that the UE transmits a UE CAPABILITY INFORMATION message after it receives a UE CAPABILITY ENQUIRY message from the SS. Test to confirm that the UE indicates an invalid message reception when invalid UE CAPABILITY ENQUIRY and UE CAPABILITY INFORMATION CONFIRM messages are received. The UE shall transmit RRC STATUS message with the correct error cause value to SS.

UE Capability in CELL_DCH state: Success after T304 timeout. Test to confirm that the UE re-transmits a UE CAPABILITY INFORMATION message until V304 is greater than N304, after the expiry of timer T304 when the UE cannot receive a UE CAPABILITY INFORMATION CONFIRM message in response to a UE CAPABILITY INFORMATION message.

UE Capability in CELL_DCH state: Failure (After N304 re-transmissions). Test to confirm that the UE stops retrying to transmit a UE CAPABILITY INFORMATION message if V304 is greater than N304. It then initiates cell update procedure.

UE Capability in CELL_FACH state: Success. Test to confirm that the UE transmits a UE CAPABILITY INFORMATION message after it receives a UE CAPABILITY ENQUIRY message from the SS. Test to confirm that the UE indicates an invalid message reception when invalid UE CAPABILITY ENQUIRY and UE CAPABILITY INFORMATION CONFIRM messages are received. The UE shall transmit RRC STATUS message with the correct error cause value to SS.

UE Capability in CELL_FACH state: Success after T304 timeout. Test to confirm that the UE re-transmits a UE CAPABILITY INFORMATION message until V304 is greater than N304, after the expiry of timer T304 when it fails to receive a downlink UE CAPABILITY INFORMATION CONFIRM message in response to the uplink UE CAPABILITY INFORMATION message sent.

UE Capability Information/ Reporting Of InterRAT Specific UE RadioAccessCapability. Test to confirm that a multi-RAT UE responds with a UE CAPABILITY INFORMATION message after it receives a UE CAPABILITY ENQUIRY message from the UTRAN and it includes the inter-RAT-specific UE radio access capability information element.

Direct Transfer in CELL DCH state (invalid message reception and no signalling connection exists). Test to confirm that the UE transmits an RRC

STATUS message on the DCCH using AM RLC if it receives a DOWNLINK DIRECT TRANSFER message with a non comprehended critical extension. Test to confirm that the UE transmits an RRC STATUS message on the DCCH using AM RLC if it receives a DOWNLINK DIRECT TRANSFER message which includes an invalid IE "CN domain identity".

Direct Transfer in CELL FACH state (invalid message reception and no signalling connection exists). Test to confirm that the UE transmits an RRC STATUS message on the DCCH using AM RLC if it receives a DOWNLINK DIRECT TRANSFER message which does not include any IEs except IE "Message Type". Test to confirm that the UE transmits an RRC STATUS message on the DCCH using AM RLC if it receives a DOWNLINK DIRECT TRANSFER message which includes an invalid IE "CN domain identity".

Measurement Report on INITIAL DIRECT TRANSFER message and UPLINK DIRECT TRANSFER message. Test to confirm that the UE reports measured results on RACH messages, if it receives IE "Intra-frequency reporting quantity for RACH reporting" and IE "Maximum number of reported cells on RACH" from System Information Block Type 11 or 12 upon a transition from idle mode to CELL_FACH state.

UPLINK Direct Transfer (RLC re-establishment). Test to confirm that the UE transmits a second UPLINK DIRECT TRANSFER message after the re-establishment of RLC on RB3 which occurs before the successful delivery of the first UPLINK DIRECT TRANSFER message.

Initial Direct Transfer: Inclusion of establishment cause. Test to confirm that, in the case the UE wants to start a new signalling connection while the UE is already in CELL_DCH state, the UE shall include the IE "Establishment cause" in the Initial Direct Transfer message.

Security mode command in CELL_DCH state (CS Domain). Test to confirm that the UE activates the new ciphering configurations after the stated activation time. Test to confirm that after the UE receives a SECURITY MODE COMMAND message, it transmits a SECURITY MODE COMPLETE message to the UTRAN using the old ciphering configuration together with the application of the new integrity protection configuration. Test to confirm that UE send SECURITY MODE FAILURE message when SS transmits a SECURITY MODE COMMAND message that causes an invalid configuration. Test to confirm that the UE sends a SECURITY MODE FAILURE message when the UE receives an invalid SECURITY MODE COMMAND message.

Security mode command in CELL_DCH state (PS Domain). Test to confirm that the UE modifies an integrity protection configuration and applies new keys on reception of a correct SECURITY MODE COMMAND message. Test to confirm that the UE modifies a ciphering configuration in the uplink and downlink and applies new keys according to transmitted activation times. Also confirms that the UE accepts a new ciphering configuration for a RB when ciphering is started for SRBs. Test to confirm that after the UE receives a SECURITY MODE

COMMAND message, it transmits a SECURITY MODE COMPLETE message to the UTRAN using the old ciphering configuration and new integrity protection configuration. Test to confirm that UE send SECURITY MODE FAILURE message when SS transmits a SECURITY MODE COMMAND message with a non comprehended critical extension. Test to confirm that the UE sends a SECURITY MODE FAILURE message when UE receives an invalid SECURITY MODE COMMAND message.

Security mode control in CELL_DCH state (CN Domain switch and new keys at RRC message sequence number wrap around). Test to confirm that the UE correctly modifies the integrity protection and ciphering configuration with a newly generated PS domain key-set for when previously using the CS domain key-set. Test to confirm that the UE can handle change of integrity protection key when the RRC message sequence number wraps around when the SECURITY MODE COMMAND is received.

Security mode control in CELL_DCH state interrupted by a cell update. Test to confirm that the UE aborts the ongoing integrity and ciphering configuration and the security mode control procedure in case it is interrupted by a cell update procedure.

Security mode command in CELL_FACH state. Test to confirm that after the UE receives a SECURITY MODE COMMAND message, it transmits a SECURITY MODE COMPLETE message to the UTRAN using the old ciphering configuration together with the application of the new integrity protection configuration. Test to confirm that the UE applies the old ciphering configuration in the downlink prior to the activation time; and uses the new ciphering configuration on and after the activation time. Test to confirm that the UE starts to cipher its uplink transmissions after the uplink activation time stated in SECURITY MODE COMPLETE message is reached. Test to confirm that the UE sends a SECURITY MODE FAILURE message when the UE receives an invalid SECURITY MODE COMMAND message.

Counter check in CELL_DCH state, with symmetric RAB. Test to confirm that the UE transmits a COUNTER CHECK RESPONSE message after it receives a COUNTER CHECK message from the SS. Test to confirm that the UE responds to the reception of an invalid downlink COUNTER CHECK message by transmitting a RRC STATUS message on the uplink DCCH, stating the correct error cause value in message.

Counter check in CELL_FACH state. Test to confirm that the UE transmits a COUNTER CHECK RESPONSE message after it receives a COUNTER CHECK message from the SS. Test to confirm that the UE responds to the reception of an invalid downlink COUNTER CHECK message by transmitting a RRC STATUS message on the uplink DCCH, stating the correct error cause value in message.

Counter check in CELL_DCH state, with asymmetric RAB. Test to confirm that the UE transmits a COUNTER CHECK RESPONSE message even if

COUNT-C does not exist for a radio bearer for a given direction for reasons given in the above section.

Signalling Connection Release Indication. Test to confirm that the UE transmits a SIGNALLING CONNECTION RELEASE INDICATION message after upper layer requests to release its signalling connection.

Signalling Connection Release Indication (RLC re-establishment): CS signalling connection release. Test to confirm that the UE re-transmits a SIGNALLING CONNECTION RELEASE INDICATION message after it re-establishes the RLC entity on signalling radio bearer RB2 if SRNS relocation occurs before the successful delivery of SIGNALLING CONECTION RELEASE INDICATION message.

Signalling Connection Release Indication (RLC re-establishment): PS signalling connection release. Test to confirm that the UE re-transmits a SIGNALLING CONNECTION RELEASE INDICATION message after it re-establishes the RLC entity on signalling radio bearer RB2 if SRNS relocation occurs before the successful delivery of SIGNALLING CONECTION RELEASE INDICATION message.

Broadcast of system information, Dynamic change of segmentation, concatenation & scheduling and handling of unsupported information blocks. Test to confirm that dynamic change of System Information is identified, new information read and used. Test to confirm that the UE can support all segment types and "all" segment combinations. Test to confirm that the UE can dynamically use different configurations. Test to confirm that the UE properly uses combinations of Default and assigned values.

Signalling Connection Release (Invalid configuration). Test to confirm that the UE ignores the SIGNALLING CONNECTION RELEASE REQUEST message which request the UE to release signalling connection of domain that contains established radio access bearers. Test to confirm that the UE transmit a RRC STATUS message to SS after detecting an invalid configuration in the received message.

Integrity Protection. Test to confirm that the UE discards any RRC messages that include wrong message authentication code, or RRC message sequence number, or do not include IE"Integrity Check Info" after integrity protection is activated.

RRC Radio Bearer Control Procedures

Radio Bearer Establishment for transition from CELL_DCH to CELL_DCH: Success. Test to confirm that the UE establishes a new radio bearer according to a RADIO BEARER SETUP message.

Radio Bearer Establishment for transition from CELL_DCH to CELL_DCH: Failure (Unsupported configuration). Test to confirm that the UE keeps its configuration and transmits a RADIO BEARER SETUP FAILURE message in case of receiving a RADIO BEARER SETUP message which includes parameters of its unsupported configuration.

Radio Bearer Establishment for transition from CELL_DCH to CELL_ DCH: Failure (Physical channel Failure and successful reversion to old configuration). Test to confirm that the UE reverts to the old configuration and transmits a RADIO BEARER SETUP FAILURE message on the DCCH using AM RLC, if the UE fails to reconfigure the radio bearer according to the RADIO BEARER SETUP message before timer T312 expires.

Radio Bearer Establishment for transition from CELL_DCH to CELL_ DCH: Failure (Invalid message reception and Invalid configuration). Test to confirm that the UE transmits a RADIO BEARER SETUP FAILURE message on the DCCH using AM RLC if it receives an invalid RADIO BEARER SETUP message which contains an unexpected critical message extension. Test to confirm that the UE transmits a RADIO BEARER SETUP FAILURE message on the DCCH using AM RLC if it receives a RADIO BEARER SETUP message including an invalid configuration.

Radio Bearer Establishment for transition from CELL_DCH to CELL_ FACH: Success. Test to confirm that the UE establishes a new radio bearer according to a RADIO BEARER SETUP message.

Radio Bearer Establishment for transition from CELL_DCH to CELL_ FACH: Success (Cell re-selection). Test to verify that the UE when receiving a RADIO BEARER SETUP message not including a value for C-RNTI initiate a cell update procedure and indicating the cause "Cell reselection". Test to confirm that the UE when the CELL UPDATE CONFIRM message does not include "RB information elements", "Transport channel information elements" nor "Physical channel information elements" but include the IE "New C-RNTI" transmit a UTRAN MOBILITY INFORMATION CONFIRM message. Test to confirm that the UE transmits RADIO BEARER SETUP COMPLETE message after it completes the cell update procedure.

Radio Bearer Establishment for transition from CELL_FACH to CELL_ DCH: Success. Test to confirm that the UE establishes a new radio bearer according to a RADIO BEARER SETUP message.

Radio Bearer Establishment for transition from CELL_FACH to CELL_ DCH: Failure (Unsupported configuration). Test to confirm that the UE keeps its configuration and transmits a RADIO BEARER SETUP FAILURE message in case of it receiving a RADIO BEARER SETUP message, which includes parameters of an unsupported configuration.

Radio Bearer Establishment for transition from CELL_FACH to CELL_ DCH: Failure (Physical channel Failure and successful reversion to old configuration). Test to confirm that the UE reverts to the old configuration and transmits a RADIO BEARER SETUP FAILURE message when the UE fails to configure the new radio bearer after it detects physical channel failure, followed by the T312 expiry.

Radio Bearer Establishment for transition from CELL_FACH to CELL_ DCH: Failure (Physical channel Failure and cell reselection). Test to confirm that the UE transmit a RADIO BEARER SETUP FAILURE message after it completes a cell update for the physical channel failure in the radio bearer establishment procedure.

Radio Bearer Establishment for transition from CELL_FACH to CELL_DCH: Failure (Incompatible simultaneous reconfiguration). Test to confirm that if the UE receives a RADIO BEARER SETUP message during a reconfiguring procedure due to a radio bearer message other than RADIO BEARER SETUP, it shall keep its configuration as if the RADIO BEARER SETUP message had not been received and complete the reconfiguration procedure according to the previously received message.

Radio Bearer Establishment for transition from CELL_FACH to CELL_FACH: Success. Test to confirm that the UE establishes a new radio access bearer according to a RADIO BEARER SETUP message.

Radio Bearer Establishment for transition from CELL_DCH to CELL_DCH: success (Subsequently received). Test to confirm that if the UE receives a new RADIO BEARER SETUP message before the UE completes the configuration of the radio bearer according to a previous RADIO BEARER SETUP message, it ignores the new RADIO BEARER SETUP message and configures according to the previous RADIO BEARER SETUP message received.

Radio Bearer Establishment for transition from CELL_FACH to CELL_DCH: Success (Subsequently received). Test to confirm that if the UE receives a new RADIO BEARER SETUP message before the UE completes the configuration of the radio bearer according to a previous RADIO BEARER SETUP message, it ignores the new RADIO BEARER SETUP message and configures according to the previous RADIO BEARER SETUP message received.

Radio Bearer Establishment for transition from CELL_DCH to CELL_FACH (Frequency band modification): Success. Test to confirm that the UE transits from CELL_DCH to CELL_FACH according to the RADIO BEARER SETUP message. Test to confirm that the UE transmits RADIO BEARER SETUP COMPLETE message on the uplink DCCH using AM RLC on a common physical channel in a different frequency.

Radio Bearer Establishment for transition from CELL_FACH to CELL_DCH (Frequency band modification): Success. Test to confirm that the UE transits from CELL_FACH to CELL_DCH according to the RADIO BEARER SETUP message. Test to confirm that the UE transmits RADIO BEARER SETUP COMPLETE message on the uplink DCCH using AM RLC on a dedicated physical channel in a different frequency.

Radio Bearer Establishment for transition from CELL_DCH to CELL_DCH (Frequency band modification): Success. Test to confirm that the UE transits from CELL_DCH to CELL_DCH according to the RADIO BEARER SETUP message. Test to confirm that the UE transmits the RADIO BEARER SETUP COMPLETE message on the uplink DCCH using AM RLC on a dedicated physical channel in a different frequency.

Radio Bearer Establishment for transition from CELL_DCH to CELL_ DCH (Inter band handover): Success. Test to confirm that the UE transits from CELL_DCH to CELL_DCH according to the RADIO BEARER SETUP message. Test to confirm that the UE transmits the RADIO BEARER SETUP COMPLETE message on the uplink DCCH using AM RLC on a dedicated physical channel in a different frequency band cell.

Radio Bearer Establishment for transition from CELL_FACH to CELL_ FACH (Frequency band modification): Success. Test to confirm that the UE transits from CELL_FACH to CELL_FACH according to the RADIO BEAR-ER SETUP message. Test to confirm that the UE transmits RADIO BEARER SETUP COMPLETE message on the uplink DCCH using AM RLC on a common physical channel in a different frequency.

Radio Bearer Establishment for transition from CELL_DCH to CELL_ DCH: Success (two radio links, start of HS-DSCH reception). Test to confirm that the UE establishes a radio bearer mapped to HS-DSCH according to the received RADIO BEARER SETUP message when having two radio links established.

Radio Bearer Establishment for transition from CELL_DCH to CELL_ DCH: Success (start of HS-DSCH reception). Test to confirm that the UE establishes a radio bearer mapped to HS-DSCH according to the received RADIO BEARER SETUP message.

Radio Bearer Establishment for transition from CELL_DCH to CELL_ DCH: Success (RB mapping for both DL DCH and HS-DSCH in cell without HS-DSCH support). Test to confirm that the UE establishes a radio bearer mapped to DCH and HS-DSCH according to the received RADIO BEARER SETUP message in a cell without HS-DSCH.

Radio Bearer Establishment for transition from CELL_DCH to CELL_ DCH: Success (Timing re-initialized hard handover to another frequency, uplink TFCS restriction and start of HS-DSCH reception). Test to confirm that the UE establishes a radio bearer mapped to HS-DSCH using uplink TFCS restriction according to the received RADIO BEARER SETUP message.

Radio Bearer Establishment for transition from CELL_DCH to CELL_ DCH: Success (Timing re-initialised hard handover to another frequency, start of HS-DSCH reception). Test to confirm that the UE establishes a radio bearer mapped to HS-DSCH and starts HS-DSCH reception in conjunction with a inter-frequency hard handover without prior measurement on the target frequency according to the received RADIO BEARER SETUP message.

Radio Bearer Establishment for transition from CELL_FACH to CELL_ DCH: Success (start of HS-DSCH reception). Test to confirm that the UE establishes a radio bearer mapped to HS-DSCH according to the received RADIO BEARER SETUP message.

Radio Bearer Establishment for transition from CELL_FACH to CELL_ DCH: Success (start of HS-DSCH reception with frequency modification).

Test to confirm that the UE establishes a radio bearer mapped to HS-DSCH according to the received RADIO BEARER SETUP message.

Radio Bearer Establishment for transition from CELL_DCH to CELL_DCH: Success (Unsynchronised RL Reconfiguration). Test to confirm that the UE establishes a new radio bearer according to a RADIO BEARER SETUP message.

Radio Bearer Establishment for transition from CELL_DCH to CELL_DCH: Success (Unsynchronised RL Reconfiguration with frequency modification). Test to confirm that the UE establishes a new radio bearer according to a RADIO BEARER SETUP message.

Radio Bearer Establishment for transition from CELL_DCH to CELL_DCH: Success (Unsynchronised RL Reconfiguration with inter band handover). Test to confirm that the UE establishes a new radio bearer according to a RADIO BEARER SETUP message in a different frequency band cell.

Radio Bearer Establishment for transition from CELL_DCH to CELL_DCH: Success (start of E-DCH transmission). Test to confirm that the UE establishes a radio bearer mapped to HS-DSCH and E-DCH according to the received RADIO BEARER SETUP message.

Radio Bearer Establishment for transition from CELL_DCH to CELL_DCH: Success (hard handover to another frequency, start of E-DCH transmission). Test to confirm that the UE performs a hard handover to another frequency, establishes a radio bearer mapped on E-DCH and starts E-DCH transmission according to the received a RADIO BEARER SETUP message.

Radio Bearer Reconfiguration from CELL_DCH to CELL_DCH: Success. Test to confirm that the UE reconfigures the radio bearers according to a RADIO BEARER RECONFIGURATION message, which indicates a change of UL scrambling code and change of RLC parameters.

Radio Bearer Reconfiguration from CELL_DCH to CELL_DCH: Failure (Unsupported configuration). Test to confirm that the UE transmits a RADIO BEARER RECONFIGURATION FAILURE message on the DCCH using AM RLC if the received RADIO BEARER RECONFIGURATION message includes unsupported configuration parameters.

Radio Bearer Reconfiguration from CELL_DCH to CELL_DCH: Failure (Physical channel failure and cell reselection). Test to confirm that the UE transmits a RADIO BEARER RECONFIGURATION FAILURE message after it completes a cell update procedure when the UE cannot reconfigure the new radio bearer and a subsequent failure to revert to the old configuration.

Radio Bearer Reconfiguration from CELL_DCH to CELL_DCH: Success (Continue and stop). Test to confirm that the UE reconfigures new radio bearer and stop the transmission and reception of the RLC entity belonging to the RB identity specified in the RADIO BEARER RECONFIGURATION message. Test to confirm that the UE reconfigures new radio bearer and restart the transmission and reception of the RLC entity belonging to the RB identity specified in the RADIO BEARER RECONFIGURATION message.

Radio Bearer Reconfiguration from CELL_DCH to CELL_FACH: Success. Test to confirm that the UE establishes the reconfigured radio bearer(s) using common physical channel, after UE receives a RADIO BEARER RECONFIGURATION message.

Radio Bearer Reconfiguration from CELL_DCH to CELL_FACH: Success (Cell re-selection). Test to confirm that the UE transmits RADIO BEARER RECONFIGURATION COMPLETE message after it completes a cell update procedure.

Radio Bearer Reconfiguration: from CELL_FACH to CELL_DCH including modification of previously signalled CELL_DCH configuration: Success. Test to confirm that the UE applies a previously signalled configuration for CELL_DCH and in addition modifies the parameters for which reconfiguration is requested in the RADIO BEARER RECONFIGURATION message that is used to initiate transition from CELL_FACH to CELL_DCH

Radio Bearer Reconfiguration from CELL_FACH to CELL_DCH: Failure (Unsupported configuration). Test to confirm that the UE transmits a RADIO BEARER RECONFIGURATION FAILURE message on the DCCH using AM RLC if the received RADIO BEARER RECONFIGURATION message includes unsupported configuration parameters.

Radio Bearer Reconfiguration from CELL_FACH to CELL_FACH: Success. Test to confirm that the UE establishes radio bearers according to a RADIO BEARER RECONFIGURATION message.

Radio Bearer Reconfiguration from CELL_FACH to CELL_FACH: Success (Cell re-selection). Test to confirm that the UE transmits RADIO BEARER RECONFIGURATION COMPLETE message in cell 2 when a cell re-selection occurs after receiving a RADIO BEARER RECONFIGURATION message.

Radio Bearer Reconfiguration from CELL_DCH to CELL_DCH: Success (Subsequently received). Test to confirm that if the UE receives a new RADIO BEARER RECONFIGURATION message before the UE configures the radio bearer according to a previous RADIO BEARER RECONFIGURATION message, it ignores the new RADIO BEARER RECONFIGURATION message and configures the radio bearer according to the previous RADIO BEARER RECONFIGURATION message received.

Radio Bearer Reconfiguration from CELL_FACH to CELL_PCH: Success. Test to confirm that the UE transmits RADIO BEARER RECONFIGURATION COMPLETE message and enters CELL_PCH state after it received a RADIO BEARER RECONFIGURATION message, which invoke the UE to transit from CELL_FACH to CELL_PCH.

Radio Bearer Reconfiguration from CELL_DCH to CELL_DCH: Success (Incompatible Simultaneous Reconfiguration). Test to confirm that the UE ignores the subsequent security reconfiguration information which is contained in the RADIO BEARER RECONFIGURATION message. Test to confirm that the UE reconfigures according to the SECURITY MODE COMMAND message. Test to confirm that the UE transmits RADIO BEARER RECONFIGURATION FAILURE message on the uplink DCCH using AM RLC. Test to confirm that the UE transmits SECURITY MODE COMPLETE message on the uplink DCCH using AM RLC.

Radio Bearer Reconfiguration for transition from CELL_DCH to CELL_ DCH (Frequency band modification): Success. Test to confirm that the UE transits from CELL_DCH to CELL_DCH according to the RADIO BEARER RECONFIGURATION message. Test to confirm that the UE transmits the RADIO BEARER RECONFIGURATION COMPLETE message on the uplink DCCH using AM RLC on a dedicated physical channel in a different frequency.

Radio Bearer Reconfiguration for transition from CELL_DCH to CELL_FACH (Transport channel type switching with frequency band modification): Success. Test to confirm that the UE transits from CELL_DCH to CELL_FACH according to the RADIO BEARER RECONFIGURATION message. Test to confirm that the UE transmits RADIO BEARER RECONFIG-URATION COMPLETE message on the uplink DCCH using AM RLC on a common physical channel in a different frequency.

Radio Bearer Reconfiguration for transition from CELL_FACH to CELL_ DCH (Frequency band modification): Success. Test to confirm that the UE transits from CELL_FACH to CELL_DCH according to the RADIO BEARER RECONFIGURATION message. Test to confirm that the UE transmits RADIO BEARER RECONFIGURATION COMPLETE message on the uplink DCCH using AM RLC on a dedicated physical channel in a different frequency.

Radio Bearer Reconfiguration for transition from CELL_FACH to CELL_ FACH (Frequency band modification): Success. Test to confirm that the UE transits from CELL_FACH to CELL_FACH according to the RADIO BEARER RECONFIGURATION message. Test to confirm that the UE transmits RADIO BEARER RECONFIGURATION COMPLETE message on the uplink DCCH using AM RLC on a common physical channel in a different frequency.

Radio Bearer Reconfiguration for transition from CELL_FACH to URA_ PCH (Frequency band modification): Success. Test to confirm that the UE transmits a RADIO BEARER RECONFIGURATION COMPLETE message on the uplink DCCH using AM RLC. Test to confirm that the UE transits from CELL_FACH to URA_PCH according to the RADIO BEARER RECONFIGU-RATION message. Test to confirm that the UE selects a common physical chan-nel in a different frequency.

Radio Bearer Reconfiguration from CELL_DCH to CELL_FACH: Suc-cessful channel switching with multiple PS RABs established. Test to con-firm that the UE transit from CELL_DCH to CELL_FACH state according to a RADIO BEARER RECONFIGURATION message when having two radio access bearers established. Test to confirm that the UE transit from CELL_FACH to CELL_DCH state according to a RADIO BEARER RECONFIGURATION message when having two radio access bearers established. Test to confirm that the UE release two radio access bearers included in a single RADIO BEARER RELEASE message.

Radio Bearer Reconfiguration for transition from CELL_DCH to CELL_DCH: Success (Start and stop of HS-DSCH reception). Test to confirm that the UE starts and stops receiving the HS-DSCH according to the received RADIO BEARER RECONFIGURATION message.

Radio Bearer Reconfiguration for transition from CELL_DCH to CELL_FACH and from CELL_FACH to CELL_DCH: Success (start and stop of HS-DSCH reception). Test to confirm that the UE transits to CELL_FACH state from CELL_DCH state and stops receiving the HS-DSCH according to the received RADIO BEARER RECONFIGURATION message. Test to confirm that the UE transits to CELL_DCH state from CELL_FACH state and starts receiving the HS-DSCH according to the received RADIO BEARER RECONFIGURATION message.

Radio Bearer Reconfiguration from CELL_DCH to CELL_DCH: Success (with active HS-DSCH reception). Test to confirm that the UE reconfigures the radio bearer while being mapped to HS-DSCH according to the received RADIO BEARER RECONFIGURATION message.

Radio Bearer Reconfiguration for transition from CELL_DCH to CELL_DCH: Success (Timing re-initialised hard handover to another frequency, start and stop of HS-DSCH reception). Test to confirm that the UE starts and stops receiving the HS-DSCH in conjunction with an inter-frequency hard handover without prior measurement on the target frequency according to the received RADIO BEARER RECONFIGURATION message.

Radio Bearer Reconfiguration for transition from CELL_DCH to CELL_FACH and from CELL_FACH to CELL_DCH: Success (frequency band modification, start and stop of HS-DSCH reception). Test to confirm that the UE transits to CELL_FACH state from CELL_DCH state in another cell and frequency and stops receiving the HS-DSCH according to the received RADIO BEARER RECONFIGURATION message. Test to confirm that the UE transits to CELL_DCH state from CELL_FACH state in another cell and frequency and starts receiving the HS-DSCH according to the received RADIO BEARER RE-CONFIGURATION message.

Radio Bearer Reconfiguration for transition from CELL_DCH to CELL_DCH: Success (Start and stop of HS-DSCH reception, during an active CS bearer). Test to confirm that the UE starts and stops receiving the HS-DSCH according to the received RADIO BEARER RECONFIGURATION message when a circuit-switched radio bearer is established and mapped to DCH.

Radio Bearer Reconfiguration for transition from CELL_DCH to CELL_DCH: Success (Timing re-initialised hard handover to another frequency, start and stop of HS-DSCH reception, during an active CS bearer). Test to confirm that the UE starts and stops receiving the HS-DSCH in conjunction with an inter-frequency hard handover without prior measurement on the target frequency according to the received RADIO BEARER RECONFIGURATION message when a circuit-switched radio bearer is established and mapped to DCH.

Radio Bearer Reconfiguration for transition from CELL_DCH to CELL_DCH: Success (Seamless SRNS relocation, without pending of ciphering, frequency band modification). Test to confirm that the UE performs a combined inter-frequency hard handover and SRNS relocation and then transmit a RADIO BEARER RECONFIGURATION COMPLETE message in the new cell. Test to confirm that the UE correctly applies integrity protection after the SRNS relocation. Test to confirm that the UE accepts a gap in the downlink RRC message sequence numbering for integrity protection on signalling radio bearer 3 after SRNS relocation. In the case that ciphering is applied by the network, to confirm that the UE restarts ciphering following a successful SRNS relocation.

Radio Bearer Reconfiguration from CELL_DCH to CELL_DCH: Success (With active E-DCH transmission). Test to confirm that the UE reconfigures radio bearer parameters during active E-DCH transmission according to the received RADIO BEARER RECONFIGURATION message.

Radio Bearer Reconfiguration for transition from CELL_FACH to CELL_DCH and CELL_DCH to CELL_FACH: Success (start and stop of E-DCH transmission). Test to confirm that the UE reconfigures radio bearer parameters, transits from CELL_FACH to CELL_DCH state, and starts E-DCH transmission according to the received RADIO BEARER RECONFIGURATION message. Test to confirm that the UE reconfigures radio bearer parameters, transits from CELL_DCH to CELL_FACH state, and stops E-DCH transmission according to the received RADIO BEARER RECONFIGURATION message.

Radio Bearer Reconfiguration for transition from CELL_DCH to CELL_DCH: Success (hard handover to another frequency, start and stop of E-DCH transmission). Test to confirm that the UE starts and stops transmitting the E-DCH, while maintaining HS-DSCH reception, in conjunction with an inter-frequency hard handover without prior measurement on the target frequency according to the received RADIO BEARER RECONFIGURATION message.

Radio Bearer Reconfiguration for transition from CELL_FACH to CELL_DCH and CELL_DCH to CELL_FACH: Success (frequency modification, start and stop of E-DCH transmission). Test to confirm that the UE reconfigures radio bearer parameters, transits from CELL_DCH to CELL_FACH state with frequency modification and stops E-DCH transmission according to the received a RADIO BEARER RECONFIGURATION message. Test to confirm that the UE reconfigures radio bearer parameters, transits from CELL_FACH to CELL_DCH state with frequency modification and starts E-DCH transmission according to the received a RADIO BEARER RECONFIGURATION message.

Radio Bearer Reconfiguration for transition from CELL_DCH to CELL_ DCH: Success (Start and stop of E-DCH transmission). Test to confirm that the UE reconfigures the radio bearer parameters and stops E-DCH transmission according to the received RADIO BEARER RECONFIGURATION message. Test to confirm that the UE reconfigures the radio bearer parameters and starts E-DCH transmission according to the received RADIO BEARER RECONFIG- URATION message.

Radio Bearer Reconfiguration for transition from CELL_DCH to CELL_ PCH: Success (stop of E-DCH transmission). Test to confirm that the UE reconfigures radio bearer parameters, transits from CELL_DCH to CELL_PCH state, and stops E-DCH transmission and HS-DSCH reception according to the received RADIO BEARER RECONFIGURATION message.

Radio Bearer Release for transition from CELL_DCH to CELL_DCH: Success. Test to confirm that the UE releases the existing radio bearer according to a RADIO BEARER RELEASE message.

Radio Bearer Release for transition from CELL_DCH to CELL_FACH: Success. Test to confirm that the UE release the existing the radio bearer according to a RADIO BEARER RELEASE message.

Radio Bearer Release for transition from CELL_DCH to CELL_FACH: Success (Cell re-selection). Test to confirm that the UE transmits a RADIO BEARER RELEASE COMPLETE message after the UE completes a cell update procedure.

Radio Bearer Release for transition from CELL_FACH to CELL_DCH: Success. Test to confirm that an UE, in state CELL_FACH, releases the radio access bearers using common physical channel. After the release, it shall access the affected radio bearers on the DPCH.

Radio Bearer Release for transition from CELL_FACH to CELL_DCH: Failure (Physical channel failure and successful reversion to old configura- tion). Test to confirm that the UE reverts to the old configuration and transmits a RADIO BEARER RELEASE FAILURE message on the DCCH using AM RLC if the UE fails to release the radio bearers in accordance with the specified settings in RADIO BEARER RELEASE message before T312 timer expires.

Radio Bearer Release for transition from CELL_FACH to CELL_FACH: Success. Test to confirm that the UE releases the existing the radio bearer(s) according to the RADIO BEARER RELEASE message.

Radio Bearer Release for transition from CELL_DCH to CELL_DCH: Success (Subsequently received). Test to confirm that if the UE receives a new RADIO BEARER RELEASE message before the UE releases the radio bearer according to a previous RADIO BEARER RELEASE message it ignore the new RADIO BEARER RELEASE message and configures according to the previous RADIO BEARER RELEASE message received.

Radio Bearer Release for transition from CELL_FACH to CELL_DCH: Success (Subsequently received). Test to confirm that if the UE receives a new

RADIO BEARER RELEASE message before the UE releases the radio bearer according to a previous RADIO BEARER RELEASE message, it ignores the new RADIO BEARER RELEASE message and configures according to the previous RADIO BEARER RELEASE message received.

Radio Bearer Release from CELL_DCH to CELL_PCH: Success. Test to confirm that the UE transmits a RADIO BEARER RELEASE COMPLETE before entering CELL_PCH state after it received a RADIO BEARER RELEASE message and released its radio access bearers.

Radio Bearer Release from CELL_DCH to URA_PCH: Success. Test to confirm that the UE transmits a RADIO BEARER RELEASE COMPLETE before entering URA_PCH state after it received a RADIO BEARER RELEASE message and released its radio bearers.

Radio Bearer Release for transition from CELL_DCH to CELL_FACH (Frequency band modification): Success. Test to confirm that the UE transits from CELL_DCH to CELL_FACH according to the RADIO BEARER RELEASE message. Test to confirm that the UE transmits RADIO BEARER RELEASE COMPLETE message on the uplink DCCH using AM RLC on a common physical channel in a different frequency.

Radio Bearer Release from CELL_DCH to CELL_PCH (Frequency band modification): Success. Test to confirm that the UE transmits RADIO BEARER RELEASE COMPLETE message on the uplink DCCH using AM RLC. Test to confirm that the UE transits from CELL_DCH to CELL_PCH according to the RADIO BEARER RELEASE message. Test to confirm that the UE releases the radio access bearer and selects a common physical channel in a different frequency indicated by SS.

Radio Bearer Release for transition from CELL_FACH to CELL_PCH: Success. Test to confirm that the UE transmits RADIO BEARER RELEASE COMPLETE message on the uplink DCCH using AM RLC. Test to confirm that the UE transits from CELL_FACH to CELL_PCH according to the RADIO BEARER RELEASE message. Test to confirm that the UE releases the radio access bearer and selects a common physical channel.

Radio Bearer Release for transition from CELL_FACH to URA_PCH: Success. Test to confirm that the UE transmits RADIO BEARER RELEASE COMPLETE message on the uplink DCCH using AM RLC. Test to confirm that the UE transits from CELL_FACH to URA_PCH according to the RADIO BEARER RELEASE message. Test to confirm that the UE releases the radio access bearer and selects a common physical channel.

Radio Bearer Release for transition from CELL_DCH to CELL_DCH (Frequency band modification): Success. Test to confirm that the UE transits from CELL_DCH to CELL_DCH according to the RADIO BEARER RELEASE message. Test to confirm that the UE transmits RADIO BEARER RELEASE COMPLETE message on the uplink DCCH using AM RLC on a dedicated physical channel in a different frequency.

Radio Bearer Release for transition from CELL_DCH to URA_PCH (Frequency band modification): Success. Test to confirm that the UE transmits a RADIO BEARER RELEASE COMPLETE message on the uplink DCCH using AM RLC. Test to confirm that the UE transits from CELL_DCH to URA_PCH according to the RADIO BEARER RELEASE message. Test to confirm that the UE releases radio access bearer, dedicated physical channel and selects a common physical channel in a different frequency.

Radio Bearer Release for transition from CELL_FACH to CELL_PCH (Frequency band modification): Success. Test to confirm that the UE transmits a RADIO BEARER RELEASE COMPLETE message on the uplink DCCH using AM RLC. Test to confirm that the UE transits from CELL_FACH to CELL_PCH according to the RADIO BEARER RELEASE message. Test to confirm that the UE releases radio access bearer and selects a common physical channel in a different frequency.

Radio Bearer Release for transition from CELL_FACH to URA_PCH (Frequency band modification): Success. Test to confirm that the UE transmits a RADIO BEARER RELEASE COMPLETE message on the uplink DCCH using AM RLC. Test to confirm that the UE transits from CELL_FACH to URA_PCH according to the RADIO BEARER RELEASE message. Test to confirm that the UE releases radio access bearer and selects a common physical channel in a different frequency.

Radio Bearer Release for transition from CELL_FACH to CELL_FACH (Frequency band modification): Success. Test to confirm that the UE transits from CELL_FACH to CELL_FACH according to the RADIO BEARER RELEASE message. Test to confirm that the UE transmits RADIO BEARER RELEASE COMPLETE message on the uplink DCCH using AM RLC on a common physical channel in a different frequency.

Radio Bearer Release for transition from CELL_DCH to CELL_DCH: Associated with signalling connection release during multi call for PS and CS services. Test to confirm that the UE releases the existing radio access bearer and signalling connection according to a RADIO BEARER RELEASE message.

Radio Bearer Release for transition from CELL_DCH to CELL_DCH: Success (stop of HS-DSCH reception). Test to confirm that the UE releases a radio bearer mapped to HS-DSCH according to the received RADIO BEARER RELEASE message.

Radio Bearer Release for transition from CELL_DCH to CELL_DCH: Success (With active HS-DSCH reception). Test to confirm that the UE releases a radio bearer according to the received RADIO BEARER RELEASE message while keeping HS-DSCH reception active for a second radio bearer mapped to HS-DSCH.

Radio Bearer Release for transition from CELL_DCH to CELL_DCH: Success (Timing re-initialised hard handover to another frequency, with active HS-DSCH reception). Test to confirm that the UE releases a radio bearer according to the received RADIO BEARER RELEASE message while keeping HS-DSCH reception active for a second radio bearer mapped to HS-DSCH.

Radio Bearer Release for transition from CELL_DCH to CELL_DCH: Success (stop of HS-DSCH reception with frequency modification). Test to confirm that the UE stops HS-DSCH reception when UE releases PS RAB according to the received RADIO BEARER RELEASE message.

Radio Bearer Release for transition from CELL_DCH to CELL_FACH: Success (stop of HS-DSCH reception with frequency modification). Test to confirm that the UE stops HS-DSCH reception when UE releases CS RAB according to the received RADIO BEARER RELEASE message.

Radio Bearer Release for transition from CELL_DCH to CELL_PCH: Success (stop of HS-DSCH reception). Test to confirm that the UE releases CS bearer and stops receiving the HS-DSCH reception according to the received RADIO BEARER RELEASE message.

Radio Bearer Release for transition from CELL_DCH to CELL_DCH: Success (frequency modification, stop of E-DCH transmission). Test to confirm that the UE releases a radio bearer mapped on E-DCH / HS-DSCH and stops E-DCH transmission / HS-DSCH reception according to the received RADIO BEARER RELEASE message with frequency modification.

Transport channel reconfiguration (Timing re-initialised hard handover with transmission rate modification) from CELL_DCH to CELL_DCH: Success. Test to confirm that the UE reconfigures the channel configuration according to a TRANSPORT CHANNEL RECONFIGURATION message, which is used to change the TFCS and the TFS while replacing the RL(s) in the active set with a set of RL(s) disjunct with the previous active set.

Transport channel reconfiguration (Transmission Rate Modification) from CELL_DCH to CELL_DCH of the same cell: Success. Test to confirm that the UE reconfigures the transport channel configuration according to a TRANSPORT CHANNEL RECONFIGURATION message, which specifies a reconfiguration by changing the TFCS. Test to confirm that the UE receives the RLC SDU and sends it according to the new UL TFCS.

Transport channel reconfiguration (Transmission Rate Modification) from CELL_DCH to CELL_DCH of the same cell: Success. Test to confirm that the UE reconfigures the transport channel configuration according to a TRANSPORT CHANNEL RECONFIGURATION message, which specifies a reconfiguration by changing the TFCS. Test to confirm that the UE receives the RLC SDU and sends it according to the new UL TFCS.

Transport channel reconfiguration from CELL_DCH to CELL_DCH: Failure (Physical channel failure and reversion to old configuration). Test to confirm that the UE reverts to the old configuration and transmits a TRANSPORT CHANNEL RECONFIGURATION FAILURE message on the DCCH using AM RLC, if the UE fails to reconfigure the new configuration according to a TRANSPORT CHANNEL RECONFIGURATION message.

Transport channel reconfiguration from CELL_DCH to CELL_DCH: Failure (Physical channel failure and cell reselection). Test to confirm that the UE transmits a TRANSPORT CHANNEL RECONFIGURATION FAILURE message after it completes a cell update procedure when the UE cannot synchronise with the SS on the new channel before T312 expires and fails to revert to the old configuration.

Transport channel reconfiguration from CELL_FACH to CELL_DCH: Success. Test to confirm that the UE reconfigures a new channel using dedicated physical channel according to a TRANSPORT CHANNEL RECONFIGURATION message.

Transport Channel Reconfiguration from CELL_DCH to CELL_DCH: Success (Subsequently received). Test to confirm that if the UE receives a TRANSPORT CHANNEL RECONFIGURATION message before the UE configures the radio bearer according to the previous TRANSPORT CHANNEL RECONFIGURATION message it ignores the second TRANSPORT CHANNEL RECONFIGURATION message and configures according to the previous TRANSPORT CHANNEL RECONFIGURATION message.

Transport Channel Reconfiguration from CELL_FACH to CELL_DCH: Success (Subsequently received). Test to confirm that if the UE receives a TRANSPORT CHANNEL RECONFIGURATION message before the UE configures the radio bearer according to the previous TRANSPORT CHANNEL RECONFIGURATION message it ignores the second TRANSPORT CHANNEL RECONFIGURATION message and configures according to the previous TRANSPORT CHANNEL RECONFIGURATION message.

Transport channel reconfiguration from CELL_DCH to CELL_DCH: Success with uplink transmission rate modification. Test to confirm that the UE transmits TRANSPORT CHANNEL RECONFIGURATION COMPLETE message on the uplink DCCH using AM RLC after reconfigure its available uplink TFC according to a TRANSPORT CHANNEL RECONFIGURATION message.

Transport channel reconfiguration from CELL_FACH to CELL_DCH (Frequency band modification): Success. Test to confirm that the UE transits from CELL_FACH to CELL_DCH according to TRANSPORT CHANNEL RECONFIGURATION message. Test to confirm that the UE transmits TRANSPORT CHANNEL RECONFIGURATION message on the uplink DCCH using AM RLC on dedicated physical channel in a different frequency.

Transport Channel Reconfiguration for transition from CELL_DCH to CELL_DCH (Frequency band modification): Success. Test to confirm that the UE transits from CELL_DCH to CELL_DCH according to the TRANSPORT CHANNEL RECONFIGURATION message. Test to confirm that the UE transmits the TRANSPORT CHANNEL RECONFIGURATION COMPLETE message on the uplink DCCH using AM RLC on a dedicated physical channel in a different frequency.

Transport Channel Reconfiguration from CELL_DCH to CELL_DCH: Success (with active HS-DSCH reception, not changing the value of TTI during UL rate modification). Test to confirm that the UE reconfigures the transport and physical channel while being mapped to HS-DSCH according to the received TRANSPORT CHANNEL RECONFIGURATION message. Test to confirm that the UE keeps the same value of TTI (transmission time interval) during the procedure.

Transport format combination control in CELL_DCH: Failure (Invalid configuration). Test to confirm that the UE transmits a TRANSPORT FORMAT COMBINATION CONTROL FAILURE message on the DCCH using AM RLC if it receives a TRANSPORT FORMAT COMBINATION CONTROL message including an invalid configuration.

Physical channel reconfiguration for transition from CELL_DCH to CELL_DCH (code modification): Success. Test to confirm that the UE reconfigures the physical channel parameters according to a PHYSICAL CHANNEL RECONFIGURATION message received from the SS. After the reconfiguration, the UE shall be able to communicate with the SS on the new physical channel.

Physical channel reconfiguration for transition from CELL_DCH to CELL_DCH (code modification): Failure (Unsupported configuration). Test to confirm that the UE keeps its configuration and transmits a PHYSICAL CHANNEL RECONFIGURATION FAILURE message on the DCCH using AM RLC if the received PHYSICAL CHANNEL RECONFIGURATION message includes unsupported configuration parameters for the UE.

Physical channel reconfiguration for transition from CELL_DCH to CELL_DCH (code modification): Failure (Incompatible simultaneous reconfiguration). Test to confirm that if the UE receives a PHYSICAL CHANNEL RECONFIGURATION message during a reconfiguring procedure due to a radio bearer message other than PHYSICAL CHANNEL RECONFIGURATION, it shall keep its configuration as if the PHYSICAL CHANNEL RECONFIGURATION message had not been received and complete the reconfiguration procedure according to the previously received message.

Physical channel reconfiguration for transition from CELL_DCH to CELL_DCH (code modification): Failure (Invalid message reception and Invalid configuration). Test to confirm that the UE transmits a PHYSICAL CHANNEL RECONFIGURATION FAILURE message on the DCCH using AM RLC if it receives an invalid PHYSICAL CHANNEL RECONFIGURATION message which does not include any IEs except IE "Message Type". Test to confirm that the UE transmits a PHYSICAL CHANNEL RECONFIGURATION FAILURE message on the DCCH using AM RLC if it receives a PHYSICAL CHANNEL RECONFIGURATION message including some IEs set to give an invalid configuration.

Physical channel reconfiguration for transition from CELL_DCH to CELL _FACH: Success. Test to confirm that the UE reconfigures a common physical channel according to the PHYSICAL CHANNEL RECONFIGURATION message, which invoke the UE to transit from CELL_DCH to CELL_FACH.

Physical channel reconfiguration for transition from CELL_DCH to CELL_FACH: Success (Cell re-selection). Test to confirm that the UE transmits a PHYSICAL CHANNEL RECONFIGURATION COMPLETE message after the UE completes a cell update procedure.

Physical channel reconfiguration for transition from CELL_FACH to CELL_DCH: Success. Test to confirm that the UE reconfigures a new physical channel according to a PHYSICAL CHANNEL RECONFIGURATION message, which invoke UE to transit from CELL_FACH to CELL_DCH.

Physical channel reconfiguration for transition from CELL_FACH to CELL_DCH: Failure (Physical channel failure and successful reversion to old configuration). Test to confirm that the UE reverts to the old configuration and transmits a PHYSICAL CHANNEL RECONFIGURATION FAILURE message on the DCCH using AM RLC if the UE fails to reconfigure the new physical channel according to a PHYSICAL CHANNEL RECONFIGURATION message before the T312 expiry.

Physical channel reconfiguration for transition from CELL_FACH to CELL_DCH: Failure (Physical channel failure and cell update). Test to confirm that the UE initiates a cell update procedure after it fails to reconfigure the new physical channel and selects another cell. Test to confirm that UE transmits a PHYSICAL CHANNEL RECONFIGURATION FAILURE message after UE completes cell update procedure.

Physical channel reconfiguration for transition from CELL_FACH to CELL_DCH: Failure (Invalid message reception and Invalid configuration). Test to confirm that the UE transmits a PHYSICAL CHANNEL RECONFIGURATION FAILURE message on the DCCH using AM RLC if the received message does not include any IEs except IE "Message Type". Test to confirm that the UE transmits a PHYSICAL CHANNEL RECONFIGURATION FAILURE message on the DCCH using AM RLC if it receives a PHYSICAL CHANNEL RECONFIGURATION message including some IEs which are set to give an invalid configuration.

Physical channel reconfiguration for transition from CELL_DCH to CELL_DCH (code modification): Success (Subsequently received). Test to confirm that if the UE receives a PHYSICAL CHANNEL RECONFIGURATION message before the UE reconfigures the radio bearer according to the previous PHYSICAL CHANNEL RECONFIGURATION message it ignores the new PHYSICAL CHANNEL RECONFIGURATION message and reconfigures according to the previous PHYSICAL CHANNEL RECONFIGURATION message.

Physical channel reconfiguration for transition from CELL_FACH to CELL_DCH: Success (Subsequently received). Test to confirm that if the UE receives a PHYSICAL CHANNEL RECONFIGURATION message before the UE reconfigures the radio bearer according to the previous PHYSICAL CHANNEL RECONFIGURATION message it ignores the new PHYSICAL CHANNEL RECONFIGURATION message and reconfigures according to the previous PHYSICAL CHANNEL RECONFIGURATION message.

Physical Channel Reconfiguration from CELL_DCH to CELL_PCH: Success. Test to confirm that the UE transmits a PHYSICAL CHANNEL RECONFIGURATION COMPLETE message and enter CELL_PCH state after it received a PHYSICAL CHANNEL RECONFIGURATION message, which invokes the UE to transit from CELL_DCH to CELL_PCH.

Physical Channel Reconfiguration from CELL_DCH to URA_PCH: Success. Test to confirm that the UE transmits a PHYSICAL CHANNEL RECONFIGURATION COMPLETE message and enter URA_PCH state after it received a PHYSICAL CHANNEL RECONFIGURATION message, which invokes the UE to transit from CELL_DCH to URA_PCH.

Physical Channel Reconfiguration from CELL_FACH to URA_PCH: Success. Test to confirm that the UE, when receiving a PHYSICAL CHANNEL RECONFIGURATION message, responds by transmitting a PHYSICAL CHANNEL RECONFIGURATION COMPLETE message on the uplink DCCH using AM RLC. Test to confirm that the response message is transmitted using the old configuration before the state transition, and that the UE enters the URA_PCH state.

Physical Channel Reconfiguration from CELL_FACH to CELL_PCH: Success. Test to confirm that the UE, when receiving a PHYSICAL CHANNEL RECONFIGURATION message, responds by transmitting a PHYSICAL CHANNEL RECONFIGURATION COMPLETE message on the uplink DCCH using AM RLC. Test to confirm that the response message is transmitted using the old configuration before the state transition, and that the UE enters the CELL_PCH state.

Physical channel reconfiguration for transition from CELL_DCH to CELL_DCH (Hard handover to another frequency with timing maintain): Success. Test to confirm that the UE transmits PHYSICAL CHANNEL RECONFIGURATION COMPLETE message on the uplink DCCH using AM RLC, on a dedicated physical channel in a different frequency band.

Physical channel reconfiguration for transition from CELL_DCH to CELL_FACH (Frequency band modification): Success. Test to confirm that the UE transits from CELL_DCH to CELL_FACH according to the PHYSICAL CHANNEL RECONFIGURATION message. Test to confirm that the UE transmits PHYSICAL CHANNEL RECONFIGURATION COMPLETE message on the uplink DCCH using AM RLC on a common physical channel in a different frequency.

Physical Channel Reconfiguration from CELL_DCH to CELL_PCH (Frequency band modification): Success. Test to confirm that the UE transmits PHYSICAL CHANNEL RECONFIGURATION COMPLETE message on the uplink DCCH using AM RLC. Test to confirm that the UE transits from CELL_DCH to CELL_PCH according to the PHYSICAL CHANNEL RECON-FIGURATION message. Test to confirm that the UE releases a dedicated physical channel and selects a common physical channel in a different frequency.

Physical channel reconfiguration from CELL_FACH to CELL_PCH: Success. Test to confirm that the UE transmits PHYSICAL CHANNEL RE-CONFIGURATION COMPLETE message on the uplink DCCH using AM RLC. Test to confirm that the UE transits from CELL_FACH to CELL_PCH according to the PHYSICAL CHANNEL RECONFIGURATION message.

Physical channel reconfiguration for transition from CELL_DCH to CELL_DCH (Downlink channeliation code modification): Success. Test to confirm that the UE change assigned downlink channelisation code by SS ac-cording to a PHYSICAL CHANNEL RECONFIGURATION message. Test to confirm that the UE response PHYSICAL CHANNEL RECONFIGURATION COMPLETE message on the uplink DCCH using AM RLC

Physical channel reconfiguration for transition from CELL_DCH to CELL_DCH (Compressed mode initiation): Success. Test to confirm that the UE activates compressed mode according to a PHYSICAL CHANNEL RE-CONFIGURATION message. Test to confirm that the UE transmits a PHYSI-CAL CHANNEL RECONFIGURATION COMPLETE message on the uplink DCCH using AM RLC. Test to confirm that the UE deactivates compressed mode according to a PHYSICAL CHANNEL RECONFIGURATION message. Test to confirm that the UE transmits a PHYSICAL CHANNEL RECONFIGURA-TION COMPLETE message on the uplink DCCH using AM RLC.

Physical channel reconfiguration for transition from CELL_DCH to CELL_DCH (Modify active set cell): Success. Test to confirm that the UE transmits PHYSICAL CHANNEL RECONFIGURATION COMPLETE mes-sage on the uplink DCCH using AM RLC on a dedicated physical channel of same frequency in another cell.

Physical channel reconfiguration transition from CELL_FACH to URA_PCH: Success. Test to confirm that the UE transmits the PHYSICAL CHANNEL RECONFIGURATION COMPLETE message on the uplink DCCH using AM RLC. Test to confirm that the UE transits from CELL_FACH to URA_PCH according to the PHYSICAL CHANNEL RECONFIGURA-TION message.

Physical channel reconfiguration for transition from CELL_DCH to URA_PCH (Frequency band modification): Success. Test to confirm that the UE transmits a PHYSICAL CHANNEL RECONFIGURATION COM-PLETE message on the uplink DCCH using AM RLC. Test to confirm that the UE transits from CELL_DCH to URA_PCH according to the PHYSICAL CHANNEL RECONFIGURATION message. Test to confirm that the UE releases the dedicated physical channel and selects a common physical channel in a different frequency.

Physical channel reconfiguration for transition from CELL_FACH to CELL_DCH (Frequency band modification): Success. Test to confirm that the UE transits from CELL_FACH to CELL_DCH according to the PHYSICAL CHANNEL RECONFIGURATION message. Test to confirm that the UE transmits PHYSICAL CHANNEL RECONFIGURATION COMPLETE message on the uplink DCCH using AM RLC on a dedicated physical channel in a different frequency.

Physical channel reconfiguration from CELL_FACH to CELL_PCH (Frequency band modification): Success. Test to confirm that the UE transmits a PHYSICAL CHANNEL RECONFIGURATION COMPLETE message on the uplink DCCH using AM RLC. Test to confirm that the UE transits from CELL_FACH to CELL_PCH according to the PHYSICAL CHANNEL RECONFIGURATION message. Test to confirm that the UE selects a common physical channel in a different frequency.

Physical channel reconfiguration for transition from CELL_FACH to URA_PCH (Frequency band modification): Success. Test to confirm that the UE transmits a PHYSICAL CHANNEL RECONFIGURATION COMPLETE message on the uplink DCCH using AM RLC. Test to confirm that the UE transits from CELL_FACH to URA_PCH according to the PHYSICAL CHANNEL RECONFIGURATION message. Test to confirm that the UE selects a common physical channel in a different frequency.

Physical channel reconfiguration for transition from CELL_FACH to CELL_FACH (Frequency band modification): Success. Test to confirm that the UE transits from CELL_FACH to CELL_FACH according to the PHYSICAL CHANNEL RECONFIGURATION message. Test to confirm that the UE transmits PHYSICAL CHANNEL RECONFIGURATION COMPLETE message on the uplink DCCH using AM RLC on a common physical channel in a different frequency.

Physical channel reconfiguration for transition from CELL_DCH to CELL_DCH (Hard handover to another frequency with timing re-initialised). Test to confirm that the UE is able to perform a hard-handover with change of frequency, with and without prior measurements on the target frequency. Test to confirm that the UE answers with a PHYSICAL CHANNEL RECONFIGURATION COMPLETE message when the procedure has been initiated with the PHYSICAL CHANNEL RECONFIGURATION message. Test to confirm that the UE stops intra-frequency measurements after the inter-frequency handover has been performed, until a MEASUREMENT CONTROL message is received from the SS. Test to confirm that the UE computes as it shall the CFN to be used after the handover. Test to confirm that the UE deactivates compressed mode (if required) when it has been ordered to do so in the PHYSICAL CHANNEL RECONFIGURATION message. Test to confirm that the UE includes the IE "COUNT-C activation time" and the IE "START list" (in the IE "Uplink counter synchronisation info") in the response message if ciphering is active for any radio bearer using RLC-TM.

Physical channel reconfiguration for transition from CELL_DCH to CELL_DCH (Hard handover to another frequency with timing re-initialised): Failure (Physical channel failure and reversion to old channel). Test to confirm that the UE reverts to the old configuration (including measurement configurations, ciphering procedures and compressed mode configurations if required) and transmits a PHYSICAL CHANNEL RECONFIGURATION FAILURE message on the DCCH using AM RLC if the UE fails to reconfigure the new physical channel according to the received PHYSICAL CHANNEL RECONFIGURATION message before timer T312 expiry.

Physical Channel Reconfiguration for transition from CELL_DCH to CELL_DCH: Success (Seamless SRNS relocation) (without pending of ciphering). Test to confirm that the UE performs a combined hard handover and SRNS relocation and then transmit a PHYSICAL CHANNEL RECONFIGURATION COMPLETE message in the new cell. In the case that ciphering is applied by the network, to confirm that the UE applies the new ciphering algorithm following a successful SRNS relocation.

Physical Channel Reconfiguration for transition from CELL_DCH to CELL_DCH: Success (serving HS-DSCH cell change without MAC-hs reset). Test to confirm that the UE changes the serving HS-DSCH cell according to the received PHYSICAL CHANNEL RECONFIGURATION message in case of no MAC-hs reset.

Physical Channel Reconfiguration for transition from CELL_DCH to CELL_DCH: Success (serving HS-DSCH cell change with MAC-hs reset). Test to confirm that the UE changes the serving HS-DSCH cell according to the received PHYSICAL CHANNEL RECONFIGURATION message, in case of MAC-hs reset.

Physical Channel Reconfiguration for transition from CELL_DCH to CELL_DCH: Success (Two radio links, change of HS-PDSCH configuration). Test to confirm that, when two radio links are used, the UE changes HS-DSCH specific uplink DPCH power control and measurement feedback configuration according to the received PHYSICAL CHANNEL RECONFIGURATION message.

Physical Channel Reconfiguration for transition from CELL_DCH to CELL_DCH: Success (change of HS-PDSCH configuration). Test to confirm that, the UE changes HS-DSCH specific HS-SCCH configuration according to the received PHYSICAL CHANNEL RECONFIGURATION message.

Physical Channel Reconfiguration for transition from CELL_DCH to CELL_DCH: Success (Timing re-initialised hard handover to another frequency, signalling only). Test to confirm that the UE makes timing re-initialised inter-frequency hard handover on a signalling only configuration without prior measurement on the target frequency according to the received PHYSICAL CHANNEL RECONFIGURATION message.

Physical Channel Reconfiguration for transition from CELL_DCH to CELL_DCH: Success (Timing re-initialized hard handover to another frequency, Serving HS-DSCH cell change). Test to confirm that the UE is able to perform a timing re-initialised hard handover to another frequency without prior measurement on the target frequency in conjunction with a serving HS-DSCH cell change according to the received PHYSICAL CHANNEL RECONFIGURATION message.

Physical Channel Reconfiguration for transition from CELL_DCH to CELL_DCH: Success (Seamless SRNS relocation with pending of ciphering). Test to confirm that the UE includes the previously received new keys from the last SECURITY MODE COMMAND in the new ciphering configuration in the case the ciphering configuration for RB2 from the last received SECURITY MODE COMMAND has not yet been applied because the activation times not having been reached.

Physical Channel Reconfiguration for transition from CELL_DCH to CELL_DCH: Failure (Radio link failure in new configuration). Test to confirm that the UE enters idle mode state when UE detects radio link failure after UE started using the new configuration but before receiving the RLC acknowledgement of the reconfiguration complete message.

Physical Channel Reconfiguration for transition from CELL_DCH to URA_PCH: Failure (Radio link failure in old configuration). Test to confirm that the UE aborts reconfiguration to URA_PCH and performs cell update when UE detects radio link failure before receiving the RLC acknowledgement of the reconfiguration complete message in the old configuration.

Physical channel reconfiguration for transition from CELL_DCH to CELL_DCH (Hard handover to another frequency with timing re-initialised. Serving HS-DSCH cell change): Failure (Physical channel failure and reversion to old channel). Test to confirm that the UE reverts to the old configuration (including measurement configurations, ciphering procedures and compressed mode configurations if required), and transmits a PHYSICAL CHANNEL RECONFIGURATION FAILURE message on the DCCH using AM RLC if the UE fails to reconfigure the new physical channel according to the received PHYSICAL CHANNEL RECONFIGURATION message before timer T312 expiry.

Physical channel reconfiguration for transition from CELL_DCH to CELL_DCH (Compressed mode initiation, with active HS-DSCH reception): Success. Test to confirm that the UE configures compressed mode according to a PHYSICAL CHANNEL RECONFIGURATION message during active HS-DSCH reception. Test to confirm that the UE activates compressed mode according to the previously stored configuration when receiving a MEASUREMENT CONTROL message during active HS-DSCH reception.

Physical Channel Reconfiguration for transition from CELL_DCH to CELL_DCH: Success (Timing re-initialized hard handover to another

frequency, serving HS-DSCH cell change, compressed mode). Test to confirm that the UE is able to perform timing re-initialised hard handover to another frequency after compressed mode measurement on the target frequency in conjunction with a serving HS-DSCH cell change according to the received PHYSICAL CHANNEL RECONFIGURATION message.

Physical Channel Reconfiguration from CELL_DCH to URA_PCH: Success (stop of HS-DSCH reception). Test to confirm that the UE transmits a PHYSICAL CHANNEL RECONFIGURATION COMPLETE message and enter URA_PCH state after it received a PHYSICAL CHANNEL RECONFIGURATION message, which invokes the UE to transit from CELL_DCH with active HS-DSCH reception to URA_PCH.

Physical Channel Reconfiguration for transition from CELL_DCH to URA_PCH: Success (frequency modification, stop of E-DCH transmission). Test to confirm that the UE transmits a PHYSICAL CHANNEL RECONFIGURATION COMPLETE message on the uplink DCCH using AM RLC; Test to confirm that the UE transits from CELL_DCH with active E-DCH transmission and HS-HSCH reception to URA_PCH state on a different frequency after it received a PHYSICAL CHANNEL RECONFIGURATION message; Test to confirm that the UE releases the dedicated physical channel and select a common physical channel on a different frequency.

Physical Channel Reconfiguration for transition from CELL_DCH to CELL_DCH: Success (serving E-DCH cell change). Test to confirm that the UE changes the serving E-DCH cell according to the received PHYSICAL CHANNEL RECONFIGURATION message.

Physical channel reconfiguration for transition from CELL_DCH to CELL_DCH: Success (Timing re-initialized hard handover to another frequency, Serving E-DCH cell change, compressed mode). Test to confirm that the UE is able to perform a timing re-initialised hard handover to another frequency after compressed mode measurement on the target frequency in conjunction with a serving E-DCH cell change according to the received PHYSICAL CHANNEL RECONFIGURATION message.

Physical Channel Reconfiguration for transition from CELL_DCH to CELL_DCH: Failure (Timing re-initialized hard handover, Serving E-DCH cell change, physical channel failure and reversion to old channel). Test to confirm that the UE reverts to the old E-DCH / HS-DSCH configuration and transmits a PHYSICAL CHANNEL RECONFIGURATION FAILURE message if the UE fails to reconfigure the new physical channel according to the received PHYSICAL CHANNEL RECONFIGURATION message before the expiry of timer T312.

RRC connection mobility procedure
Cell Update: cell reselection in CELL_FACH. Test to confirm that the UE executes a cell update procedure after the successful reselection of another

UTRA cell. Test to confirm that the UE sends the correct uplink response message when executing cell update procedure due to cell reselection.

Cell Update: cell reselection in CELL_FACH (Cells belong to different frequency bands). Test to confirm that the UE executes a cell update procedure after the successful reselection of another UTRA cell in different frequency band. Test to confirm that the UE sends the correct uplink response message when executing cell update procedure due to cell reselection

Cell Update: cell reselection in CELL_PCH. Test to confirm that the UE, in CELL_PCH state, executes a cell update procedure after the successful reselection of another UTRA cell.

Cell Update: periodical cell update in CELL_FACH. Test to confirm that the UE executes a periodical cell update procedure following the expiry of timer T305.

Cell Update: periodical cell update in CELL_PCH. Test to confirm that the UE, in CELL_PCH state, executes a cell update procedure after the expiry of timer T305.

Cell Update: UL data transmission in URA_PCH. Test to confirm that the UE executes a cell update procedure when the UE transmits uplink data if the UE is in URA_PCH state.

Cell Update: UL data transmission in CELL_PCH. Test to confirm that the UE executes a cell update procedure when the UE transmits uplink data if the UE is in CELL_PCH state.

Cell Update: re-entering of service area after T305 expiry and being out of service area. Test to confirm that the UE performs a cell search after experiencing an "out of service area" condition. Test to confirm that the UE initiates cell updating procedure if it manages to re-enter the service area.

Cell Update: expiry of T307 after T305 expiry and being out of service area. Test to confirm that the UE moves to idle mode after the expiry of T307, indicating that it is out of service area when attempting to perform a periodic cell updating procedure.

Cell Update: Success after T302 time-out. Test to confirm that the UE repeats the transmission of CELL UPDATE message after failing to receive any response from the SS before T302 timer expires.

Cell Update: Failure (After Maximum Re-transmissions). Test to confirm that the UE repeats the cell update procedure upon the expiry of timer T302 and moves to idle state when its internal counter V302 is greater than N302.

Cell Update: Reception of Invalid CELL UPDATE CONFIRM Message. Test to confirm that the UE retransmits a CELL UPDATE message when it receives an invalid CELL UPDATE CONFIRM message, before the number of retransmissions has reached the maximum allowed value.

Cell Update: Incompatible simultaneous reconfiguration. Test to confirm that the UE retransmits a CELL UPDATE message when it receives a CELL UPDATE CONFIRM message that includes "Physical channel information elements" and UE's variable ORDERED_RECONFIGURATION is set to TRUE because of an ongoing Reconfiguration procedure, before the number of retransmissions has reached the maximum allowed value.

Cell Update: Unrecoverable error in Acknowledged Mode RLC SRB. Test to confirm that the UE reports the occurrence of an unrecoverable error in a C-plane AM RLC entity by initiating cell update procedure. Test to confirm that UE enters idle mode state after receiving RRC CONNECTION RELEASE message on the downlink CCCH.

Cell Update: Failure (UTRAN initiate an RRC connection release procedure on CCCH). Test to confirm that the UE moves to idle state upon the reception of a RRC CONNECTION RELEASE message on CCCH.

Cell Update: Radio Link Failure (T314>0, T315=0), CS RAB established. Test to confirm that the UE shall try to find a new cell after detecting that a radio link failure has occurred. Test to confirm that the UE performs a cell selection procedure when it fails to configure the physical channel(s) indicated in the CELL UPDATE CONFIRM message.

Cell Update: Reception of CELL UPDATE CONFIRM Message that causes invalid configuration. Test to confirm that the UE retransmits a CELL UPDATE message when it receives a CELL UPDATE CONFIRM message that will trigger an invalid configuration in the UE, if the number of retransmissions has not reached the maximum allowed value.

Cell Update: Cell reselection to cell of another PLMN belonging to the equivalent PLMN list. Test to confirm that the UE executes a cell update procedure after a successful reselection to another UTRA cell with a PLMN identity different from the original cell but with a PLMN identity that is part of the equivalent PLMN list in the UE. Test to confirm that the UE sends the correct uplink response message when executing cell update procedure due to cell reselection. Test to confirm that the UE refrains from executing a cell update procedure to a better UTRA cell with another PLMN identity when that PLMN identity is not part of the equivalent PLMN list in the UE.

Cell update: Restricted cell reselection to a cell belonging to forbidden LA list (Cell_FACH). Test to confirm that the UE executes a cell update procedure after a successful reselection of another UTRA cell with a LA identity that is not part of the list of LAs stored in the UE as "forbidden location areas for roaming". Test to confirm that if the UE get a release message and is moved to idle mode, performs a registration update where the LA list is updated and the UE again enters connected mode, that the UE refrains from selecting that same UTRA cell if that is part of the forbidden LA list.

Cell Update: HCS cell reselection in CELL_FACH. Test to confirm that the UE can read HCS related SIB information and act upon all HCS parameters in CELL_FACH state. Test to confirm that the UE executes a cell update procedure after the successful reselection of another UTRA cell in CELL_FACH state. Test to confirm that the UE sends the correct uplink response message when executing cell update procedure due to cell reselection.

Cell Update: HCS cell reselection in CELL_PCH. Test to confirm that the UE can read HCS related SIB information and act upon all HCS parameters in CELL_PCH state. Test to confirm that the UE executes a cell update procedure after the successful reselection of another UTRA cell in CELL_PCH state. Test to confirm that the UE sends the correct uplink response message when executing cell update procedure due to cell reselection.

CELL UPDATE: Radio Link Failure (T314=0, T315=0). Test to confirm that the UE releases all resources and enters idle mode when there is a radio link failure.

Cell Update: Radio Link Failure (T314>0, T315=0), PS RAB established. Test to confirm that the UE shall indicate to the non-access stratum the release of radio access bearer which is associated with T315 and try to find a new cell after detecting that a radio link failure has occurred.

Cell Update: Radio Link Failure (T314=0, T315>0), CS RAB. Test to confirm that the UE release radio access bearer which is associated with T314 and try to find a new cell after detecting that a radio link failure has occurred.

Cell Update: Radio Link Failure (T314=0, T315>0), PS RAB. Test to confirm that the UE release radio access bearer which is associated with T314 and try to find a new cell after detecting that a radio link failure has occurred.

Cell Update: Radio Link Failure (T314>0, T315>0), CS RAB. Test to confirm that the UE shall indicate to the non-access stratum the release of radio access bearer which is associated with T314 and try to find a new cell after detecting that a radio link failure has occurred. Test to confirm that the UE enters idle mode after T314 expires and T302 and T315 are not running.

Cell Update: Radio Link Failure (T314>0, T315>0), PS RAB. Test to confirm that the UE shall indicate to the non-access stratum the release of radio access bearer which is associated with T315 and try to find a new cell after detecting that a radio link failure has occurred. Test to confirm that the UE shall indicate to the non-access stratum the release of radio access bearer which is associated with T314 and try to find a new cell after detecting that a radio link failure has occurred. (This test purpose is only applicable when CS RAB is set up in the initial condition.). Test to confirm that the UE enters idle mode after T315 expires and T302 and T314 are not running.

Cell Update: re-entering of service area from URA_PCH after T316 expiry but before T317 expiry. Test to confirm that the UE executes a cell update procedure when the UE re-enters the service area before the expiry of timer T317, after expiry of T316.

Cell Update: Transition from URA_PCH to CELL_DCH, start of HS-DSCH reception. Test to confirm that the UE enters the CELL_DCH state after it receives a CELL UPDATE CONFIRM message with a physical channel configuration causing it to start HS-DSCH reception.

Cell Update: Transition from CELL_PCH to CELL_DCH, start of HS-DSCH reception, frequency band modification. Test to confirm that the UE enters the CELL_DCH state after it receives a CELL UPDATE CONFIRM message with a physical channel configuration causing it to start HS-DSCH reception on a different cell and frequency. Test to confirm that the UE enters CELL_PCH state on another frequency and stops HS-DSCH reception when it receives a PHYSICAL CHANNEL RECONFIGURATION message.

Cell Update: Transition from CELL_DCH to CELL_FACH, stop of HS-DSCH reception. Test to confirm that the UE stops HS-DSCH reception after a radio link failure in CELL_DCH during HS-DSCH reception.

Cell Update: Transition from CELL_DCH to CELL_DCH, with active HS-DSCH reception. Test to confirm that the UE keeps the RB mapping option for HS-DSCH reception after a radio link failure in CELL_DCH during HS-DSCH reception.

Cell Update: Transition from CELL_DCH to CELL_FACH (stop of HS-DSCH reception with frequency modification). Test to confirm that the UE stops HS-DSCH reception after a radio link failure in CELL_DCH during HS-DSCH reception.

Cell Update: Transition from CELL_DCH to CELL_DCH (with active HS-DSCH reception and frequency modification). Test to confirm that the UE keeps the RB mapping option for HS-DSCH reception after a radio link failure in CELL_DCH during HS-DSCH reception.

Cell Update: state specific handling of Treselection and Qhyst for cell reselection in CELL_FACH. Test to confirm that the UE uses the correct SIB 4 IEs to perform cell reselection calculation in CELL_FACH.

Cell Update: state specific handling of Treselection and Qhyst for cell reselection in CELL_PCH. Test to confirm that the UE uses the correct SIB 4 IEs to perform cell reselection calculation in CELL_PCH.

Cell update: Transition from CELL_PCH to CELL_DCH, inclusion of establishment cause. Test to confirm that, in the case the Cell Update procedure is initiated by a UE in CELL_PCH state in order to transmit the Initial Direct Transfer message, the UE shall include the IE "Establishment cause" in the Cell Update message. Test to confirm that the IE "Establishment Cause" is not included in the following Initial Direct Transfer message.

Cell Update: Transition from URA_PCH to CELL_DCH: Success (start of E-DCH transmission). Test to confirm that the UE enters the CELL_DCH state after it receives a CELL UPDATE CONFIRM message with a physical channel configuration causing it to start E-DCH transmission and HS_DSCH reception. Test to confirm that the UE executes a cell update procedure when the UE transmits uplink data if the UE is in CELL_PCH state.

Cell Update: Transition from CELL_PCH to CELL_DCH: Success (frequency modification, start of E-DCH transmission). Test to confirm that the UE enters the CELL_DCH state after it receives a CELL UPDATE CONFIRM message with a physical channel configuration causing it to start E-DCH transmission and HS-DSCH reception on a different cell and frequency. Test to confirm that the UE enters CELL_PCH state on another frequency and stops E-DCH transmission and HS-DSCH reception when it receives a RADIO BEARER RE-CONFIGURATION message.

Cell Update: Radio Link Failure, with active E-DCH transmission. Test to confirm that the UE detects the radio link failure condition when the F-DPCH physical channel is established. Test to confirm that the UE stops the E-DCH transmission and performs a Cell Update procedure after radio link failure. Test to confirm that the UE keeps the radio bearer mapping option and transport channel configuration for E-DCH after the radio link failure. Test to confirm that the UE resumes the E-DCH transmission after the Cell Update procedure.

URA Update: Change of URA. Test to confirm that the UE executes an URA update procedure after the successful change of URA. Test to confirm that the UE performs an URA update procedure after it detects that SIB 2 is not broadcasted. Test to confirm that the UE performs an URA update procedure after it detects a confirmation error of URA identity list.

URA Update: Change of URA (Cells belong to different frequency bands). Same test purpose as in previous test except that the cells belong to different frequency bands

URA Update: Periodical URA update and Reception of Invalid message. Test to confirm that the UE executes a URA update procedure after the expiry of timer T305. Test to confirm that the UE handles an invalid URA UPDATE CONFIRM message correctly when executing the URA update procedure.

URA Update: loss of service after expiry of timers T307 and T305. Test to confirm that the UE moves to idle mode after the expiry of timer T307, following an expiry of timer T305 when it discovers that it is out of service area.

URA Update: Success after Confirmation error of URA-ID list. Test to confirm that the UE retries to perform the URA update procedure following a confirmation error of URA-ID list.

URA Update: Failure (V302 is greater than N302: Confirmation error of URA-ID list). Test to confirm that the UE make repeated attempts to perform the URA update procedure following a detection of a confirmation error of URA-ID list. It then moves to idle state when internal counter V302 is greater than N302.

URA Update: Success after T302 timeout. Test to confirm that the UE attempts to repeat the URA update procedure upon the expiry of timer T302.

URA Update: Failure (UTRAN initiate an RRC connection release procedure on CCCH). Test to confirm that the UE moves to idle state upon the reception of RRC CONNECTION RELEASE message on downlink CCCH during a URA update procedure.

URA Update: Reception of URA UPDATE CONFIRM message that causes invalid configuration. Test to confirm that the UE retransmits a URA UPDATE message when it receives a URA UPDATE CONFIRM message that will trigger an invalid configuration in the UE, if the number of retransmissions has not reached the maximum allowed value.

URA Update: Cell reselection to cell of another PLMN belonging to the equivalent PLMN list. Test to confirm that the UE executes a URA update procedure after a successful reselection of another UTRA cell with a URA identity that is not the URA of the UE and with a PLMN identity different from the original cell but with a PLMN that is part of the equivalent PLMN list in the UE. Test to confirm that the UE refrains from executing a URA update procedure to a better UTRA cell with another PLMN identity when that PLMN identity is not part of the equivalent PLMN list in the UE.

Restricted cell reselection to a cell belonging to forbidden LA list (URA_ PCH). Test to confirm that the UE refrains from selects a UTRA cell and performs a URA update if that cell has a LA identity that is part of the list of LAs stored in the UE as "forbidden location areas for roaming".

URA Update: Change of URA due to HCS Cell Reselection. Test to confirm that the UE can read HCS related SIB information and act upon all HCS parameters in URA_PCH state. Test to confirm that the UE executes an URA update procedure after the successful change of URA due to HCS Cell Reselection in URA_PCH state. Test to confirm UE responds correctly when it re-selects to a new cell while waiting from URA UPDATE CONFIRM message from SS.

UTRAN Mobility Information: Success. Test to confirm that the UE starts to use the new identities after it receives a UTRAN MOBILITY INFORMATION message from the SS.

UTRAN Mobility Information: Failure (Invalid message reception). Test to confirm that the UE ignore the erroneous UTRAN MOBILITY INFORMATION message and report this event to the UTRAN by sending UTRAN MOBILITY INFORMATION FAILURE message, stating the appropriate failure cause and information.

UTRAN MOBILITY INFORMATION: Seamless SRNS relocation in CELL_DCH (without pending of ciphering). Test to confirm that the UE sends calculated START values for each CN domain to SS after a successful SRNS relocation. In the case that ciphering is applied by the network, to confirm that the UE restarts ciphering following a successful SRNS relocation. Test to confirm that the UE correctly applies integrity protection after the SRNS relocation.

UTRAN Mobility Information: Shared Network. Test to confirm that the UE reacts on the IE "Primary PLMN identity" in message UTRAN MOBILITY INFORMATION, and forwards this IE to NAS. Test to confirm that the UE sets the IE "PLMN Identity" in the INITIAL DIRECT TRANSFER message to the correct PLMN information received in UTRAN MOBILITY INFORMATION message.

Active set update in soft handover: Radio Link addition. Test to confirm that the UE continues to communicate with the SS on both the additional radio link and an already existing radio link after the radio link addition.

Active set update in soft handover: Radio Link removal. Test to confirm that the UE continues to communicate with the SS on the remaining radio link after radio link removal on the active set. Test to confirm that the UE is not using the removed radio link to communicate with the SS.

Active set update in soft handover: Combined radio link addition and removal. Test to confirm that the UE continues to communicate with the SS on the added radio link and removes radio link which exists prior to the execution of active set update procedure.

Active set update in soft handover: Invalid Configuration. Test to confirm that the UE transmits an ACTIVE SET UPDATE FAILURE message on the DCCH using AM RLC, if the received ACTIVE SET UPDATE message includes a radio link which is specified in both IE "Radio Link Addition Information" and IE "Radio Link Removal Information".

Active set update in soft handover: Reception of an ACTIVE SET UPDATE message in wrong state. Test to confirm that the UE transmit an ACTIVE SET UPDATE FAILURE message when it receives an ACTIVE SET UPDATE message in any state other then CELL_DCH.

Active set update in soft handover: Invalid Message Reception. Test to confirm that the UE retains its active set list and transmits an ACTIVE SET UPDATE FAILURE message when it receives an invalid ACTIVE SET UPDATE message.

Active set update in soft handover: Radio Link addition in multiple radio link environments. Test to confirm that the UE communicates with the SS on all radio link in the active set and keeps the connection when some of the radio links are faded out.

Active set update in soft handover: Radio Link removal (stop of HS-PDSCH reception). Test to confirm that the UE continues to communicate with the SS on the remaining radio link after radio link removal on the active set. Test to confirm that UE removes the HS-PDSCH configuration when the serving HS-DSCH radio link is removed. Test to confirm that the UE is not using the removed radio link to communicate with the SS.

Active set update in soft handover: Radio Link addition and serving HS-DSCH / E-DCH cell change. Test to confirm that the UE performs a radio link addition with serving HS-DSCH / E-DCH serving cell change according to the received ACTIVE SET UPDATE message.

ISHO from UTRAN/To GSM/Speech/Success. To test that the UE supporting both GSM and UTRAN hands over from a UTRAN serving cell to the indicated channel of GSM target cell when the UE is in the speech call active state and receives an HANDOVER FROM UTRAN COMMAND.

ISHO from UTRAN/To GSM/Data/Same data rate/Success. To test that the UE hands over to the indicated channel of same data rate in the GSM target cell when it is in the data call active state in the UTRAN serving cell and receives an HANDOVER FROM UTRAN COMMAND.

ISHO from UTRAN/To GSM/Data/Same data rate/Extended Rates/ Success. To test that the UE hands over to the indicated channel of same data rate in the GSM target cell when it is in the data call active state in the UTRAN serving cell and receives an HANDOVER FROM UTRAN COMMAND.

ISHO from UTRAN/To GSM/Data/Data rate down grading/Success. To test that the UE hands over to the indicated channel of lower data rate in the GSM target cell when it is in the data call active state in the UTRAN serving cell and receives an HANDOVER FROM UTRAN COMMAND.

ISHO from UTRAN/To GSM/Data/Data rate down grading/Extended Rates/Success. To test that the UE hands over to the indicated channel of lower data rate in the GSM target cell when it is in the data call active state in the UTRAN serving cell and receives an HANDOVER FROM UTRAN COMMAND.

ISHO from UTRAN/To GSM/Speech/Establishment/Success. To test that the UE hands over to the indicated channel in the GSM target cell when it is in the call establishment phase in the UTRAN serving cell and receives an HANDOVER FROM UTRAN COMMAND.

ISHO from UTRAN/To GSM/Speech/Failure. To test that the UE reactivates the old configuration and uses this to transmit a HANDOVER FROM UTRAN FAILURE message to the network including IE "Inter-RAT Handover failure cause" which is set to "physical channel failure", when it receives an HANDOVER FROM UTRAN COMMAND and the connection to GSM for handover can not be established. Test to confirm that after the handover failure the UE resumes previously configured compressed mode patterns and measurements.

ISHO from UTRAN/To GSM/Speech/Failure (L2 Establishment). To Test that the UE shall keep its old configuration and transmit a HANDOVER FROM UTRAN FAILURE message, which is set to "physical channel failure" in IE "Inter_RAT HO failure cause", when it receives a HANDOVER FROM UTRAN COMMAND and the connection to GSM for handover cannot be established due to failure in L2 establishment.

ISHO from UTRAN/To GSM/Speech/Failure (L1 Synchronization). To test that the UE reactivates its old configuration and transmit a HANDOVER FROM UTRAN FAILURE message, which is set to "physical channel failure" in IE "Inter-RAT Handover failure cause", when it receives a HANDOVER FROM UTRAN COMMAND and the connection to GSM for handover cannot be established due to failure in L1 Synchronization.

ISHO from UTRAN/To GSM/Speech/Failure (Invalid Inter-RAT message). To Test that the UE shall keep its old configuration and transmit a HANDOVER FROM UTRAN FAILURE message, which is set to "Inter-RAT protocol error" in IE "Inter_RAT HO failure cause", when it receives a Handover From UTRAN message, with the IE "Inter-RAT message" received within the HANDOVER FROM UTRAN COMMAND message not including a valid inter RAT handover message in accordance with the protocol specifications for the target RAT.

ISHO from UTRAN/To GSM/Speech/Failure (Unsupported configuration). To test that the UE shall keep its old configuration and transmit a HANDOVER FROM UTRAN FAILURE message, which is set to "configuration unacceptable" in IE "Inter-RAT Handover failure cause", when it receives a HANDOVER FROM UTRAN COMMAND message, with the IE "GSM message" containing a HANDOVER COMMAND message including a configuration not supported by the UE.

ISHO from UTRAN/To GSM/Speech/Failure (Reception by UE in CELL_FACH). The UE shall keep its old configuration when the UE receives a HANDOVER FROM UTRAN COMMAND message when in CELL_FACH state and then transmit a HANDOVER FROM UTRAN FAILURE message on the DCCH using AM RLC, which sets value "protocol error" in IE "Inter_RAT HO failure cause" and is set to "Message not compatible with receiver state" in IE "Protocol error cause".

ISHO from UTRAN/To GSM/Speech/Failure (Invalid message reception). The UE shall keep its old configuration when the UE receives a Handover From UTRAN message, that cause the variable PROTOCOL_ERROR_REJECT to be set to TRUE. It shall then transmit a HANDOVER FROM UTRAN FAILURE message on the uplink DCCH. The IE "Protocol error information" shall contain an IE "Protocol error cause" set to " Message extension not comprehended ".

ISHO from UTRAN/To GSM/Speech/Failure (Physical channel Failure and Reversion Failure). The UE shall perform a cell update when the UE fails to revert to the old configuration after the detection of physical channel failure in the target RAT cell as given in the HANDOVER FROM UTRAN COMMAND message. After the UE completes the cell update procedure, the UE shall transmit a HANDOVER FROM UTRAN FAILURE message on the DCCH using AM RLC, including IE "failure cause" set to "physical channel failure".

ISHO from UTRAN/To GSM/ success / call under establishment. To test that the UE supporting both GSM and UTRAN performs handover from UTRAN to the indicated channel of GSM target cell when the UE receives a HANDOVER FROM UTRAN COMMAND in call establishment phase. To test that the UE continues the call in the GSM cell, after successful completion of the Handover.

ISHO from UTRAN/To GSM/Speech/Success (stop of HS-DSCH reception). To test that the UE supporting both GSM and UTRAN hands over from a UTRAN serving cell to the indicated channel of GSM target cell when the UE is in the speech call active state, active PS RAB with HS-DSCH reception and receives an HANDOVER FROM UTRAN COMMAND. Test to confirm that UE stops HS-DSCH reception after receiving the HANDOVER FROM UTRAN COMMAND. For the UEs supporting compressed mode, to verify that the HS-DSCH reception has no impact on the GSM cells measurement when GSM compressed mode is activated.

ISHO from UTRAN/To GSM/Speech/Failure(stop of HS-DSCH reception). To test that the UE reactivates the old configuration and uses this to transmit a HANDOVER FROM UTRAN FAILURE message to the network including IE "Inter-RAT Handover failure cause" which is set to "physical channel failure", when it receives an HANDOVER FROM UTRAN COMMAND and the connection to GSM for handover can not be established. Test to confirm that UE stops using the HS-PDSCH configuration after receiving the HANDOVER FROM UTRAN COMMAND.

ISHO from UTRAN/To GSM/Simultaneous CS and PS domain services/Success/TBF Establishment Success. To test that in UTRAN cell when UE (not supporting DTM) is in speech call active state and PS data call is established, UE performs handover to GSM RAT after receiving HANDOVER FROM UTRAN COMMAND.

ISHO from UTRAN/To GSM/DTM Support/Simultaneous CS and PS domain services/Success. Test to confirm that in UTRAN cell when UE (supporting DTM) is in speech call active state and PS data call is established, UE performs handover to GSM RAT after receiving HANDOVER FROM UTRAN COMMAND.

Cell reselection if cell becomes barred or S<0; UTRAN to GPRS (CELL_ FACH). Test to confirm that the UE performs reselection from UTRAN to GPRS in the state CELL_FACH on the following occasions: Serving cell becomes barred and S<0 for serving cell. Test to confirm when the UE has succeeded in reselecting a cell in the target radio access technology and has initiated the establishment of a connection, it shall release all UTRAN specific resources.

Cell reselection if cell becomes barred or S<0; UTRAN to GPRS (URA_ PCH). Test to confirm that the UE performs reselection from UTRAN to GPRS in the state URA_PCH on the following occasions: Serving cell becomes barred and S<0 for serving cell.

Cell reselection if cell rank changes; UTRAN to GPRS (UE in CELL_ FACH fails to complete an inter-RAT cell reselection). Test to confirm if the inter-RAT cell reselection fails before the UE in CELL_FACH succeeds in initiating the establishment of a connection to the GPRS cell, the UE shall: resume the connection to UTRAN using the resources used before initiating the inter-RAT cell reselection procedure.

Cell reselection if S<0; UTRAN to GPRS (UE in CELL_PCH fails to complete an inter-RAT cell reselection). Test to confirm if the inter-RAT cell reselection fails before the UE in CELL_PCH succeeds in initiating the establishment of a connection to the GPRS cell, the UE shall: resume the connection to UTRAN using the resources used before initiating the inter-RAT cell reselection procedure.

Successful Cell Reselection with RAU – Qoffset value modification; UTRAN to GPRS (CELL_FACH). Test to confirm that the UE performs reselection correctly considering the Qoffset value broadcast in SIB 11.

Inter-RAT cell change order from UTRAN/To GPRS/CELL_DCH/ Success. To test that the UE shall be able to receive a CELL CHANGE ORDER FROM UTRAN message in CELL_DCH state and perform a cell change to another RAT, even if no prior UE measurements have been performed on the target cell. The UE regards the procedure as completed when it has received a successful response from the target RAT, e.g. in case of GSM when it received the response to a (PACKET) CHANNEL REQUEST in the new cell.

Inter-RAT cell change order from UTRAN/To GPRS/CELL_FACH/ Success. To test that the UE shall be able to receive a CELL CHANGE ORDER FROM UTRAN message in CELL_FACH state and perform a cell change to another RAT, even if no prior UE measurements have been performed on the target cell. The UE regards the procedure as completed when it has received a successful response from the target RAT, e.g. in case of GSM when it received the response to a CHANNEL REQUEST in the new cell.

Inter-RAT cell change order from UTRAN/To GPRS/CELL_DCH/ Failure (T309 expiry). Test to confirm that when UE received CELL CHANGE ORDER FROM UTRAN message in CELL_DCH state and if the establishment of the connection to the other RAT failed due to expiry of timer T309 prior to the successful establishment of a connection to the target RAT: a. revert back to the UTRA configuration; b. establish the UTRA physical channel(s) used at the time for reception of CELL CHANGE ORDER FROM UTRAN; c. transmit the CELL CHANGE ORDER FROM UTRAN FAILURE message and set the IE "Inter-RAT change failure" to "physical channel failure".

Inter-RAT cell change order from UTRAN/To GPRS/CELL_DCH/ Failure (Physical channel Failure and Reversion Failure). Test to confirm that when UE received CELL CHANGE ORDER FROM UTRAN message in CELL_DCH state and if the establishment of the connection to the other RAT failed due to other reasons e.g. (random) access failure, rejection due to lack of resources: a. revert back to the UTRA configuration; b. if the UE does not succeed in establishing the UTRA physical channel(s): - perform a cell update procedure with cause "Radio link failure"; c. when the cell update procedure is completed successfully, it transmits the CELL CHANGE ORDER FROM UTRAN FAILURE message and set the IE "Inter-RAT change failure" to "physical channel failure".

Inter-RAT cell change order from UTRAN/To GPRS/CELL_FACH/ Failure (T309 expiry). Test to confirm that when UE received CELL CHANGE ORDER FROM UTRAN message in CELL_FACH state and if the establishment of the connection to the other RAT failed due to expiry of timer T309 prior to the successful establishment of a connection to the target RAT: a. revert to the cell it was camped on at the reception of the CELL CHANGE ORDER FROM UTRAN message; b. transmit the CELL CHANGE ORDER FROM UTRAN FAILURE message and set the IE "Inter-RAT change failure" to "physical channel failure".

Inter-RAT cell change order from UTRAN/To GPRS/CELL_FACH/ Failure (Physical channel Failure and Reversion Failure). Test to confirm that when UE received CELL CHANGE ORDER FROM UTRAN message in

CELL_FACH state and if the establishment of the connection to the other RAT failed due to other reasons e.g. (random) access failure, rejection due to lack of resources: a. revert to the cell it was camped on at the reception of the CELL CHANGE ORDER FROM UTRAN message; b. if the UE is unable to return to this cell: - select a suitable UTRA cell; c. initiate the cell update procedure using the cause "cell re-selection"; d. when the cell update procedure is completed successfully, it transmits the CELL CHANGE ORDER FROM UTRAN FAILURE message and set the IE "Inter-RAT change failure" to "physical channel failure".

Inter-RAT cell change order from UTRAN/To GPRS/ Failure (Unsupported configuration). Test to confirm if the UTRAN instructs the UE to perform a non-supported cell change order or to use a non-supported configuration, the UE shall: a. Transmit a CELL CHANGE ORDER FROM UTRAN FAILURE message, setting the IE "Inter-RAT change failure" to "configuration unacceptable"; b. Resume normal operation

Inter-RAT cell change order from UTRAN/To GPRS/ Failure (Invalid Inter-RAT message). Test to confirm that the UE shall keep its old configuration and transmit a CELL CHANGE ORDER FROM UTRAN FAILURE message, with the "Inter-RAT change failure" set to "protocol error", when it receives a CELL CHANGE ORDER FROM UTRAN message, not including a valid message in accordance with the protocol specifications for the target RAT.

Inter-RAT Cell Change Order from UTRAN to GPRS/CELL_DCH/ Success (stop of HS-DSCH reception). To test that the UE shall be able to receive a CELL CHANGE ORDER FROM UTRAN message in CELL_DCH state when Radio bearers are mapped to HSDSCH channels and perform a cell change to another RAT, even if no prior UE measurements have been performed on the target cell and HS-PDSCH channels are active. The UE regards the procedure as completed when it has received a successful response from the target RAT, e.g. in case of GSM when it received the response to a (PACKET) CHANNEL REQUEST in the new cell.

Inter-RAT Cell Change Order from UTRAN to GPRS/CELL_DCH/ Failure (Physical channel Failure). Test to confirm that when UE received CELL CHANGE ORDER FROM UTRAN message in CELL_DCH state and if the establishment of the connection to the other RAT failed due to expiry of timer T309 prior to the successful establishment of a connection to the target RAT: a. revert back to the UTRA configuration b. establish the UTRA physical channel(s) used at the time for reception of CELL CHANGE ORDER FROM UTRAN; c. transmit the CELL CHANGE ORDER FROM UTRAN FAILURE message and set the IE "Inter-RAT change failure" to "physical channel failure".

Inter-RAT cell change order from UTRAN/To GPRS/CELL_FACH/ No RAB established/Success. To test that the UE shall be able to receive a CELL CHANGE ORDER FROM UTRAN message in CELL_FACH state and perform a cell change to another RAT, when no RABs are established.

Inter-RAT cell change order from UTRAN/To GPRS/CELL_DCH/ Network Assisted Cell Change/Success. To test that the UE shall be able to receive a CELL CHANGE ORDER FROM UTRAN message which includes a correct and consistent set of SI or PSI messages in the IE "GERAN System Information" in CELL_DCH state and perform a cell change to another RAT using this as the system information to begin access on the target GERAN cell, even if no prior UE measurements have performed on the target cell. The UE regards the procedure as completed when it has received a successful response from the target RAT, e.g. in case of GSM when it received the response to a (PACKET) CHANNEL REQUEST in the new cell.

Inter-RAT cell change order from UTRAN/To GPRS/CELL_DCH/ Network Assisted Cell Change with Invalid SI/Success. To test that the UE shall be able to receive a CELL CHANGE ORDER FROM UTRAN message which includes an incorrect set of SI or PSI messages in the IE "GERAN System Information" in CELL_DCH state and perform a cell change to another RAT, ignoring the IE "GERAN System Information", even if no prior UE measurements have been performed on the target cell. The UE regards the procedure as completed when it has received a successful response from the target RAT, e.g. in case of GSM when it received the response to a (PACKET) CHANNEL REQUEST in the new cell.

Inter-RAT Cell Change Order from UTRAN to GPRS/CELL_DCH/ Success (stop of E-DCH transmission). To test that the UE is able to receive a CELL CHANGE ORDER FROM UTRAN message in CELL_DCH state when Radio bearers are mapped to the E-DCH channel and perform a cell change to another RAT, even if no prior UE measurements have been performed on the target cell and E-DPDCH channels are active.

RRC Measurement Control and Report

Measurement Control and Report: Intra-frequency measurement for transition from idle mode to CELL_DCH state (FDD). Test to confirm that the UE continues to monitor intra-frequency measurement quantity of the cells listed in System Information Block type 11 or 12 messages, after it has entered CELL_DCH state from idle mode. When the intra-frequency measurement reporting criteria specified in System Information Block type 11 or 12 messages have been met, it shall report the measurements using MEASUREMENT REPORT message(s). Test to confirm that the UE terminates monitoring and reporting activities for the cells listed in "intra-frequency cell info list" IE in System Information Block type 11 or 12 messages, after it has received a MEASUREMENT CONTROL message that specifies the measurement type to be "intra-frequency measurement" with the same measurement identity as in System Information Block Type 11 or 12 messages. Test to confirm that the UE reconfigures the monitoring and reporting activities based on the last MEASUREMENT CONTROL message received.

Measurement Control and Report: Inter-frequency measurement for transition from idle mode to CELL_DCH state (FDD). Test to confirm that the UE stops monitoring the list of cells assigned in the IE "inter-frequency cell info" in System Information Block type 11 messages, after it enters CELL_DCH state from idle mode. Test to confirm that the UE, which requires compressed mode, starts to perform inter-frequency measurement and related reporting activities, when it receives a MEASUREMENT CONTROL message with the "DPCH compressed mode status info" IE indicating that a stored compressed mode pattern sequence be simultaneously activated. Test to confirm that the UE, which does not require compressed mode, starts to perform inter-frequency measurement and related reporting activities when it receives a MEASUREMENT CONTROL message without IE "DPCH compressed mode status info". Test to confirm that the UE excludes the IE "Measured Results" for any cells in the MEASUREMENT REPORT messages, after it receives a MEASUREMENT CONTROL message with "Reporting cell status" IE omitted.

Measurement Control and Report: Inter-band measurement for transition from idle mode to CELL_DCH state (FDD). Same test purpose as above except that the cells belong to different frequency bands.

Measurement Control and Report: Intra-frequency measurement for transition from idle mode to CELL_FACH state (FDD). Test to confirm that the UE begins or continues to monitor cells listed in IE "intra-frequency cell info list" of System Information Block type 11 or 12 messages after it has entered CELL_FACH state from idle mode. Test to confirm that the UE applies the reporting criteria stated in "intra-frequency measurement reporting criteria" IE in System Information Block Type 11 or 12 in a subsequent transition to CELL_DCH state. Test to confirm that the UE reports measured results on RACH messages, if it receives IE "Intra-frequency reporting quantity for RACH reporting" and IE "Maximum number of reported cells on RACH" from System Information Block Type 11 or 12 upon a transition from idle mode to CELL_FACH state.

Measurement Control and Report: Inter-frequency measurement for transition from idle mode to CELL_FACH state (FDD). Test to confirm that the UE begins to monitor the list of cells assigned in the IE "inter-frequency cell info list" in System Information Block type 11 or 12 messages, after it enters CELL_FACH state from idle mode. However, it shall not transmit any MEASUREMENT REPORT messages to report measured results for inter-frequency cells.

Measurement Control and Report: Intra-frequency measurement for transition from CELL_DCH to CELL_FACH state (FDD). Test to confirm that the UE stops performing intra-frequency measurement reporting specified in a MEASUREMENT CONTROL message, when it moves from CELL_DCH state to CELL_FACH state. Test to confirm that the UE reads the System Information Block type 11 or 12 messages when it enters CELL_FACH state from CELL_DCH state, and starts to monitor the cells listed in the IE "intra-frequency cell info list". Test to confirm that the UE performs measurements on uplink RACH transmissions and appends the measured results in RACH messages, when it receives IE "intra-frequency reporting quantity for RACH reporting" and IE "Maximum number of reported cells on RACH" in the System Information Block type 11 or 12 messages. Test to confirm that the UE applies the reporting criteria in IE "intra-frequency reporting criteria" in System Information Block Type 11 or 12 messages following a state transition from CELL_FACH to CELL_DCH, if no intra-frequency measurements applicable to CELL_DCH are stored.

Measurement Control and Report: Inter-frequency measurement for transition from CELL_DCH to CELL_FACH state (FDD). Test to confirm that UE ceases inter-frequency type measurement reporting assigned in MEASUREMENT CONTROL message when moving from CELL_DCH state to CELL_FACH. Test to confirm that the UE begins to monitor the cells listed in "inter-frequency cell info" received in System Information Block type 11 or 12 messages, following a state transition from CELL_DCH state to CELL_FACH state.

Measurement Control and Report: Intra-frequency measurement for transition from CELL_FACH to CELL_DCH state (FDD). Test to confirm that UE retrieves stored measurement control information for intra-frequency measurement type with "measurement validity" assigned to "CELL_DCH", after it enters CELL_DCH state from CELL_FACH state. Test to confirm that the UE continues to monitor the neighbouring cells listed "intra-frequency cell info" IE in the System Information Block type 11 or 12 messages, if no intra-frequency measurements applicable to CELL_DCH are stored. Test to confirm that the UE transmits MEASUREMENT REPORT messages if reporting criteria stated in IE "intra-frequency measurement reporting criteria" in System Information Block type 11 or 12 messages are fulfilled. Test to confirm that a MEASUREMENT CONTROL message received in CELL_DCH state overrides the measurement and associated reporting contexts maintained in the UE by virtue of System Information Block type 11 or 12. Test to confirm that the UE delete all measurements of type intra-frequency upon cell reselection while in CELL_FACH.

Measurement Control and Report: Inter-frequency measurement for transition from CELL_FACH to CELL_DCH state (FDD). Test to confirm that the UE resumes inter-frequency measurements and reporting stored for which the measurement control information has IE "measurement validity" assigned to the value "CELL_DCH", after it re-enters CELL_DCH state from CELL_FACH state. Test to confirm that the UE resumes inter-frequency measurement and reporting activities after it has received a MEASUREMENT CONTROL message specifying that a stored compressed mode pattern sequence be re-activated.

Measurement Control and Report: Unsupported measurement in the UE. Test to confirm that the UE transmits a MEASUREMENT CONTROL FAILURE message, with the value "unsupported measurement" in IE "failure cause" when the SS instructs the UE to perform an unsupported measurement by sending a MEASUREMENT CONTROL message. Test to confirm that the UE retains its existing valid measurement configuration, after receiving a MEASURE-MENT CONTROL message containing an unsupported measurement.

Measurement Control and Report: Failure (Invalid Message Reception). Test to confirm that the UE continues its ongoing processes and procedures after it has received an invalid MEASUREMENT CONTROL message. Test to confirm that the UE transmits MEASUREMENT CONTROL FAILURE message, after it has received an invalid MEASUREMENT CONTROL message.

Measurement Control and Report: Cell forbidden to affect reporting range (FDD). Test to confirm that the UE reports the triggering of event 1A to the SS, if a primary CPICH currently measured by the UE enters the reporting range. Test to confirm that the UE reports the triggering of event 1B to the SS, if a primary CPICH currently measured by the UE leaves the reporting range. Test to confirm that the UE use the forbidden cell indicated in the MEASUREMENT CONTROL message to affect the reporting range. Test to confirm that the UE ignores that a primary CPICH is forbidden to affect the reporting range when (a) the primary CPICH concerned is included in active set and (b) all cells in the active set are defined as primary CPICHs forbidden to affect the reporting range.

Measurement Control and Report: Traffic volume measurement for transition from idle mode to CELL_FACH state. Test to confirm that after a state transition from idle mode to CELL_FACH state, the UE shall begin a traffic volume type measurement, as specified in System Information Block type 11 or 12 messages on BCCH. Test to confirm that in CELL_FACH state, the UE shall send a MEASUREMENT REPORT message when reporting criteria is satisfied. During CELL_FACH state, if the UE receives a MEASUREMENT CONTROL message, it shall perform the measurement and reporting tasks based on the MEASUREMENT CONTROL message received.

Measurement Control and Report: Traffic volume measurement for transition from idle mode to CELL_DCH state. Test to confirm that after a state transition from idle mode to CELL_DCH state, the UE begin a traffic volume type measurement, as specified in System Information Block type 11 or 12 messages on BCCH. When entering CELL_DCH state, the UE shall send a MEASUREMENT REPORT message when reporting criteria is satisfied. During CELL_DCH state, if the UE receives a MEASUREMENT CONTROL message, it shall perform the measurement and reporting tasks based on the MEASUREMENT CONTROL message received.

Measurement Control and Report: Traffic volume measurement for transition from CELL_FACH state to CELL_DCH state. Test to confirm that the UE performs traffic volume measurements and the associated reporting when it enters CELL_DCH state from CELL_FACH state, and that such measurement contexts (and optionally, the reporting context) valid for CELL_DCH state have been previously stored. Test to confirm that the UE shall continue to perform traffic volume measurement listed in the System Information Block type 11 or 12 messages, if no previously assigned measurements are present. The UE shall transmit MEASUREMENT REPORT messages if reporting conditions stated in System Information Block type 11 or 12 messages have been satisfied.

Measurement Control and Report: Traffic volume measurement for transition from CELL_DCH to CELL_FACH state. The UE shall performs traffic volume measurements and the associated reporting when it enters CELL_FACH state from CELL_DCH state, and that such measurement contexts (and optionally, the reporting context) valid for CELL_FACH state have been previously stored. The UE shall perform traffic volume measurement listed in the System Information Block type 11 or 12 messages, if no previously assigned measurements are present. The UE shall transmit MEASUREMENT REPORT messages if reporting conditions has been satisfied.

Measurement Control and Report: Quality measurements. Test to confirm that the UE performs quality measurement as specified in MEASUREMENT CONTROL message received. In CELL_DCH state, the UE shall send MEASUREMENT REPORT message when the reporting criteria is fulfilled for any ongoing quality measurement.

Measurement Control and Report: Intra-frequency measurement for events 1C and 1D. Test to confirm that the UE sends MEASUREMENT REPORT message if event 1C is configured, and number of cells in active set is greater than or equal to 'Replacement activation threshold' parameter, and if monitored or detected primary CPICH on same frequency becomes better than a primary CPICH in active set. Test to confirm that the UE does not send MEASUREMENT REPORT message indicating event 1C if number of cells in active set is less than 'Replacement activation threshold' parameter, and if monitored or detected primary CPICH on same frequency becomes better than a primary CPICH in active set. Test to confirm that the UE stops periodic reporting of event 1C if the cell that triggered event 1C is added into active set. Test to confirm that from Rel-5 onwards the UE sends a MEASUREMENT REPORT message as soon as event 1D is configured, when there is more than one cell in the active set. Test to confirm that the UE sends MEASUREMENT REPORT message if event 1D is configured and intra-frequency measurement indicates change in best cell.

Measurement Control and Report: Inter-frequency measurement for event 2A. Test to confirm that the UE sends MEASUREMENT REPORT message if event 2A is configured, and if any of the non- used frequencies quality estimate becomes better than the currently used frequency quality estimate. Test to confirm that the UE does not send MEASUREMENT REPORT message indicating event 2A if hysteresis condition is not fulfilled. Test to confirm that the UE does not send MEASUREMENT REPORT message indicating event 2A if time to trigger condition is not fulfilled.

Measurement Control and Report: Inter-band measurement for event 2A. Same test purpose as above except that the cells belong to different frequency bands.

Measurement Control and Report: Inter-frequency measurement for events 2B and 2E. Test to confirm that the UE sends MEASUREMENT REPORT message when event 2E is configured and the estimated quality of a non-used frequency is below the value of the IE "Threshold non-used frequency". This MEASUREMENT REPORT message shall contain at least the best primary CPICH info (for FDD) on the non-used frequency that triggered the event. Test to confirm that the UE sends MEASUREMENT REPORT message when event 2B is configured and estimated quality of the currently used frequency is below the value of the IE "Threshold used frequency" and the estimated quality of a non-used frequency is above the value of the IE "Threshold non-used frequency". This MEASUREMENT REPORT message shall contain at least the best primary CPICH info (for FDD) on the non-used frequency that triggered the event.

Measurement Control and Report: Inter-band measurement for events 2B and 2E. Same test as above except that the cells belong to different frequency bands.

Measurement Control and Report: Measurement for events 2D and 2F. Test to confirm that the UE sends MEASUREMENT REPORT message when event 2F is configured and estimated quality of the currently used frequency is above the value of the IE "Threshold used frequency". Test to confirm that the UE sends MEASUREMENT REPORT message when event 2D is configured and estimated quality of the currently used frequency is below the value of the IE "Threshold used frequency".

Measurement Control and Report: UE internal measurement for events 6A and 6B. Test to confirm that the UE performs UE internal measurements and reporting for events 6A and 6B, when requested by the UTRAN to do so in the MEASUREMENT CONTROL message.

Measurement Control and Report: UE internal measurement for events 6F (FDD) and 6G. Test to confirm that the UE performs UE internal measurements and reporting for events 6F and 6G, when requested by the UTRAN to do so in the MEASUREMENT CONTROL message.

Measurement Control and Report: Event based Traffic Volume measurement in CELL_FACH state. Test to confirm that in CELL_FACH state when event 4a triggered at TVM set up UE sends Measurement Report with correct measurement identity and indication of UL transport channel type, radio bearer identities and corresponding RLC buffer payloads in number of bytes. Test to confirm that in CELL_FACH state when event 4a triggerreds after TVM set up UE sends Measurement Report with correct measurement identity and indication of UL transport channel type, radio bearer identities and corresponding RLC buffer payloads in number of bytes. Test to confirm that the UE sends MEASUREMENT REPORT message, with measurement report in IE "Measurement results on RACH" as specified in System Information Block type 12.

Measurement Control and Report: Event based Traffic Volume measurement in CELL_DCH state. Test to confirm that in CELL_DCH state when event 4a or 4b triggered at setup TVM UE sends RRC: Measurement Report with correct measurement identity and indication of uplink transport channel type and identity, radio bearer identities and corresponding RLC buffer payloads in number of bytes. Test to confirm that in CELL_DCH state when event 4a or 4b triggered after setup TVM UE sends RRC: Measurement Report with correct measurement identity and indication of uplink transport channel type and identity, radio bearer identities and corresponding RLC buffer payloads in number of bytes.

Measurement Control and Report: Inter-RAT measurement in CELL_DCH state. Purpose of this test is to verify that UE is capable to perform GSM RSSI and GSM Initial BSIC identification measurements.

Measurement Control and Report: Inter-RAT measurement, event 3a. Test to confirm that the UE starts compressed mode and inter-RAT measurements when so required by the network in a MEASUREMENT CONTROL message. Test to confirm that the UE sends MEASUREMENT REPORT message if event 3a is configured, if the quality of the currently used UTRAN frequency is below a given threshold and the estimated quality of the other system is above a certain threshold. Test to confirm that the hysteresis and time to trigger behaviours for event 3a are correctly implemented. Test to confirm that the UE verifies the BSIC of the cell triggering the event if so required by UTRAN and if the proper compressed mode patterns have been configured in the UE by UTRAN. Test to confirm that the content of the MEASUREMENT REPORT sent by the UE is according to what was required by UTRAN.

Measurement Control and Report: Inter-RAT measurement, event 3b. Test to confirm that the UE sends MEASUREMENT REPORT message if event 3b is configured, if the estimated quality of the other system is below a given threshold. Test to confirm that the hysteresis and time to trigger behaviours for event 3b are correctly implemented. Test to confirm that the UE updates the list of inter-RAT cells it stores according to what is ordered in the MEASUREMENT CONTROL messages received from UTRAN.

Measurement Control and Report: Inter-RAT measurement, event 3c. Test to confirm that the UE sends MEASUREMENT REPORT message if event 3c is configured, and if the quality of the other system becomes better than the given threshold for event 3c. Test to confirm that no other UE MEASUREMENT REPORT message is sent by the UE for a cell that has already triggered event 3c as long as the hysteresis condition for triggering once again event 3c has not been fulfilled.

Measurement Control and Report: Inter-RAT measurement, event 3d. Test to confirm that the UE sends MEASUREMENT REPORT message if event 3d is configured, and if the best cell changes in the other system. Test to confirm that no other UE MEASUREMENT REPORT message is sent by the UE for a cell that has already triggered event 3d as long as the hysteresis condition for triggering once again event 3d has not been fulfilled.

Measurement Control and Report: UE internal measurement, event 6c. Test to confirm that the UE sends a measurement report for event 6c when the

UE Tx power reaches its minimum value when event 6c has been configured in the UE through a MEASUREMENT CONTROL message.

Measurement Control and Report: UE internal measurement, event 6d. Test to confirm that the UE sends a measurement report for event 6d when the UE Tx power reaches its maximum value when event 6d has been configured in the UE through a MEASUREMENT CONTROL message.

Measurement Control and Report: UE internal measurement, event 6e. Test to confirm that the UE sends a measurement report for event 6e when the UE RSSI reaches the UE's dynamic receiver range when event 6e has been configured in the UE through a MEASUREMENT CONTROL message.

Measurement Control and Report: Inter-RAT measurement, event 3C, in CELL_DCH state using sparse compressed mode pattern. This test is only applicable to UEs supporting both FDD and GSM, and which require compressed mode to perform the GSM related measurements. Test to confirm that the UE performs Inter-RAT measurement using a sparse compressed mode pattern as specified in the MEASUREMENT CONTROL message. Test to confirm that the UE send MEASUREMENT REPORT message when event 3C is triggered, and if the quality of the other system becomes better than the given threshold for event 3c. Test to confirm that no other UE MEASUREMENT REPORT message is sent by the UE for a cell that has already triggered event 3c as long as the hysteresis condition for triggering once again event 3c has not been fulfilled.

Measurement Control and Report: Additional Measurements list. Test to confirm that the UE reports measured results for a referenced additional measurement. Test to confirm that the UE transmits MEASUREMENT REPORT messages for a measurement, also if this measurement is referenced as an additional measurement by another measurement.

Measurement Control and Report: Change of Compressed Mode Method. Test to confirm that the UE supports change of compressed mode method included in a RADIO BEARER SETUP message. Test to confirm that the UE supports change of compressed mode method included in a RADIO BEARER RELEASE message.

Measurement Control and Report: Compressed Mode Reconfiguration. Test to confirm that the UE supports de- activation of compressed mode included in a RADIO BEARER SETUP message. Test to confirm that the UE supports reconfiguration of transport channel parameters (rate reduction PS RAB) and change of compressed mode method included in a TRANSPORT CHANNEL RECONFIGURATION message. Test to confirm that the UE supports change of compressed mode included in a RADIO BEARER RELEASE message. Test to confirm that the UE supports reconfiguration of transport channel parameters (rate increase PS RAB) without performing hard handover included in a TRANSPORT CHANNEL RECONFIGURATION message.

Measurement Control and Report: Event triggered periodic measurement for event 1B (FDD). Test to confirm that the UE reverts to periodical measurement reporting of event 1B, after event 1B has been initially triggered, when primary CPICH currently measured by the UE leaves the reporting range. Test to confirm that the event triggered periodic measurement reporting is terminated after UE no longer finds any monitored cell within the removal range. Test to confirm that the event triggered periodic measurement reporting is terminated after UTRAN has removed cells from the active set so that there is no longer the minimum amount of active cells for event 1B to be triggered. Test to confirm that the event triggered periodic measurement is no longer triggered if the reporting interval is set to zero. Test to confirm that the event-triggered periodic measurement reporting is terminated after UE has sent the maximum number of MEASUREMENT REPORT messages (defined by the "amount of reporting" parameter).

Measurement Control and Report: Combined Inter-frequency measurement for event 2b and Inter-RAT measurement, event 3a (FDD). Test to confirm that the UE sends MEASUREMENT REPORT message when event 2b and 3a are configured and estimated quality of the currently used frequency is below the value of the IE "Threshold used frequency" and the estimated quality of a non-used frequency is above the value of the IE "Threshold non-used frequency". This MEASUREMENT REPORT message shall contain at least the best primary CPICH info of the non-used frequency that triggered the event. Test to confirm that the UE sends MEASUREMENT REPORT message when event 2b and 3a are configured and estimated quality of the currently used UTRAN frequency is below the value of the IE "Threshold own system" and the estimated quality of the other system is above the value of the IE "Threshold other system". This MEASUREMENT REPORT message shall contain the inter-RAT cell id of the GSM cell that triggered the event.

REFERENCES

[1] 3GPP TS 34.123-1 V6.5.0, Technical Specification, 3rd Generation Partnership Project; Technical Specification Group Radio Access Network; User Equipment (UE) conformance specification; Part 1: **Protocol conformance specification**

[2] 3GPP TS 34.123-2 V6.5.0, Technical Specification, 3rd Generation Partnership Project; Technical Specification Group Radio Access Network; User Equipment (UE) conformance specification; Part 2: **Implementation Conformance Statement (ICS) proforma specification**

[3] 3GPP TS 26.132 V6.0.0, Technical Specification, 3rd Generation Partnership Project; Technical Specification Group Services and System Aspects; **Speech and video telephony terminal acoustic test specification**

[4] 3GPP TS 31.121 V6.6.0, Technical Specification, 3rd Generation Partnership Project; Technical Specification Group Core Network and Terminals; **UICC-terminal interface; Universal Subscriber Identity Module (USIM) application test specification**

[5] 3GPP TS 31.124 V6.7.0, Technical Specification, 3rd Generation Partnership Project; Technical Specification Group Core Network and Terminals; **Mobile Equipment (ME) conformance test specification; Universal Subscriber Identity Module Application Toolkit (USAT) conformance test specification**

[6] 3GPP TS 34.131 V6.0.1, Technical Specification, 3rd Generation Partnership Project; Technical Specification Group Terminals; **Test Specification for C-language binding to (Universal) Subscriber Interface Module ((U)SIM) Application Programming Interface (API)**

[7] 3GPP TS 34.171 V6.5.0, Technical Specification, 3rd Generation Partnership Project; Technical Specification Group Radio Access Network; **Terminal conformance specification; Assisted Global Positioning System (A-GPS); Frequency Division Duplex (FDD)**

[8] 3GPP TS 34.229-1 V6.0.0, Technical Specification, 3rd Generation Partnership Project; Technical Specification Group Radio Access Network; **Internet Protocol (IP) multimedia call control protocol based on Session Initiation Protocol (SIP) and Session Description Protocol (SDP); User Equipment (UE) conformance specification; Part 1: Protocol conformance specification**

[9] 3GPP TS 51.010-1 V7.4.0 (2006-12), Technical Specification, 3rd Generation Partnership Project; Technical Specification Group, GSM/EDGE Radio Access Network, Digital cellular telecommunications system (Phase 2+); **Mobile Station (MS) conformance specification; Part 1: Conformance specification**

Chapter Two

UMTS Conformance Protocol Testing – Part Two

MORE UE PROTOCOL TESTS

MBMS, IMS , AGPS, USIM, USAT, etc. are the type of testing that are used throughout the initial stages of development to ensure the accuracy of the related protocol implementation. Capable SS's to perform the above have became very complicated. In general, protocol conformance testing is appreciated in the telecommunications world. At ETSI, protocol conformance testing specifications have dominated the testing activities and will continue to do so in the future.

Variety of other areas that need to be addressed when test program is to be arranged; protocols, RF performance, USIM, audio performance and RF. Protocol testing determines the overall operation of the UE. If the protocol software operates incorrectly, then the UE will not operate properly on a network. There even have been instances when the incorrect operation of UE has been the cause of problems with a network. Due the complexity of the protocols that are used, this testing can be very involved and requires the use of sophisticated network simulators. These system simulators emulate a variety of network entities, such as Node-Bs, RNCs. Scenarios from registering to terminating a call and all the different forms of handovers can be simulated, i.e. any situation that can be encountered in real operation can be fully tested.

Conformance testing is able to determine whether the behaviour of an implementation conforms to the requirements laid out in its base specification including the full range of error and exception conditions which can only be induced or replicated by dedicated test equipment. It exercises most, if not all, of the possible ways of achieving each of a component's function. On the other hand, conformance testing does not prove end-to-end interoperability of functions between the two similar communicating systems, and does not exercise all system components and their interfaces together to determine whether the implementation works in a real-life environment. Also it does not prove the operation of proprietary features, functions, interfaces, and systems that are not in the public domain. However, these proprietary facilities may be exercised indirectly as part of the configuration or execution of the conformance tests.

The testers, which in a real testing environment may be distributed, execute test programs or scripts which are called Test Cases. The entire set of Test Cases is known as a Test Suite. ETSI develops Abstract Test Suites (ATS) written in the standardized testing language TTCN which can be compiled and run on a variety of real test systems.

In each case there may be different conformance test suites for the different protocols (components) that make up the products. At no time does an individual test suite test the product as a complete system.

MBMS

MBMS Session Start in Idle mode. Test to confirm that the UE receives the MBMS info in idle mode state and to verify that the UE starts the reception of MBMS services at MCCH acquisition.

MBMS session start in CELL_FACH state. Test to confirm that the UE receives the MBMS information on MCCH in CELL_FACH state. Test to confirm that the UE correctly handles the notification procedure after receiving the MBMS MODIFIED SERVICES INFORMATION message via MCCH if no ongoing MBMS p-t-m session (MICH supported/not supported by the UE). Test to confirm that the UE starts the reception of MBMS services according to notification via MICH/MCCH when the UE is in CELL_FACH state.

MBMS session start in CELL_DCH state, MCCH notification. Test to confirm that the UE handles reading of MCCH during compressed mode measurements in CELL_DCH state. Test to confirm that the UE correctly handles the notification procedure after receiving the MBMS MODIFIED SERVICES INFORMATION message via MCCH in CELL_DCH state. Test to confirm that the UE starts the p-t-m reception of MBMS services according to notification via MCCH when the UE is in CELL_DCH state.

MBMS session start at MCCH acquisition in CELL_DCH (for a non-MBMS service) when entering into an MBMS cell (UE capable of MBMS p-t-m reception in CELL_DCH). Test to confirm that the UE receives the MBMS information on MCCH in CELL_DCH state. Test to confirm that the UE, when entering into an MBMS cell in CELL_DCH state, starts the p-t-m reception of MBMS services according to the information on MCCH.

MBMS Session Start (Frequency Layer Convergence)/Session Stop (Frequency Layer Dispersion) in Idle mode. Test to confirm that in idle mode, the UE at session start re-selects to the preferred Frequency Layer for the MBMS services it has joined. Test to confirm that at session stop the UE re-select to the frequency where it has previously camped.

MBMS session stop with frequency layer dispersion - no previous frequency layer available (URA_PCH). Test to confirm UE performs frequency layer dispersion procedure at session stop in URA_PCH state, when the IE 'MBMS dispersion indicator' is set to TRUE and no suitable cell in the previously stored frequency is found.

MBMS session stop with frequency layer dispersion - no previous frequency layer available (CELL_FACH). Test to confirm UE performs frequency layer dispersion procedure at session stop in CELL_FACH state, when the IE 'MBMS dispersion indicator' is set to TRUE and no suitable cell in the previously stored frequency is found.

MBMS Counting in Idle Mode. Test to confirm that the UE starts the MBMS counting procedure in idle mode upon receiving an MBMS MODIFIED SERVICES message including IE "MBMS required UE action" and a ACCESS INFORMATION message including IE "Access probability factor – Idle" set to 0 (corresponding to the actual probability factor value 1).

MBMS No Counting in CELL_DCH. Test to confirm that the UE in CELL_DCH state (with the probability factor set to 1) does not perform the MBMS counting procedure.

MBMS Counting in Cell_PCH. Test to confirm that the UE correctly handles the counting procedure upon receiving an MBMS MODIFIED SERVICES message including IE "MBMS required UE action" and a ACCESS INFORMATION message including IE "Access probability factor – Connected" set to 1 for the selected services.

MBMS serving cell reselection in CELL_FACH during ongoing session. Test to confirm that a UE in CELL_FACH state performs an MBMS Serving Cell Reselection during an ongoing Session. Test to confirm that the UE is able of read MCCH in combination with FACH measurements.

Mobility Management

TMSI reallocation. TMSI Reallocation procedure is to assign a new temporary identity for the UE. If the message is not understood by the UE, the network could not establish a link to the UE. As this is a common MM procedure, it can be initiated at any time. This test is to verify that the UE is able to receive and acknowledge a new TMSI by means of an explicit TMSI reallocation procedure. Also to verify that the UE has stored the TMSI in a non-volatile memory.

Authentication accepted. To check that a UE correctly responds to an AUTHENTICATION REQUEST message by sending an AUTHENTICATION RESPONSE message with the RES information field set to the same value as the one produced by the authentication algorithm in the network. To check that a UE indicates in a PAGING RESPONSE message the ciphering key sequence number which was allocated to it through the authentication procedure.

Authentication rejected by the network. To check that, after reception of an AUTHENTICATION REJECT message, the UE: does not perform normal location updating; does not perform periodic location updating; does not respond to paging with TMSI; rejects any request from CM entity for MM connection except for emergency call; does not perform IMSI detach if deactivated. To check that, after reception of an AUTHENTICATION REJECT message the UE, if it supports emergency speech call, accepts a request for an emergency call by sending a RRC CONNECTION REQUEST message with the establishment cause set to "emergency call" and includes an IMEI as UE identity in the CM SERVICE REQUEST message. To check that, after reception of an AUTHENTICATION REJECT message and after having been deactivated and reactivated, the UE performs location updating using its IMSI as UE identity and indicates deleted LAI and CKSN.

Authentication rejected by the UE (MAC code failure). To check that a UE shall correctly respond to an AUTHENTICATION REQUEST message, with a MAC code failure in the AUTN parameter, by sending an AUTHENTICATION FAILURE message with the reject cause 'MAC failure'. To check that upon reception of an IDENTITY REQUEST message, requesting for IMSI, the UE identifies itself by sending an IDENTITY RESPONSE message including the IMSI to the network. To check that upon receiving the second AUTHENTICATION REQUEST message from the network, the UE shall stop the timer T3214, if running, and then process the challenge information as normal. To check that upon successfully validating the network (an AUTHENTICATION REQUEST that contains a valid MAC is received), the UE sends the AUTHENTICATION RESPONSE message to the network.

Authentication rejected by the UE (SQN failure). To check that a UE shall correctly respond to an AUTHENTICATION REQUEST message, with an SQN failure in the AUTN parameter, by sending an AUTHENTICATION FAILURE message with the reject cause 'Synch failure'. To check that upon successfully validating the network (a second AUTHENTICATION REQUEST is received which contains a valid SQN) while T3216 is running, the UE shall send the AUTHENTICATION RESPONSE message to the network.

Authentication rejected by the UE / fraudulent network. To test UE treating a cell as barred: when the UE receives the second or third AUTHENTICATION REQUEST message with invalid MAC value during the T3214 is running. when the timer T3214 has expired.

Identification. The purpose of this procedure is to check that the UE gives its identity as requested by the network. If this procedure does not work, it will not be possible for the network to rely on the identity claimed by the UE.

General Identification. Test to confirm that the UE sends identity information as requested by the system in the following cases: IMSI and TMSI are requested in non-security mode, IMEI is requested in security mode. Test to confirm that the UE sends its IMEI, when requested to do so, in non- security mode. Test to confirm that the UE sends its IMEISV, when requested to do so, in non- security mode.

Handling of IMSI shorter than the maximum length. To check that the UE behaves correctly when activated with an IMSI of length less than the maximum length. In this condition, the UE shall: - perform location updating; - answer to paging with IMSI; - give the correct IMSI when asked by an IDENTITY

REQUEST; - attempt CM connection establishment when requested to; - attempt IMSI detach when needed; - erase its TMSI when the IMSI is sent by the network in a LOCATION UPDATING ACCEPT or a TMSI REALLOCATION COMMAND message.

Location updating. This procedure is used to register the UE in the network. If it is not performed correctly, no call can be established.

Location updating / accepted. To test the behaviour of the UE if the network accepts the location updating of the UE. For the network response three different cases are identified: TMSI is allocated; location updating accept contains neither TMSI nor IMSI; location updating accept contains IMSI.

Location updating / rejected / IMSI invalid. To test the behaviour of the UE if the network rejects the location updating of the UE with the cause "IMSI unknown in HLR", "illegal MS" or "Illegal ME".

Location updating / rejected / PLMN not allowed. To test the behaviour of the UE if the network rejects the location updating of the UE with the cause "PLMN not allowed".

Location updating / rejected / location area not allowed. To test the behaviour of the UE if the network rejects the location updating of the UE with the cause "Location Area not allowed". To test that the UE deletes the list of forbidden LAs after switch off (power off).

Location updating / rejected / roaming not allowed in this location area. To test that on receipt of a rejection using the Roaming cause code, the UE ceases trying to update on that cell, that this situation continues for at least one periodic location interval period, and that the corresponding list is re-set by switching off the UE or removing its power source. To test that if no cell is available, the UE rejects a request from CM entity other than for emergency calls. To test that at least 6 entries can be held in the list of "forbidden location areas for roaming" (the requirement in is to store at least 10 entries. This is not fully tested by the third procedure). To test that if a cell of the Home PLMN is available then the UE returns to it in preference to any other available cell. To test that if the USIM is removed the list of "forbidden location areas for roaming" is cleared.

Location updating / rejected / No Suitable Cells In Location Area. To test the behaviour of the UE if the network rejects the location updating of the UE with the cause "No Suitable Cells In Location Area".

Location updating / abnormal cases / attempt counter less or equal to 4, LAI different. Test to confirm that the UE performs normal location updating procedures when its attempt counter is smaller than 4. To check that the UE does not perform the IMSI detach procedure when "idle not updated". Test to confirm that when "idle not updated" the UE can perform an emergency call. Test to confirm that when "idle not updated" the UE uses requests from CM layer other than emergency call as triggering of a normal location updating procedure. Test to confirm that the UE performs a normal location updating procedure if it enters a new cell while being "idle not updated".

Location updating / abnormal cases / attempt counter equal to 4. Test to confirm that the UE performs normal location updating procedures after T3212 expiry, when its attempt counter has reached value 4 and that the UE reset its attempt counter after a timer T3212 expiry. Test to confirm that the UE still follows the MM IDLE state and ATTEMPTING TO UPDATE substate requirements after its attempt counter has reached value 4. Test to confirm that the attempt counter is reset in the cases where it has to be done.

Location updating / abnormal cases / attempt counter less or equal to 4, stored LAI equal to broadcast LAI. Test to confirm that in the case when the attempt counter is smaller than 4 and the broadcast LAI is equal to the stored LAI, the UE is in the MM IDLE state and NORMAL SERVICE substate. Test to confirm that timer T3211 is stopped after a MM connection establishment. Test to confirm that the UE uses the T3211 timer. and that it enters the MM IDLE state and NORMAL SERVICE substate when its attempt counter reaches value 4 even in the case where the stored LAI is equal to the broadcast LAI.

Location updating / abnormal cases / Failure due to non-integrity protection. Test to confirm that the UE ignores NAS signalling messages when the security mode procedure is not activated.

Location updating /abnormal cases / CS domain is changed from barred to unbarred because of domain specific access control. To test the behaviour of the UE if the CS domain is barred and if it is then changed from barred to unbarred because of domain specific access control in a network that operates mode II.

Location updating / release / expiry of T3240. Test to confirm that the UE aborts the RR-connection at the expiry of timer T3240.

Location updating / periodic spread. To check that when the location updating timer is reduced, the timer running in the UE is started with a value depending on the current timer value and the new broadcasted T3212 value. Test to confirm that when the UE is reactivated in the same cell (as the one in which it was deactivated), IMSI attach being forbidden, the UE starts the timer T3212 with a value between zero and the broadcasted value.

Location updating / periodic normal / test 1. Test to confirm that the UE stops and resets the timer T3212 of the periodic location updating procedure when: - the first MM-message is received in the case of MM-connection establishment, security mode being not set; - the UE has responded to paging and the first correct L3 message that is not an RRC message is received.

Location updating / periodic normal / test 2. Test to confirm that the UE stops and resets the timer T3212 of the periodic location updating procedure when a LOCATION UPDATING ACCEPT message is received.

Location updating / periodic search for HPLMN or higher priority PLMN / UE waits time T. Test to confirm that if a UE is camped on a VPLMN it will perform a search for higher priority networks (e.g. HPLMN) with a periodicity of T, which is the Search Period stored in the USIM. This test will confirm that, if a cell from a new PLMN becomes available, within a time T the UE will perform a location updating on it only if the following requirements are met: - The PLMN of this new cell if from the same country as the VPLMN, and - This PLMN is the HPLMN stored in the USIM, or has a higher priority than the serving VPLMN or any PLMN from the country of the VPLMN that is stored in the equivalent PLMN list.

Location updating / periodic search for HPLMN or higher priority PLMN / UE in manual mode. Test to confirm that no Search for HPLMN or Higher Priority PLMN is performed when the UE is not in automatic mode.

Location updating / periodic search for HPLMN or higher priority PLMN / UE waits at least two minutes and at most T minutes. Test to confirm that the UE waits at least 2 minutes and at most T minutes before attempting its first Search for HPLMN or higher priority PLMN.

Location updating/periodic search of the higher priority PLMN, VPLMN in a foreign country – higher priority/UE is in automatic mode. Test to confirm that the UE selects the highest priority network if the HPLMN/ higher priority PLMN Search is performed, when a UE is receiving foreign country's VPLMN and UE is in automatic mode.

Location updating/periodic search of the higher priority PLMN, VPLMN in a foreign country – lower priority/UE is in automatic mode. Test to confirm that the UE remains on the highest priority network if the HPLMN/higher priority PLMN Search is performed, when a UE is receiving foreign country's VPLMN and UE is in automatic mode.

Location updating/periodic search of the higher priority PLMN, VPLMN in a foreign country – List of EPLMN contain HPLMN /UE is in automatic mode. Test to confirm that, in automatic mode, when registered on a VPLMN of a country different to it's HPLMN, the MS only selects the highest priority network available from upon those of the same country as the serving PLMN. It also verifies that the MS does not take into account PLMNs, including the HPLMN, which are included in the Equivalent PLMN list.

Location updating / inter-working of attach and periodic. To check that if the PLU timer expires while the UE is out of coverage, the UE informs the network of its return to coverage. To check that the PLU timer is not disturbed by cells of forbidden PLMNs. To check that if the PLU timer does not expire while out of coverage and if the UE returns to the LA where it is updated, the UE does not inform the network of its return to coverage.

Location Updating / accept with replacement or deletion of Equivalent PLMN list. Test to confirm that the UE replaces its stored equivalent PLMN list if the equivalent PLMN list is contained in the LOCATION UPDATING ACCEPT message received from the network during a location updating procedure. Test to confirm that the UE deletes its stored equivalent PLMN list if no equivalent PLMN list is contained in the LOCATION UPDATING ACCEPT message received from the network during a location updating procedure.

Location Updating after UE power off. Test to confirm that the UE stores the equivalent PLMN list at UE switch off and uses the stored equivalent PLMN list after UE switch on.

Location Updating / Accept, Interaction between Equivalent PLMNs and Forbidden PLMNs. Test to confirm that the UE shall not select a forbidden PLMN even though it is included in the equivalent PLMN list provided by the network because forbidden PLMNs shall not be stored in the UE's equivalent PLMN list.

MM connection / establishment in security mode. Test to confirm that the UE can correctly set up an MM connection in an origination and interpret security mode setting as acceptance of its CM service request.

MM connection / establishment rejected. Test to confirm that the UE stops timer T3230, informs the requesting CM sublayer entity and returns to the previous state.

MM connection / establishment rejected cause 4. Test to confirm that the UE can correctly set up an MM connection in a UE Originating CM connection attempt and send a CM SERVICE REQUEST message with CKSN information element as stored in the USIM and UE Identity information element set to TMSI. Test to confirm that the UE, when receiving a CM SERVICE REJECT message with reject cause "IMSI unknown in VLR" shall wait for the network to release the RRC connection. Test to confirm that the UE shall then perform a normal location updating procedure.

MM connection / expiry T3230. Test to confirm that at T3230 expiry, the UE aborts the MM-connection establishment.

MM connection / abortion by the network / cause #6. To check that upon reception of an ABORT message with cause #6 during call establishment: - the UE does not send any layer 3 message; - after reception of an ABORT message and after having been deactivated and reactivated, the UE performs location updating using its IMSI as UE identity and indicates deleted LAI and CKSN; - the UE does not perform location updating, does not answer to paging with TMSI, rejects any request for UE originating call except emergency call, does not perform IMSI detach; - the UE accepts a request for emergency call.

MM connection / abortion by the network / cause not equal to #6. To check that when multiple MM connections are established, the UE releases all MM connections upon reception of an ABORT message, in the case when the two MM connections are established for a UE terminating call and a non call related supplementary service operation. To check that the TMSI is not deleted from UE after reception of ABORT message with cause another than #6.

MM connection / follow-on request pending / test 1. To check that when the network does not include the follow on proceed IE in a LOCATION UP-DATING ACCEPT message, a UE that has a CM application request pending does not attempt to establish a new MM connection on that RRC connection.

MM connection / follow-on request pending / test 2. To check that when the network includes the follow on proceed IE in a LOCATION UPDATING ACCEPT message, a UE that supports the follow on request procedure and that has a CM application request pending establishes successfully a new MM connection on that RRC connection.

MM connection / follow-on request pending / test 3. To check that a UE that has no CM application request pending sets the follow on proceed IE to No follow-on request pending in a LOCATION UPDATING REQUEST message. To check that when the network includes the follow on proceed IE in a LOCATION UPDATING ACCEPT message, a UE that has no CM application request pending does not attempt to establish a new MM connection on that RRC connection. To check that the UE accepts establishment by the network of a new MM connection on the existing RRC connection.

MM connection / abnormal cases / CS domain barred because of domain specific access control. To test the behaviour of the UE if the CS domain is changed from unbarred to barred because of domain specific access control in a network that operates mode II.

CS Call Control
Outgoing Call
Outgoing call / U0 null state / MM connection requested. Test to confirm that upon initiation of an outgoing basic call by user the UE initiates establishment of an MM connection, using as first MM message a CM SERVICE REQUEST message with CM service type "UE originating call establishment or packet mode connection establishment" or " Emergency call establishment ".

Outgoing call / U0.1 MM connection pending / CM service rejected. Test to confirm that a CC entity of the UE in CC-state U0.1, "MM-connection pending", upon the UE receiving a CM SERVICE REJECT message, returns to CC state U0, "Null".

Outgoing call / U0.1 MM connection pending / CM service accepted. Test to confirm that a CC entity of the UE in CC-state U0.1, "MM connection pending", after completion of the security mode control procedure, sends a SETUP message specifying the Called party BCD number that was entered into the UE and then enters CC state U1, "Call initiated".

Outgoing call / U0.1 MM connection pending / lower layer failure. Test to confirm that the UE with a CC entity in state U0.1, "MM connection pending", aborts MM connection establishment, stops timer T3230 and returns to idle mode in case an RR connection failure occurs.

Outgoing call / U1 call initiated / receiving CALL PROCEEDING. Test to confirm that a CC entity of the UE in CC-state U1, "Call initiated", upon receipt of a CALL PROCEEDING message, enters CC state U3, "UE originating call proceeding".

Outgoing call / U1 call initiated / rejecting with RELEASE COMPLETE. Test to confirm that a CC entity of the UE in CC-state U1, "Call initiated", upon receipt of a RELEASE COMPLETE message with valid cause value, enters CC state U0, "Null". Test to confirm that in returning to idle mode, the CC entities relating to the seven UE originating transaction identifiers are in state U0, "Null".

Outgoing call / U1 call initiated / T303 expiry. Test to confirm that a CC entity of the UE in CC-state U1, "Call initiated", upon expiry of T303 sends a DISCONNECT message to its peer entity and enters state U11, "Disconnect request".

Outgoing call / U1 call initiated / lower layer failure. Test to confirm that after the UE with a CC entity in state U1 "Call initiated", has detected a lower layer failure and has returned to idle mode, the CC entity is in state U0, "Null".

Outgoing call / U1 call initiated / receiving ALERTING. Test to confirm that a CC entity of the UE in CC-state U1, "Call initiated", upon receipt of an ALERTING message, enters CC state U4, "Call delivered".

Outgoing call / U1 call initiated / entering state U10. Test to confirm that a CC entity of the UE in CC-state U1, "Call initiated", upon receipt of a CONNECT message, sends a CONNECT ACKNOWLEDGE message to its peer entity and enters CC state U10, "Active".

Outgoing call / U1 call initiated / unknown message received. Test to confirm that a CC entity of the UE in CC-state U1, "Call initiated", upon receipt of a message with message type not defined for the protocol discriminator from its peer entity returns a STATUS message.

Outgoing call / U3 UE originating call proceeding / ALERTING received. Test to confirm that a CC-entity of the UE in CC-state U3, "UE Originating Call Proceeding", upon receipt of a ALERTING message enters CC-state U4, "Call Delivered".

Outgoing call / U3 UE originating call proceeding / CONNECT received. Test to confirm that a CC-entity of the UE in CC-state U3, "UE Originating Call Proceeding", upon receipt of a CONNECT message returns a "CONNECT ACKNOWLEDGE" message to its peer entity and enters the CC state U10, "Active".. Test to confirm that the UE stops locally generated alerting indication, if any.

Outgoing call / U3 UE originating call proceeding / PROGRESS received without in band information. Test to confirm that a CC-entity of the UE in CC-state U3, "UE Originating Call Proceeding", upon receipt of a PROGRESS message with valid cause values stays in CC-state U3. Test to confirm that after receipt of the PROGRESS message timer T310 is stopped.

Outgoing call / U3 UE originating call proceeding / PROGRESS with in band information. Test to confirm that a CC-entity of the UE in CC-state U3, "UE Originating Call Proceeding", upon receipt of a PROGRESS message indicating in-band announcement through-connects the traffic channel for speech, if

DTCH is in speech mode. If DTCH is not in a speech mode, the UE does not through-connect the DTCH. Test to confirm that after receipt of the PROGRESS message, T310 is stopped.

Outgoing call / U3 UE originating call proceeding / DISCONNECT with in band tones. Test to confirm that a CC-entity of the UE in CC-state U3, "UE Originating Call Proceeding", upon receipt of a DISCONNECT with progress indicator #8 through-connects the speech channel to make in-band announcements available, if traffic channel is in speech mode. If DTCH is not in speech mode, the UE sends a RELEASE message.

Outgoing call / U3 UE originating call proceeding / DISCONNECT without in band tones. Test to confirm that a CC-entity of the UE in CC-state U3, "UE Originating Call Proceeding", upon receipt of a DISCONNECT without progress indicator returns a RELEASE message and enters the CC-state U19, "Release Request".

Outgoing call / U3 UE originating call proceeding / RELEASE received. Test to confirm that a CC-entity of the UE in CC-state U3, "UE Originating Call Proceeding", upon receipt of a RELEASE will return a RELEASE COMPLETE and enter the CC-state U0, "Null". Test to confirm that the UE on returning to the idle mode releases the MM-connection and that the CC-entities relating to the seven UE originating transaction identifiers are in CC-state U0, "Null".

Outgoing call / U3 UE originating call proceeding / termination requested by the user. Test to confirm that a CC-entity of the UE in CC-state U3, "UE Originating Call Proceeding", upon request by the user to terminate will send a DISCONNECT message and enter the CC-state U11, "Disconnect Request".

Outgoing call / U3 UE originating call proceeding / traffic channel allocation. Test to confirm that a CC-entity of the UE in CC-state U3, "UE Originating Call Proceeding", when a traffic channel is allocated by the network performing the radio bearer establishment procedure, stays in CC-state U3.

Outgoing call / U3 UE originating call proceeding / timer T310 time-out. Test to confirm that a CC-entity of the UE in CC-state U3, "UE Originating Call Proceeding" will, upon expiry of timer T310, initiate call release by sending DISCONNECT and enter the CC-state U11, "Disconnect Request".

Outgoing call / U3 UE originating call proceeding / lower layer failure. Test to confirm that a CC-entity of the UE in CC-state U3, "UE Originating Call Proceeding" having detected a lower layer failure and having returned to idle mode, the CC entity is in state U0, "Null".

Outgoing call / U3 UE originating call proceeding / unknown message received. Test to confirm that a CC-entity of the UE in CC-state U3, "UE Originating Call Proceeding" having received an unknown message from its peer entity returns a STATUS message.

Outgoing call / U3 UE originating call proceeding / Internal alerting indication. When the call control entity of the UE in the "UE originating call proceeding" state receives an ALERTING message then it enters "call delivered" state and, for speech calls, if the user connection is not attached to the radio path, the UE generates internally an alerting indication.

Outgoing call / U4 call delivered / CONNECT received. Test to confirm that a CC-entity of the UE in CC-state U4, "Call Delivered", upon receipt of the CONNECT message returns a CONNECT ACKNOWLEDGE to its peer entity and enters the CC-state U10, "Active".

Outgoing call / U4 call delivered / termination requested by the user. Test to confirm that a CC-entity of the UE in CC-state U4, "Call Delivered", upon request by the user to terminate will send a DISCONNECT message and enter the CC-state U11, "Disconnect Request".

Outgoing call / U4 call delivered / DISCONNECT with in band tones. Test to confirm that a CC-entity of the UE in CC-state U4, "Call Delivered", upon receipt of a DISCONNECT with a progress indicator indicating in-band information, through-connects the speech channel to make in-band announcements available, if traffic channel is in speech mode. If DTCH is not in speech mode, the UE shall send a RELEASE message.

Outgoing call / U4 call delivered / DISCONNECT without in band tones. Test to confirm that a CC-entity of the UE in CC-state U4, "Call Delivered", upon receipt of a DISCONNECT without progress indicator, returns a RELEASE message and enters the CC-state U19, "Release Request".

Outgoing call / U4 call delivered / RELEASE received. Test to confirm that a CC-entity of the UE in CC-state U4, "Call Delivered", upon receipt of the RELEASE message will respond with the RELEASE COMPLETE message and enter the CC-state U0, "Null".

Outgoing call / U4 call delivered / lower layer failure. Test to confirm that a CC-entity of the UE in CC-state U4, "Call Delivered" having detected a lower layer failure and has returned to idle mode, the CC-entity is in CC-state U0, "Null".

Outgoing call / U4 call delivered / traffic channel allocation. Test to confirm that a CC-entity of the UE in CC-state U4, "Call Delivered", when a traffic channel is allocated by the network performing the radio bearer establishment procedure, stays in CC-state U4.

Outgoing call / U4 call delivered / unknown message received. Test to confirm that a CC-entity of the UE in CC-state U4, "Call Delivered", having received an unknown message from its peer entity returns a STATUS message.

U10 active / termination requested by the user. Test to confirm that the a CC-entity of the UE in CC-state U10, "Active", upon request by the user to terminate will send a DISCONNECT message and enter the CC-state U11, "Disconnect Request".

U10 active / RELEASE received. Test to confirm that the a CC-entity of the UE in CC-state U10, "Active", upon receive of the RELEASE will respond with the RELEASE COMPLETE message and enter the CC-state U0, "Null"

U10 active / DISCONNECT with in band tones. Test to confirm that a CC-entity of the UE in CC-state U10, "Active", upon receipt of a DISCONNECT message with a Progress Indicator indicating in-band information, through-connects the speech channel to make in-band announcements available, if traffic channel is in speech mode. If DTCH is not in speech mode, the UE sends a RELEASE message.

U10 active / DISCONNECT without in band tones. Test to confirm that a CC-entity of the UE in CC-state U10, "Active", upon receipt of a DISCONNECT message without progress indicator, returns a RELEASE message and enters the CC-state U19, "Release Request".

U10 active / RELEASE COMPLETE received. Test to confirm that a CC entity of the UE in CC-state U10, "Active" upon receipt of a RELEASE COMPLETE message with valid cause value, enters CC state U0, "Null". Test to confirm that in returning to idle mode, the CC entities relating to the seven UE originating transaction identifiers are in state U0, "Null".

U10 active / SETUP received. Test to confirm that a User Equipment that has a call established and receives a SETUP message answers either with a CALL CONFIRMED message with cause "user busy" if it supports call waiting, or with a RELEASE COMPLETE message with cause "user busy" otherwise. Test to confirm that after having sent this message, the UE is still in state U10 for the established call.

U11 disconnect request / clear collision. Test to confirm that the a CC-entity of the UE in CC-state U11, "Disconnect Request", upon receipt of a DISCONNECT message, returns to its peer entity the RELEASE message and enters the CC-state U19, "Release Request".

U11 disconnect request / RELEASE received. Test to confirm that the a CC-entity of the UE in CC-state U11, "Disconnect Request", upon receipt of the RELEASE message shall return RELEASE COMPLETE and enter the CC-state U0, "Null".

U11 disconnect request / timer T305 time-out. Test to confirm that the CC-entity of the UE in CC-state U11, "Disconnect Request" shall on expiry of T305, proceed with the connection release procedure by sending the RELEASE message to its peer entity and enters the CC-state U19, "Release Request".

U11 disconnect request / lower layer failure. Test to confirm that the a CC-entity of the UE in CC-state U11, "Disconnect Request" having detected a lower layer failure returns to the idle mode. The CC entity is thus in state U0, "Null".

U11 disconnect request / unknown message received. Test to confirm that a CC-entity of the UE in CC-state U11, "Disconnect Request", having received an unknown message from its peer entity returns a STATUS message.

U12 disconnect indication / call releasing requested by the user. Test to confirm that a CC-entity of the UE in CC-state U12, "Disconnect Indication" being in network initiated call release phase, shall, upon receiving a call release request from the user sends a RELEASE to its peer entity and enters CC-state U19, "Release Request"

U12 disconnect indication / RELEASE received. Test to confirm that a CC-entity of the UE in CC-state U12, "Disconnect Indication", upon receipt of a RELEASE message returns to its peer entity the RELEASE COMPLETE message and enters the CC-state U0, "Null". Test to confirm that the UE on returning to the idle mode releases the MM-connection and that the CC-entities relating to the seven UE originating transaction identifiers are in CC-state U0, "Null".

U12 disconnect indication / lower layer failure. Test to confirm that a CC-entity of the UE in CC-state U12, "Disconnect Indication" having detected a lower layer failure returns to idle mode. The CC-entity is thus in state U0, "Null".

U12 disconnect indication / unknown message received. Test to confirm that a CC-entity of the UE in CC-state U12, "Disconnect Indication" having received an unknown message from its peer entity returns a STATUS message.

Outgoing call / U19 release request / timer T308 time-out. Test to confirm that a CC-entity of the UE in CC-state U19, "Release Request" will, upon the first expiry of timer T308 send the RELEASE message to its peer entity and remain in the CC-state U19.

Outgoing call / U19 release request / 2nd timer T308 time-out. Test to confirm that a CC-entity of the UE in CC-state U19, "Release Request", upon the 2nd expiry of the timer T308, enters the CC-state U0, "Null". Test to confirm that subsequently the UE proceeds with releasing the MM-connection and enters the idle mode with the CC entities relating to the seven UE originating transaction identifiers in state U0, "Null".

Outgoing call / U19 release request / RELEASE received. Test to confirm that a CC-entity of the UE in CC-state U19, "Release Request", upon receipt of a RELEASE, shall release the MM-connection and enters the CC-state U0, "Null" with the CC entities relating to the seven UE originating transaction identifiers in state U0, "Null".

Outgoing call / U19 release request / RELEASE COMPLETE received. Test to confirm that a CC-entity of the UE in CC-state U19, "Release Request", upon receipt of a RELEASE COMPLETE, shall release the MM-connection and enters the CC-state U0, "Null" with the CC entities relating to the seven UE originating transaction identifiers in state U0, "Null".

Outgoing call / U19 release request / lower layer failure. Test to confirm that a CC-entity of the UE in CC-state U19, "Release Request", having detected a lower layer failure, returns to the idle mode, the CC entity is in state U0, "Null".

Establishment of an incoming call / Initial conditions

Incoming call / U0 null state / SETUP received with a non supported bearer capability. Test to confirm that a CC entity of the UE, upon receipt of SETUP containing one bearer capability and this bearer capability is not supported, returns a RELEASE COMPLETE with correct cause value to its peer entity, and returns to the idle mode. Test to confirm that the CC-entities relating to the seven UE terminating transaction identifiers are then in the state U0, "Null".

Incoming call / U6 call present / automatic call rejection. Test to confirm that a CC entity of the UE in CC-state U6, "Call Present", shall upon receipt of a rejection indication of the incoming call from the user, shall send RELEASE COMPLETE with the appropriate cause value to its peer entity and enter the CC-state U0, "Null". The CC entities relating to the seven UE terminating transaction identifiers are then in state U0, "Null".

Incoming call / U9 UE terminating call confirmed / alerting or immediate connecting. Test to confirm that a CC entity in CC-state U9, "UE Terminating Call Confirmed", (if signalled by the network in previous SETUP message that it may alert) will either send a ALERTING message to its peer entity and enter state U7, or send a CONNECT message to its peer entity and enter U8.

Incoming call / U9 UE terminating call confirmed / DTCH assignment. Test to confirm that a CC-entity of the UE in CC-state U9, "UE Terminating Call Confirmed", when a traffic channel is allocated by the network performing the radio bearer establishment procedure, shall sends an ALERTING message and enters state U7.

Incoming call / U9 UE terminating call confirmed / DISCONNECT received. Test to confirm that a CC-entity of the UE in CC-state U9, "UE Terminating Call Confirmed", upon receipt of a DISCONNECT returns a RELEASE message and enters the CC-state U19, "Release Request".

Incoming call / U9 UE terminating call confirmed / RELEASE received. Test to confirm that a CC-entity of the UE in CC-state U9, "UE Terminating Call Confirmed", upon receipt of a RELEASE will return a RELEASE COMPLETE and enter the CC-state U0, "Null". Test to confirm that the UE on returning to the idle mode releases the MM-connection and that the CC-entities relating to the seven UE terminating transaction identifiers are in CC-state U0, "Null".

Incoming call / U9 UE terminating call confirmed / lower layer failure. Test to confirm that a CC entity of the UE in CC-state U9, "UE Terminating Call Confirmed", having detected a lower layer failure returns to idle mode, the CC entity is in state U0, "Null".

Incoming call / U9 UE terminating call confirmed / unknown message received. Test to confirm that a CC-entity of the UE in CC-state U9, "UE Terminating Call Confirmed" having received an unknown message from its peer entity returns a STATUS message.

Incoming call / U7 call received / call accepted. Test to confirm that a CC entity of a UE in CC-state U7, "Call Received", upon a user accepting the incoming call, shall send a CONNECT message to its peer entity and enter the CC-state U8, "Connect Request".

Incoming call / U7 call received / termination requested by the user. Test to confirm that a CC entity of a UE in CC-state U7, "Call Received", upon request by the user to terminate will send a DISCONNECT message and enter the CC-state U11, "Disconnect Request".

Incoming call / U7 call received / DISCONNECT received. Test to confirm that a CC entity of a UE in CC-state U7, "Call Received", upon receipt of a DISCONNECT with a progress indicator indicating in-band information from network, if a DTCH was not assigned, returns a RELEASE message and enters the CC-state U19, "Release Request".

Incoming call / U7 call received / RELEASE received. Test to confirm that a CC entity of a UE in CC-state U7, "Call Received", upon receipt of a RELEASE will return a RELEASE COMPLETE and enter the CC-state U0, "Null". Test to confirm that the UE on returning to the idle mode releases the MM-connection and that the CC-entities relating to the seven UE terminating transaction identifiers are in CC-state U0, "Null".

Incoming call / U7 call received / lower layer failure. Test to confirm that a CC entity of a UE in CC-state U7, "Call Received", having detected a lower layer failure returns to idle mode, the CC entity is in state U0, "Null".

Incoming call / U7 call received / unknown message received. Test to confirm that a CC entity of a UE in CC-state U7, "Call Received", having received an unknown message from its peer entity returns a STATUS message.

Incoming call / U7 call received / DTCH assignment. Test to confirm that a CC entity of a UE in CC-state U7, "Call Received", when a traffic channel is allocated by the network performing the radio bearer establishment procedure, stays in CC-state U7.

Incoming call / U7 call received / RELEASE COMPLETE received. Test to confirm that a CC entity of the UE in CC-state U7, "Call received", upon receipt of a RELEASE COMPLETE message with valid cause value, enters CC state U0, "Null". Test to confirm that in returning to idle mode, the CC entities relating to the seven UE terminating transaction identifiers are in state U0, "Null".

Incoming call / U8 connect request / CONNECT acknowledged. Test to confirm that a CC entity of a UE in CC-state U8, "Connect Request", upon receipt of CONNECT ACKNOWLEDGE shall enter the CC-state U10, " Active".

Incoming call / U8 connect request / timer T313 time-out. Test to confirm that a CC entity of a UE in CC-state U8, "Connect Request", having waited for a reasonable length of time (e.g. expiry of timer T313) without receiving the appropriate protocol message to complete the incoming call, shall initiate the clearing of that incoming call by sending the CC message DISCONNECT and enter the CC-state U11, "Disconnect Request".

Incoming call / U8 connect request / termination requested by the user. Test to confirm that a CC entity of a UE in CC-state U8, "Connect Request", upon request by the user to terminate will send a DISCONNECT message and enter the CC-state U11, "Disconnect Request".

Incoming call / U8 connect request / DISCONNECT received with in-band information. Test to confirm that a CC entity of a UE in CC-state U8, "Connect Request", upon receipt of a DISCONNECT with progress indicator #8 enters CC-state U12, if the traffic channel is in speech mode, and that the UE sends a RELEASE message and enters CC-state U19 if the DTCH is not in speech mode.

Incoming call / U8 connect request / DISCONNECT received without in-band information. Test to confirm that a CC entity of a UE in CC-state U8,

"Connect Request", upon receipt of a DISCONNECT without progress indicator, returns a RELEASE message and enters the CC-state U19, "Release Request".

Incoming call / U8 connect request / RELEASE received. Test to confirm that a CC entity of a UE in CC-state U8, "Connect Request", upon receipt of a RELEASE will return a RELEASE COMPLETE and enter the CC-state U0, "Null". Test to confirm that the UE on returning to the idle mode releases the MM-connection and that the CC-entities relating to the seven UE terminating transaction identifiers are in CC-state U0, "Null".

Incoming call / U8 connect request / lower layer failure. Test to confirm that a CC entity of a UE in CC-state U8, "Connect Request", having detected a lower layer failure returns to idle mode, the CC entity is in state U0, "Null".

Incoming call / U8 connect request / DTCH assignment. Test to confirm that a CC entity of a UE in CC-state U8, "Connect Request", when a traffic channel is allocated by the network performing the radio bearer establishment procedure, stays in the CC-state U8.

Incoming call / U8 connect request / unknown message received. Test to confirm that a CC entity of a UE in CC-state U8, "Connect Request", having received an unknown message from its peer entity returns a STATUS message.

In call functions

In-call functions / DTMF information transfer / basic procedures. Test to confirm that an UE supporting the UE originating DTMF protocol control procedure, having a CC entity for speech in state U10, "Active": when made to send a DTMF tone, sends a START DTMF message. Test to confirm that an UE supporting the UE originating DTMF protocol control procedure, having a CC entity for speech in state U10, "Active": when made to send a DTMF tone (the corresponding IA5 character being selected from among the ones supported), sends a START DTMF message specifying the correct IA5 character in the "keypad information" field of the keypad facility information element and to verify that acknowledgement send by the SS is used in the UE to generate a feedback indication for a successful transmission, if applicable. Test to confirm that the UE will send a STOP DTMF message to the network. Test to confirm that the state U10 of the UE CC entity has remained unchanged throughout the test procedure.

In-call functions / User notification / UE terminated. Test to confirm that a CC entity of a UE in CC-state U10, "active", upon receiving of a NOTIFY message remains in the active state.

In-call functions / channel changes / a successful channel change in active state/ Hard handover. Test to confirm that the UE being in the active state after having successful completed a physical channel reconfiguration remains in the active state.

In-call functions / channel changes / an unsuccessful channel change in active mode/Hard handover. Test to confirm that the UE, when returning to the old channel after physical channel reconfiguration failure, will remain in the active state.

Session Management

PDP context activation

Initiated by the UE, Attach initiated by context activation/QoS Offered by Network is the QoS Requested. Test to check that the UE initiates a PS attach, if one is not already active, when PDP context activation is requested. To test the behaviour of the UE when SS responds to the PDP context activation request with the requested QoS.

Attach initiated by context activation/QoS Offered by Network is the QoS Requested/Correct handling of QoS extensions for rates above 8640 kbps. To check that the UE initiates a PS attach, if one is not already active, when PDP context activation is requested. To check that the UE performs correct handling of QoS extensions for rates above 8640 kbps. To check the UE successfully completes the PDP context activation when the SS responds to the PDP context activation request with the requested QoS.

PDP context activation requested by the network, successful and unsuccessful. To test behaviour of the UE upon receipt of a PDP context activation request from the SS: a) When UE supports PDP context activation requested by the network. b) When UE supporting PDP context activation requested by the network, receives REQUEST PDP CONTEXT ACTIVATION message with transaction identifier relating to an already active PDP context. c) When UE does not support PDP context activation requested by the network

Abnormal Cases, T3380 Expiry. To test the behaviour of the UE when the SS does not reply to PDP CONTEXT ACTIVATION REQUEST.

Abnormal Cases, Collision of UE initiated and network requested PDP context activation. To test the behaviour of the UE when there is a collision between an UE initiated and network requested PDP context activation detected by the UE.

Abnormal Cases, Network initiated PDP context activation request for an already activated PDP context (on the UE side). To test the behaviour of the UE when it detects a network initiated PDP context activation for the PDP context already activated on the UE side.

Secondary PDP context activation procedures, Successful Secondary PDP Context Activation Procedure Initiated by the UE, QoS Offered by Network is the QoS Requested. To test the behaviour of the UE when SS responds to a Secondary PDP context activation request with the requested QoS.

Secondary PDP context activation procedures, Successful Secondary PDP Context Activation Procedure Initiated by the UE, QoS Offered by Network is a lower QoS, LLC SAPI rejected by the UE. Test to confirm the behaviour of the UE when the network responds to the ACTIVATE SECONDARY PDP CONTEXT REQUEST message with a negotiated LLC SAPI which is not supported by the UE.

Unsuccessful Secondary PDP Context Activation Procedure Initiated by the UE. To test the behaviour of the UE when network rejects the UE initiated Secondary PDP context activation.

Abnormal cases, T3380 Expiry. To test the behaviour of the UE when the SS does not reply to ACTIVATE SECONDARY PDP CONTEXT REQUEST message.

PDP context modification procedure

Network initiated PDP context modification. To test behaviour of the UE upon receipt of a MODIFY PDP CONTEXT REQUEST message from SS.

UE initiated PDP Context Modification accepted by network. To test the behaviour of the UE upon receipt of a MODIFY PDP CONTEXT ACCEPT message from the SS with -Requested QoS;

UE initiated PDP Context Modification not accepted by the network. To test the behaviour of the UE upon receipt of a MODIFY PDP CONTEXT REJECT message from the System Simulator.

Abnormal cases, T3381 Expiry. To test the behaviour of the UE when SS does not reply to MODIFY PDP CONTEXT REQUEST message.

Collision of UE and network initiated PDP context modification procedures. To test behaviour of the UE when it identifies collision of the UE and SS initiated PDP context modification with the same TI.

PDP context deactivation procedures

PDP context deactivation initiated by the UE. To test the behaviour of the UE upon receipt of a DEACTIVATE PDP CONTEXT ACCEPT message from the SS in PDP context deactivation procedure initiated by the UE. To test the behaviour of the UE upon receipt of a session management message (except REQUEST PDP CONTEXT ACTIVATION or SM-STATUS) specifying a transaction identifier which is not recognised as relating to an active context or to a context that is in the process of activation or deactivation.

PDP context deactivation initiated by the network. To test the behaviour of the UE upon receipt of a DEACTIVATE PDP CONTEXT REQUEST message from the SS.

Abnormal cases, T3390 Expiry. To test the behaviour of the UE when the SS does not reply to a DEACTIVATE PDP CONTEXT REQUEST message from the UE.

Collision of UE and network initiated PDP context deactivation requests. To test the behaviour of the UE when there is a collision between an UE initiated and network initiated context deactivation.

Error cases. To test the behaviour of the UE when messages with unknown or unforeseen transaction identifiers or non-semantical mandatory information element errors occur.

MBMS Context Activation

MBMS Context Activation requested by the network, Successful and Unsuccessful procedure. To test behaviour of the UE upon receipt of a MBMS context activation request from the SS: a) When UE supports MBMS context activation requested by the network, b) When UE supporting MBMS context activation requested by the network, receives REQUEST MBMS CONTEXT ACTIVATION message with transaction identifier relating to an already active MBMS or PDP context

Abnormal Cases, T3380 Expiry. To test the behaviour of the UE when the SS does not reply to MBMS CONTEXT ACTIVATION REQUEST.

MBMS Context deactivation

MBMS Context deactivation requested by the network, Successful. Test to confirm that the UE correctly handle a MBMS context deactivation procedure requested by the network, upon reception of a DEACTIVATE PDP CONTEXT REQUEST. Test to confirm that the UE correctly handle a MBMS context deactivation procedure locally, when the linked PDP context is deactivated.

MBMS Service request procedure not accepted by the network. Test to confirm that the UE correctly handles a SERVICE REJECT message with SM cause "#40 No PDP context activated" received from the network and deactivates the MBMS context locally.

MBMS Service Request procedure collision with Routing Area Update. Test to confirm that the UE correctly aborts a SERVICE REQUEST procedure if a ROUTING AREA UPDATE procedure is initiated before the SERVICE REQUEST procedure has been completed.

PS attach procedure

This procedure is used to indicate for the network that the IMSI is available for traffic by establishment of a GMM context.

Normal PS attach

The normal PS attach procedure is a GMM procedure used by PS UEs of UE operation mode A or C to IMSI attach for PS services only.

Normal PS attach / accepted. To test the behaviour of the UE if the network accepts the PS attach procedure. The following cases are identified: 1) P-TMSI / P-TMSI signature is allocated; 2) P-TMSI / P-TMSI signature is reallocated; 3) Old P-TMSI / P-TMSI signature is not changed.

Normal PS attach / rejected / IMSI invalid / illegal UE. To test the behaviour of the UE if the network rejects the PS attach procedure of the UE with the cause 'illegal MS.

Normal PS attach / rejected / IMSI invalid / PS services not allowed. To test the behaviour of the UE if the network rejects the PS attach procedure of the UE with the cause 'GPRS services not allowed' (no valid PS-subscription for the IMSI).

Normal PS attach / rejected / PLMN not allowed. To test the behaviour of the UE if the network rejects the PS attach procedure of the UE with the cause 'PLMN not allowed'.

Normal PS attach / rejected / roaming not allowed in this location area. This is to 1) test that on receipt of a rejection using the 'roaming not allowed in this location area' cause code, the UE ceases trying to attach on that location area. Successful PS attach procedure is possible in other location areas, 2) test that if the UE is switched off or the USIM is removed the list of 'forbidden location areas for roaming' is cleared, 3) test that at least 6 entries can be held in the list of 'forbidden location areas for roaming' (the requirement in 3GPP TS 24.008 is to store at least 10 entries. This is not fully tested by the third procedure), 4 test that if a cell of the Home PLMN is available then the UE returns to it in preference to any other available cell.

Normal PS attach / rejected / No Suitable Cells In Location Area. To test the behaviour of the UE if the network rejects the PS attach procedure of the UE with the cause 'No Suitable Cells In Location Area'.

Normal PS attach / rejected / Location area not allowed. To test the behaviour of the UE if the network rejects the PS attach procedure of the UE with the cause 'Location area not allowed'.

Normal PS attach / rejected / PS services not allowed in this PLMN. To test the behaviour of the UE if the network rejects the PS attach procedure of the UE with the cause 'GPRS services not allowed in this PLMN'.

Normal PS attach / abnormal cases / access barred due to access class control. To test the behaviour of the UE in case of access class control (access is granted). To test the behaviour of the UE in case of access class control (Cell is changed).

PS attach / abnormal cases / change of routing area. To test the behaviour of the UE in case of procedure collision.

PS attach / abnormal cases / power off. To test the behaviour of the UE in case of procedure collision.

PS attach / abnormal cases / PS detach procedure collision. To test the behaviour of the UE in case of procedure collision.

PS attach / abnormal cases / Failure due to non-integrity protection. Test to confirm that the UE ignores NAS signalling messages when the security mode procedure is not activated.

PS attach / accepted / follow-on request pending indicator set. To test the behaviour of the UE if the follow-on request pending indicator can be set on during the attach procedure. 1) follow-on request pending indicator may be set to indicate further signalling messages from the UE, 2) follow-on request pending indicator not set, no further signalling messages expected from the UE.

PS attach / abnormal cases / access barred due to domain specific access restriction for PS domain. To test the behaviour of the UE in case of domain specific access control for PS domain.

Combined PS attach

Combined PS attach / PS and non-PS attach accepted. To test the behaviour of the UE if the network accepts the PS attach procedure. The following cases are identified: 1) P-TMSI / P-TMSI signature is allocated; 2) P-TMSI / P-TMSI signature is reallocated; 3) Old P-TMSI / P-TMSI signature is not changed; 4) UE terminating CS call is allowed with IMSI; 5) UE terminating CS call is not allowed with TMSI.

Combined PS attach / PS only attach accepted. 1) To test the behaviour of the UE if the network accepts the PS attach procedure with indication PS only, GMM cause 'IMSI unknown in HLR'. 2) To test the behaviour of the UE which does not support an automatic MM IMSI attach if the network accepts the PS attach procedure with indication PS only, GMM cause 'MSC temporarily not reachable', 'Network failure' or 'Congestion'. 3) To test the behaviour of the UE which supports an automatic MM IMSI attach if the network accepts the PS attach procedure with indication PS only, GMM cause 'MSC temporarily not reachable', 'Network failure' or 'Congestion'.

Combined PS attach / PS attach while IMSI attach. To test the behaviour of the UE if PS attach performed while IMSI attached.

Combined PS attach / rejected / IMSI invalid / illegal ME. To test the behaviour of the UE if the network rejects the combined PS attach procedure of the UE with the cause 'Illegal ME'.

Combined PS attach / rejected / PS services and non-PS services not allowed. To test the behaviour of the UE if the network rejects the combined PS attach procedure of the UE with the cause 'GPRS services and non-GPRS services not allowed'.

Combined PS attach / rejected / PS services not allowed. To test the behaviour of the UE if the network rejects the PS attach procedure of the UE with the cause 'GPRS services not allowed'.

Combined PS attach / rejected / location area not allowed. To test the behaviour of the UE if the network rejects the combined PS attach procedure with the cause 'Location Area not allowed'. To test that the UE deletes the list of forbidden LAs when power is switched off.

Combined PS attach / rejected / No Suitable Cells In Location Area. To test the behaviour of the UE if the network rejects the combined PS attach procedure with the cause 'No Suitable Cells In Location Area'.

Combined PS attach / rejected / Roaming not allowed in this location area. To test the behaviour of the UE if the network rejects the PS attach procedure of the UE with the cause 'Roaming not allowed in this location area'.

Combined PS attach / rejected / PS services not allowed in this PLMN. To test the behaviour of the UE if the network rejects the PS attach procedure of the UE with the cause 'GPRS services not allowed in this PLMN'.

Combined PS attach / abnormal cases / attempt counter check / miscellaneous reject causes. To test the behaviour of the UE with respect to the attempt counter.

Combined PS attach / abnormal cases / PS detach procedure collision. To test the behaviour of the UE in case of procedure collision.

PS detach procedure

UE initiated PS detach procedure

PS detach / power off / accepted. To test the behaviour of the UE for the detach procedure.

PS detach / accepted. To test the behaviour of the UE for the detach procedure, including treatment of P-TMSI signature.

PS detach / abnormal cases / attempt counter check / procedure timeout. To test the behaviour of the UE with respect to the attempt counter.

PS detach / abnormal cases / GMM common procedure collision. To test the behaviour of the UE in case of procedure collision.

PS detach / power off / accepted / PS/IMSI detach. To test the behaviour of the UE for the detach procedure.

PS detach / accepted / PS/IMSI detach. To test the behaviour of the UE for the detach procedure.

PS detach / accepted / IMSI detach. To test the behaviour of the UE for the detach procedure.

PS detach / abnormal cases / change of cell into new routing area. To test the behaviour of the UE in case of procedure collision.

PS detach / abnormal cases / PS detach procedure collision. To test the behaviour of the UE in case of procedure collision.

Network initiated PS detach procedure

PS detach / re-attach not required / accepted. To test the behaviour of the UE for the detach procedure.

PS detach / rejected / IMSI invalid / PS services not allowed. To test the behaviour of the UE if the network orders a PS detach procedure with the cause 'GPRS services not allowed' (no valid PS-subscription for the IMSI).

PS detach / IMSI detach / accepted. To test the behaviour of the UE for the detach procedure.

PS detach / re-attach requested / accepted. To test the behaviour of the UE for the detach procedure in case automatic re-attach.

PS detach / rejected / location area not allowed. To test the behaviour of the UE if the network orders the PS detach procedure with the cause 'Location Area not allowed'.

PS detach / rejected / No Suitable Cells In Location Area. To test the behaviour of the UE if the network sends the DETACH REQUEST messages with the cause 'No Suitable Cells In Location Area'.

PS detach / rejected / Roaming not allowed in this location area. To test the behaviour of the UE if the network orders the PS detach procedure with the cause ' Roaming not allowed in this location area '.

PS detach / rejected / PS services not allowed in this PLMN. To test the behaviour of the UE if the network initiates a PS detach procedure with the cause "GPRS services not allowed in this PLMN". To test the behaviour of the UE operating in UE operation mode A in network operation mode I if the network initiates a PS detach procedure with the cause "GPRS services not allowed in this PLMN"

Routing area updating procedure

This procedure is used to update the actual routing area of an UE in the network.

Normal routing area updating

The routing area updating procedure is a GMM procedure used by PS UEs of UE operation mode A or C that are IMSI attached for PS services only.

Routing area updating / accepted. To test the behaviour of the UE if the network accepts the routing area updating procedure. The following cases are identified: To test the behaviour of the UE if the UE enters the new PLMN.

Routing area updating / accepted / Signalling connection re-establishment. To test the behaviour of the UE if the UE receives a RRC CONNECTION RELEASE message with cause = "Directed signalling connection re-establishment".

Handling of MBMS context status information in ROUTING AREA UPDATE procedure. Test to confirm that the UE includes correctly the MBMS context status information in the ROUTING AREA UPDATE REQUEST message. Test to confirm that if the MBMS context status information element is included in the ROUTING AREA UPDATE ACCEPT message, then the UE deactivates all those MBMS contexts locally (without peer to peer signalling between the UE and network) which are not in SM state PDP-INACTIVE in the UE, but are indicated by the network as being in state PDP-INACTIVE. Test to confirm that if no MBMS context status information element is included, then the UE deactivates all MBMS contexts locally which are not in SM state PDP-INACTIVE in the MS.

Routing area updating / rejected / IMSI invalid / illegal ME. To test the behaviour of the UE if the network rejects the routing area updating procedure of the UE with the cause 'Illegal ME'.

Routing area updating / rejected / UE identity cannot be derived by the network. To test the behaviour of the UE if the network rejects the routing area updating procedure of the UE with the cause 'MS identity cannot be derived by the network'.

Routing area updating / rejected / location area not allowed. To test the behaviour of the UE if the network rejects the routing area updating procedure of the UE with the cause 'Location Area not allowed'. To test that the UE deletes the list of forbidden LAs when power is switched off.

Routing area updating / rejected / No Suitable Cells In Location Area. To test the behaviour of the UE if the network rejects the routing area updating procedure with the cause 'No Suitable Cells In Location Area'.

Routing area updating / rejected / PS services not allowed in this PLMN. To test the behaviour of the UE if the network rejects the routing area updating procedure of the UE with the cause 'GPRS services not allowed in this PLMN'.

Routing area updating / rejected / Roaming not allowed in this location area. To test that on receipt of a rejection using the 'Roaming not allowed in this location area' cause code, the UE ceases trying a routing area updating procedure on that location area. Successful routing area updating procedure is possible in other location areas. To test that if the UE is switched off or the USIM is removed the list of 'forbidden location areas for roaming' is cleared.

Routing area updating / abnormal cases / attempt counter check / miscellaneous reject causes. To test the behaviour of the UE with respect to the attempt counter.

Routing area updating / abnormal cases / change of cell into new routing area. To test the behaviour of the UE in case of procedure collision.

Routing area updating / abnormal cases / P-TMSI reallocation procedure collision. To test the behaviour of the UE in case of procedure collision.

Combined routing area updating / combined RA/LA accepted. To test the behaviour of the UE if the network accepts the combined routing area updating procedure. The following cases are identified: 1) P-TMSI / P-TMSI signature is reallocated. 2) Old P-TMSI / P-TMSI signature is not changed. 3) UE terminating CS call is allowed with IMSI. 4) UE terminating CS call is allowed with TMSI.

Combined routing area updating / UE in CS operation at change of RA. To test the behaviour of the UE if the routing area is changed during an ongoing circuit

Combined routing area updating / RA only accepted. To test the behaviour of the UE if the network accepts the routing area updating procedure with indication RA only, GMM cause 'IMSI unknown in HLR'. To test the behaviour of the UE if the network accepts the routing area updating procedure with indication RA only, GMM cause 'MSC temporarily not reachable', 'Network failure' or 'Congestion'.

Combined routing area updating / rejected / PLMN not allowed. To test the behaviour of the UE if the network rejects the combined routing area updating procedure of the UE with the cause 'PLMN not allowed'.

Combined routing area updating / rejected / roaming not allowed in this location area. To test that on receipt of a rejection using the 'Roaming not allowed in this location area' cause code, the UE ceases trying a routing area updating procedure on that location area. Successful combined routing area updating procedure is possible in other location areas. To test that if the UE is switched off or the USIM is removed the list of 'forbidden location areas for roaming' is cleared.

Combined routing area updating / rejected / No Suitable Cells In Location Area. To test the behaviour of the UE if the network rejects a combined routing area updating procedure of the UE with the cause 'No Suitable Cells In

Location Area'. To test that the UE deletes the list of forbidden LAs when power is switched off'.

Combined routing area updating / rejected / Location area not allowed. To test the behaviour of the UE if the network rejects the routing area updating procedure of the UE with the cause 'GPRS services not allowed in this PLMN'.

Combined routing area updating / rejected / PS services not allowed in this PLMN. To test the behaviour of the UE if the network rejects the routing area updating procedure of the UE with the cause 'GPRS services not allowed in this PLMN'.

Combined routing area updating / abnormal cases / access barred due to access class control. To test the behaviour of the UE in case of access class control (access is granted). To test the behaviour of the UE in case of access class control (cell is changed).

Combined routing area updating / abnormal cases / attempt counter check / procedure timeout. To test the behaviour of the UE with respect to the attempt counter.

Combined routing area updating / abnormal cases / change of cell into new routing area. To test the behaviour of the UE in case of procedure collision.

Combined routing area updating / abnormal cases / PS detach procedure collision. To test the behaviour of the UE in case of procedure collision.

Combined routing area updating / abnormal cases / access barred due to domain specific access restriction for CS domain. To test the behaviour of the UE in case of domain specific access control for CS domain.

Combined routing area updating / abnormal cases / access barred due to domain specific access restriction for PS domain. To test the behaviour of the UE in case of domain specific access class control for PS domain.

Periodic routing area updating / accepted. To test the behaviour of the UE with respect to the periodic routing area updating procedure.

Periodic routing area updating / accepted / T3312 default value. To test the behaviour of the UE with respect to the periodic routing area updating procedure.

Periodic routing area updating / no cell available / network mode I. To test the behaviour of the UE with respect to the periodic routing area updating procedure.

Periodic routing area updating / no cell available. To test the behaviour of the UE with respect to the periodic routing area updating procedure.

P-TMSI reallocation

UE should be checked that is able to receive and acknowledge a new P-TMSI by means of an explicit P-TMSI reallocation procedure. Also it should be able to store the P-TMSI in a non-volatile memory. The implicit reallocation procedure is tested in the attach procedure.

PS authentication

Test of authentication

The purpose of this procedure is to verify the user identity. A correct response is essential to guarantee the establishment of the connection. If not, the connection will drop.

Authentication accepted. To test the behaviour of the UE if the network accepts the authentication and ciphering procedure.

Authentication rejected by the network. To test the behaviour of the UE if the network rejects the authentication and ciphering procedure.

Authentication rejected by the UE, GMM cause 'MAC failure'. To test the behaviour of the UE when considers the MAC code (supplied by the core network in the AUTN parameter) to be invalid.

Authentication rejected by the UE, GMM cause 'Synch failure'. To test the behaviour of the UE, when the UE considers the SQN (supplied by the core network in the AUTN parameter) to be out of range.

Authentication rejected by the UE / fraudulent network. To test UE treating a cell as barred when: 1) the network sends the third AUTHENTICATION & CIPHERING REQUEST message with invalid MAC code during the timer T3318 is running, 2) timer T3318 has expired.

Identification procedure

The purpose of this procedure is to check that the UE gives its identity as requested by the network. If this procedure does not work, it will not be possible for the network to rely on the identity claimed by the UE.

General Identification. Test to confirm that the UE sends identity information as requested by the system. The following identities can be requested: IMSI, IMEI and IMEISV.

GMM READY timer handling

Test to verify that READY timer value received in UTRA can be used in GSM.

Service Request procedure (UMTS Only)

Service Request Initiated by UE Procedure. To test the behaviour of the UE if the UE initiates the CM layer service (e.g. SM or SMS) procedure.

Service Request Initiated by Network Procedure. To test the behaviour of the UE if the UE receives the paging request for PS domain service from the network.

Service Request / rejected / Illegal MS. To test the behaviour of the UE if the network rejects the service request procedure with the cause "Illegal MS".

Service Request / rejected / PS services not allowed. To test the behaviour of the UE if the network rejects the service request procedure with the cause "GPRS services not allowed in this PLMN".

Service Request / rejected / MS identity cannot be derived by the network. To test the behaviour of the UE if the network rejects the service request procedure with the cause "MS identity cannot be derived by the network".

Service Request / rejected / PLMN not allowed. To test the behaviour of the UE if the network rejects the service request procedure with the cause "PLMN not allowed".

Service Request / rejected / No PDP context activated. To test the behaviour of the UE if the network rejects the service request procedure with the cause "No PDP context activated".

Service Request / rejected / No Suitable Cells In Location Area. To test the behaviour of the UE if the network rejects the service request procedure with the cause "No Suitable Cells In Location Area".

Service Request / rejected / Roaming not allowed in this location area. To test the behaviour of the UE if the network rejects the service request procedure with the cause "Roaming not allowed in this location area".

Service Request / Abnormal cases / Access barred due to access class control. To test the behaviour of the UE in case of access class control (access is granted).

Service Request / Abnormal cases / Routing area update procedure is triggered. To test the behaviour of the UE in case of collision between Routing area update procedure and Service request procedure.

Service Request / Abnormal cases / Power off. To test the behaviour of the UE in case of collision between Service request procedure and "powered off".

Service Request / Abnormal cases / Service request procedure collision. To test the behaviour of the UE in case of collision between Service request procedure and PS detach procedure.

Service Request / RAB re-establishment / UE initiated / Single PDP context. Test to confirm that the UE initiates a Service request procedure due to uplink data transmission with one preserved PDP context with traffic class "Background class" after normal RRC connection release as well as when radio coverage is lost. Test to confirm that the radio access bearer can be re-established for the preserved PDP context, initiated by the UE.

Service Request / RAB re-establishment / UE initiated / multiple PDP contexts. Test to confirm that the UE initiates a Service request procedure due to uplink data transmission with two PDP contexts with different traffic classes are activated, when one is of traffic class "background class" and the other is of traffic class "interactive class", after normal RRC connection release. Test to confirm that the radio access bearers can be re-established with a single radio bearer establishment procedure for the preserved PDP contexts, when initiated by the UE.

Service Request / RAB re-establishment / Network initiated / single PDP context. Test to confirm that the radio access bearers can be re-established for the preserved PDP context with traffic class "Background class", when initiated from the network, after normal RRC connection release.

Service Request / abnormal cases / access barred due to domain specific access restriction for PS domain. To test the behaviour of the UE in case of domain specific access control for PS domain.

MBMS SERVICE REQUEST /counting / MBMS multicast service. To test the behaviour of the UE that, when MBMS Modified services Information is sent with "MBMS required UE action" IE set to "Acquire counting info", then UE initiates a Service Request procedure.

MBMS SERVICE REQUEST / point to point RBs / MBMS multicast service. To test the behaviour of the UE that, when MBMS Modified services Information is sent with "MBMS required UE action" IE set to "Rrequest PTP RB", then UE initiates a Service Request procedure.

Handling of MBMS context status information in SERVICE REQUEST and SERVICE ACCEPT messages. Test to confirm that the UE includes correct MBMS context status information in SERVICE REQUEST message. Test to confirm that if the MBMS context status information element is included in the SERVICE ACCEPT message, then the UE deactivates all those MBMS contexts locally (without peer to peer signalling between the UE and network) which are not in SM state PDP-INACTIVE in the UE, but are indicated by the network as being in state PDP-INACTIVE. Test to confirm that if no MBMS context status information element is included, then the UE deactivates all MBMS contexts locally which are not in SM state PDP-INACTIVE in the UE.

Emergency Call

Emergency call / with USIM / accept case. Test to confirm that an UE supporting speech in the state "MM idle", when made to call the emergency call number, sends a CM SERVICE REQUEST message specifying the correct CKSN and TMSI, with CM service type IE "emergency call establishment". Test to confirm that after security mode setting by the SS, the UE sends an EMERGENCY SETUP message. Test to confirm that, the SS having sent a CALL PROCEEDING message and then an ALERTING message the correct performance of a connect procedure and that the UE has through connected the DTCH in both directions. Test to confirm that the call is cleared correctly.

Emergency call / without USIM / accept case. Test to confirm that the UE in the "MM idle, no IMSI" state (no USIM inserted) when made to call the emergency call number, sends a CM SERVICE REQUEST message in which the ciphering key sequence number IE indicates "no key is available", the CM service type IE indicates "emergency call establishment" and the UE identity IE specifies the IMEI of the UE. Test to confirm that after receipt of a CM SERVICE ACCEPT message without security mode procedure applied from the SS, the UE sends an EMERGENCY SETUP message. Test to confirm that the SS having sent a CALL PROCEEDING message and then an ALERTING message the correct performance of a connect procedure and that the UE has through connected the DTCH in both directions. Test to confirm that the call is cleared correctly.

Emergency call / without USIM / reject case. Test to confirm that the UE in the "MM idle, no IMSI" state (no USIM inserted) when made to call the emergency call number, sends a CM SERVICE REQUEST message in which the ciphering key sequence number IE indicates "no key is available", the CM service type IE indicates "emergency call establishment", and the UE identity IE specifies the IMEI of the UE. Test to confirm that after receipt of a CM SERVICE REJECT message from the SS, the UE abandons the emergency call establishment.

Radio Bearers Services
Combinations on DPCH
Conversational / speech / UL:12.2 DL:12.2 kbps / CS RAB + UL:3.4 DL:3.4 kbps SRBs for DCCH. Test to confirm radio bearer establishment and correct data transfer for reference radio bearer configuration.

Conversational / speech / UL:(12.2 7.95 5.9 4.75) DL:(12.2 7.95 5.9 4.75) kbps / CS RAB + UL:3.4 DL:3.4 kbps SRBs for DCCH. Test to verify establishment and data transfer of reference radio bearer configuration.

Conversational / speech / UL:(12.2 7.4 5.9 4.75) DL:(12.2 7.4 5.9 4.75) kbps / CS RAB + UL:3.4 DL:3.4 kbps SRBs for DCCH + DL:0.15 kbps SRB#5 for DCCH. Test to verify establishment and data transfer of reference radio bearer configuration.

Conversational / speech / UL:10.2 DL:10.2 kbps / CS RAB + UL:3.4 DL: 3.4 kbps SRBs for DCCH. Test to confirm radio bearer establishment and correct data transfer for reference radio bearer configuration.

Conversational / speech / UL:(10.2, 6.7, 5.9, 4.75) DL:(10.2, 6.7, 5.9, 4.75) kbps / CS RAB + UL:3.4 DL:3.4 kbps SRBs for DCCH. Test to confirm radio bearer establishment and correct data transfer for reference radio bearer configuration.

Conversational / speech / UL:7.95 DL:7.95 kbps / CS RAB + UL:3.4 DL:3.4 kbps SRBs for DCCH. Test to confirm radio bearer establishment and correct data transfer for reference radio bearer configuration.

Conversational / speech / UL:7.4 DL:7.4 kbps / CS RAB+ UL:3.4 DL:3.4 kbps SRBs for DCCH. Test to confirm radio bearer establishment and correct data transfer for reference radio bearer configuration.

Conversational / speech / UL:(7.4, 6.7, 5.9, 4.75) DL:(7.4, 6.7, 5.9, 4.75) kbps / CS RAB + UL:3.4 DL:3.4 kbps SRBs for DCCH. Test to confirm radio bearer establishment and correct data transfer for reference radio bearer configuration.

Conversational / speech / UL:6.7 DL:6.7 kbps / CS RAB + UL:3.4 DL:3.4 kbps SRBs for DCCH. Test to confirm radio bearer establishment and correct data transfer for reference radio bearer configuration.

Conversational / speech / UL:5.9 DL:5.9 kbps / CS RAB + UL:3.4 DL:3.4 kbps SRBs for DCCH. Test to confirm radio bearer establishment and correct data transfer for reference radio bearer configuration.

Conversational / speech / UL:5.15 DL:5.15 kbps / CS RAB + UL:3.4 DL:3.4 kbps SRBs for DCCH. Test to confirm radio bearer establishment and correct data transfer for reference radio bearer configuration.

Conversational / speech / UL:4.75 DL:4.75 kbps / CS RAB + UL:3.4 DL:3.4 kbps SRBs for DCCH. Test to confirm radio bearer establishment and correct data transfer for reference radio bearer configuration.

Conversational / unknown / UL:28.8 DL:28.8 kbps / CS RAB + UL:3.4 DL:3.4 kbps SRBs for DCCH . Test to confirm radio bearer establishment and correct data transfer for reference radio bearer configuration.

Conversational / unknown / UL:64 DL:64 kbps / CS RAB + UL:3.4 DL:3.4 kbps SRBs for DCCH. Test to confirm radio bearer establishment and correct data transfer for reference radio bearer configuration for 20 ms TTI case.

Conversational / unknown / UL:64 DL:64 kbps / CS RAB / 40 ms TTI. Test to confirm radio bearer establishment and correct data transfer for reference radio bearer configuration for 40 ms TTI case.

Conversational / unknown / UL:32 DL:32 kbps / CS RAB + UL:3.4 DL:3.4 kbps SRBs for DCCH. Conversational / unknown / UL:32 DL:32 kbps / CS RAB / 20 ms TTI. Test to confirm radio bearer establishment and correct data transfer for reference radio bearer configuration for 20 ms TTI case.

Conversational / unknown / UL:32 DL:32 kbps / CS RAB + UL:3.4 DL:3.4 kbps SRBs for DCCH. Conversational / unknown / UL:32 DL:32 kbps / CS RAB / 40 ms TTI. Test to confirm radio bearer establishment and correct data transfer for reference radio bearer configuration for 40 ms TTI case.

Streaming / unknown / UL:14.4/DL:14.4 kbps / CS RAB + UL:3.4 DL:3.4 kbps SRBs for DCCH. Test to confirm radio bearer establishment and correct data transfer for reference radio bearer configuration.

Streaming / unknown / UL:28.8/DL:28.8 kbps / CS RAB + UL:3.4 DL:3.4 kbps SRBs for DCCH. Test to confirm radio bearer establishment and correct data transfer for reference radio bearer configuration.

Streaming / unknown / UL:57.6/DL:57.6 kbps / CS RAB + UL:3.4 DL:3.4 kbps SRBs for DCCH. Test to confirm radio bearer establishment and correct data transfer for reference radio bearer configuration.

Interactive or background / UL:32 DL:8 kbps / PS RAB + UL:3.4 DL:3.4 kbps SRBs for DCCH. Interactive or background / UL:32 DL:8 kbps / PS RAB / (TC,10 ms TTI). Test to verify establishment and data transfer of reference radio bearer configuration for the turbo channel coding and uplink 10 ms TTI case.

Interactive or background / UL:32 DL:8 kbps / PS RAB + UL:3.4 DL:3.4 kbps SRBs for DCCH. Interactive or background / UL:32 DL:8 kbps / PS RAB / (CC, 10 ms TTI). Test to verify establishment and data transfer of reference radio bearer configuration for the convolutional channel coding and uplink 10 ms TTI case.

Interactive or background / UL:8 DL:8 kbps / PS RAB + UL:3.4 DL:3.4 kbps SRBs for DCCH / TC. Test to verify establishment and data transfer of reference radio bearer configuration for a turbo coding case. Test to verify establishment and data transfer of reference radio bearer configuration.

Interactive or background / UL:32 DL:32 kbps / PS RAB + UL:3.4 DL:3.4 kbps SRBs for DCCH. Test to verify establishment and data transfer of reference radio bearer configuration. Test to verify establishment and data transfer of reference radio bearer configuration.

Interactive or background / UL:32 DL:32 kbps / PS RAB (20 ms TTI) + UL:3.4 DL:3.4 kbps SRBs for DCCH. Test to verify establishment and data transfer of reference radio bearer configuration.

Interactive or background / UL:32 DL: 64 kbps / PS RAB + UL: 3.4 DL:3.4 kbps SRBs for DCCH. Interactive or background / UL:32 DL: 64 kbps / PS RAB / (TC, 10 ms TTI). Test to verify establishment and data transfer of reference radio bearer configuration for the uplink turbo channel coding and 10 ms TTI case.

Interactive or background / UL:32 DL: 64 kbps / PS RAB + UL:3.4 DL:3.4 kbps SRBs for DCCH Interactive or background / UL:32 DL:64 kbps / PS RAB / (CC, 10 ms TTI). Test to verify establishment and data transfer of reference radio bearer configuration for the uplink convolutional channel coding and 10 ms TTI case. Test to verify establishment and data transfer of reference radio bearer configuration for the uplink convolutional channel coding and 20 ms TTI case. Test to verify establishment and data transfer of reference radio bearer configuration.

Interactive or background / UL:64 DL:128 kbps / PS RAB + UL:3.4 DL:3.4 kbps SRBs for DCCH. Test to verify establishment and data transfer of reference radio bearer configuration.

Interactive or background / UL:64 DL:144 kbps / PS RAB + UL:3.4 DL: 3.4 kbps SRBs for DCCH. Test to verify establishment and data transfer of reference radio bearer configuration.

Interactive or background / UL:64 DL:256 kbps / PS RAB + UL:3.4 DL: 3.4 kbps SRBs for DCCH. Interactive or background / UL:64 DL:256 kbps / PS RAB / 10 ms TTI. Test to verify establishment and data transfer of reference radio bearer configuration for the downlink 10 ms TTI case.

Interactive or background / UL:64 DL:384 kbps / PS RAB + UL:3.4 DL: 3.4 kbps SRBs for DCCH. Interactive or background / UL:64 DL:384 kbps / PS RAB / 10 ms TTI. Test to verify establishment and data transfer of reference radio bearer configuration for the 10 ms TTI case.

Interactive or background / UL:128 DL:384 kbps / PS RAB + UL:3.4 DL: 3.4 kbps SRBs for DCCH. Test to verify establishment and data transfer of reference radio bearer configuration for the 10 ms TTI case.

Interactive or background / UL:128 DL:384 kbps / PS RAB / 20 ms TTI. Test to verify establishment and data transfer of reference radio bearer configuration for the 20 ms TTI case.

Interactive or background / UL:384 DL:384 kbps / PS RAB + UL:3.4 DL:3.4 kbps SRBs for DCCH. Test to verify establishment and data transfer of reference radio bearer configuration for the 10 ms TTI case.

Interactive or background / UL:384 DL:384 kbps / PS RAB / 20 ms TTI. Test to verify establishment and data transfer of reference radio bearer configuration for the 20 ms TTI case

Interactive or background / UL:64 DL:2048 kbps / PS RAB + UL:3.4 DL:3.4 kbps SRBs for DCCH. Test to verify establishment and data transfer of reference radio bearer configuration for the 10 ms TTI case.

Interactive or background / UL:64 DL:2048 kbps / PS RAB / 20 ms TTI. Test to verify establishment and data transfer of reference radio bearer configuration for the 20 ms TTI case.

Conversational / speech / UL:12.2 DL:12.2 kbps / CS RAB + Interactive or background / UL:32 DL:8 kbps / PS RAB + UL:3.4 DL:3.4 kbps SRBs for DCCH. Test to verify establishment and data transfer of reference radio bearer configuration for the turbo channel coding and 20 ms TTI case.

Conversational / speech / UL:12.2 DL:12.2 kbps / CS RAB + Interactive or background / UL:32 DL:8 kbps / PS RAB / (TC, 10 ms TTI). Test to verify establishment and data transfer of reference radio bearer configuration for the turbo channel coding and 10 ms TTI case.

Conversational / speech / UL:12.2 DL:12.2 kbps / CS RAB + Interactive or background / UL:32 DL:8 kbps / PS RAB / (CC, 20 ms TTI). Test to verify establishment and data transfer of reference radio bearer configuration for the convolutional channel coding and 20 ms TTI case.

Conversational / speech / UL:12.2 DL:12.2 kbps / CS RAB + Interactive or background / UL:32 DL:8 kbps / PS RAB / (CC, 10 ms TTI). Test to verify establishment and data transfer of reference radio bearer configuration for the convolutional channel coding and 10 ms TTI case.

Conversational / speech / UL:12.2 DL:12.2 kbps / CS RAB + Interactive or background / UL:0 DL:0 kbps / PS RAB + UL:3.4 DL:3.4 kbps SRBs for DCCH. Test to verify establishment and data transfer of reference radio bearer configuration.

Conversational / speech / UL:12.2 DL:12.2 kbps / CS RAB + Interactive or background / UL:8 DL:8 kbps / PS RAB + UL:3.4 DL:3.4 kbps SRBs for DCCH. Test to verify establishment and data transfer of reference radio bearer configuration.

Conversational / speech / UL:12.2 DL:12.2 kbps / CS RAB + Interactive or background / UL:32 DL:32 kbps / PS RAB + UL:3.4 DL:3.4 kbps SRBs for DCCH. Test to verify establishment and data transfer of reference radio bearer configuration.

Conversational / speech / UL:12.2 DL:12.2 kbps / CS RAB + Interactive or background / UL:64 DL:64 kbps / PS RAB + Interactive or background /

UL:64 DL:64 kbps / PS RAB + UL:3.4 DL:3.4 kbps SRBs for DCCH. Test to verify establishment and data transfer of reference radio bearer configuration.

Conversational / speech / UL:(12.2 7.95 5.9 4.75) DL:(12.2 7.95 5.9 4.75) kbps / CS RAB + Interactive or background / UL:0 DL:0 kbps / PS RAB + UL:3.4 DL:3.4 kbps SRBs for DCCH. Test to verify establishment and data transfer of reference radio bearer configuration.

Conversational / speech / UL:(12.2 7.95 5.9 4.75) DL:(12.2 7.95 5.9 4.75) kbps / CS RAB + Interactive or background / UL:8 DL:8 kbps / PS RAB + UL:3.4 DL:3.4 kbps SRBs for DCCH. Test to verify establishment and data transfer of reference radio bearer configuration.

Conversational / speech / UL:(12.2 7.95 5.9 4.75) DL:(12.2 7.95 5.9 4.75) kbps / CS RAB + Interactive or background / UL:16 DL:16 kbps / PS RAB + UL:3.4 DL:3.4 kbps SRBs for DCCH. Test to verify establishment and data transfer of reference radio bearer configuration.

Conversational / speech / UL:(12.2 7.95 5.9 4.75) DL:(12.2 7.95 5.9 4.75) kbps / CS RAB + Interactive or background / UL:32 DL:32 kbps / PS RAB + UL:3.4 DL:3.4 kbps SRBs for DCCH. Test to verify establishment and data transfer of reference radio bearer configuration.

Conversational / speech / UL:(12.2 7.95 5.9 4.75) DL:(12.2 7.95 5.9 4.75) kbps / CS RAB + Interactive or background / UL:64 DL:64 kbps / PS RAB + UL:3.4 DL:3.4 kbps SRBs for DCCH. Test to verify establishment and data transfer of reference radio bearer configuration.

Conversational / speech / UL:(12.2 7.95 5.9 4.75) DL:(12.2 7.95 5.9 4.75) kbps / CS RAB + Interactive or background / UL:64 DL:128 kbps / PS RAB + UL:3.4 DL:3.4 kbps SRBs for DCCH. Test to verify establishment and data transfer of reference radio bearer configuration.

Conversational / speech / UL:12.2 DL:12.2 kbps / CS RAB + Interactive or background / UL:32 DL:64 kbps / PS RAB+ UL:3.4 DL: 3.4 kbps SRBs for DCCH. Test to verify establishment and data transfer of reference radio bearer configuration for the uplink turbo channel coding and 10 ms TTI case.

Conversational / speech / UL:12.2 DL:12.2 kbps / CS RAB + Interactive or background / UL:32 DL:64 kbps / PS RAB / (TC, 20 ms TTI). Test to verify establishment and data transfer of reference radio bearer configuration for the uplink turbo channel coding and 20 ms TTI case.

Conversational / speech / UL:12.2 DL:12.2 kbps / CS RAB + Interactive or background / UL:32 DL:64 kbps / PS RAB / (CC, 10 ms TTI). Test to verify establishment and data transfer of reference radio bearer configuration for the uplink convolutional channel coding and 10 ms TTI case.

Conversational / speech / UL:12.2 DL:12.2 kbps / CS RAB + Interactive or background / UL:32 DL:64 kbps / PS RAB / (CC, 20 ms TTI). Test to verify establishment and data transfer of reference radio bearer configuration for the uplink convolutional channel coding and 20 ms TTI case.

Conversational / speech / UL:12.2 DL:12.2 kbps / CS RAB + Interactive or background / UL:64 DL:64 kbps / PS RAB+ UL:3.4 DL: 3.4 kbps SRBs for DCCH. Test to verify establishment and data transfer of reference radio bearer configuration.

Conversational / speech / UL:12.2 DL:12.2 kbps / CS RAB + Interactive or background / UL:64 DL:128 kbps / PS RAB + UL:3.4 DL:3.4 kbps SRBs for DCCH. Test to verify establishment and data transfer of reference radio bearer configuration.

Conversational / speech / UL:12.2 DL:12.2 kbps / CS RAB + Interactive or background / UL:64 DL:256 kbps / PS RAB + UL:3.4 DL:3.4 kbps SRBs for DCCH. Test to verify establishment and data transfer of reference radio bearer configuration for the downlink 10 ms TTI case.

Conversational / speech / UL:12.2 DL:12.2 kbps / CS RAB + Interactive or background / UL:64 DL:256 kbps / PS RAB / 20 ms TTI. Test to verify establishment and data transfer of reference radio bearer configuration for the downlink 20 ms TTI case.

Conversational / speech / UL:12.2 DL:12.2 kbps / CS RAB + Interactive or background / UL:64 DL:384 kbps / PS RAB + UL:3.4 DL:3.4 kbps SRBs for DCCH. Test to verify establishment and data transfer of reference radio bearer configuration for the downlink 10 ms TTI case.

Conversational / speech / UL:12.2 DL:12.2 kbps / CS RAB + Interactive or background / UL:64 DL:384 kbps / PS RAB / 20 ms TTI. Test to verify establishment and data transfer of reference radio bearer configuration for the downlink 20 ms TTI case.

Conversational / speech / UL:12.2 DL:12.2 kbps / CS RAB + Interactive or background / UL:128 DL:2048 kbps / PS RAB + UL:3.4 DL:3.4 kbps SRBs for DCCH. Test to verify establishment and data transfer of reference radio bearer configuration for the downlink 10 ms TTI case.

Conversational / speech / UL:12.2 DL:12.2 kbps / CS RAB + Interactive or background / UL:128 DL:2048 kbps / PS RAB / 20 ms TTI. Test to verify establishment and data transfer of reference radio bearer configuration for the downlink 20 ms TTI case.

Conversational / speech / UL:12.2 DL:12.2 kbps / CS RAB + Streaming / unknown / UL:57.6 DL:57.6 kbps / CS RAB + UL:3.4 DL:3.4 kbps SRBs for DCCH. Test to verify establishment and data transfer of reference radio bearer configuration.

Conversational / speech / UL:12.2 DL:12.2 kbps / CS RAB + Conversational / unknown / UL:64 DL:64 kbps / CS RAB + UL:3.4 DL:3.4 kbps SRBs for DCCH. Test to verify establishment and data transfer of reference radio bearer configuration for the 20 ms TTI case.

Conversational / speech / UL:12.2 DL:12.2 kbps / CS RAB + Conversational / unknown / UL:64 DL:64 kbps / CS RAB / 40 ms TTI. Test to verify establishment and data transfer of reference radio bearer configuration for the 40 ms TTI case.

Conversational / speech / UL:(12.2 7.95 5.9 4.75) DL(12.2 7.95 5.9 4.75) kbps / CS RAB + Conversational / unknown / UL:64 DL:64 kbps / CS RAB+ UL:3.4 DL: 3.4 kbps SRBs for DCCH (20ms TTI). Test to verify establishment and data transfer of reference radio bearer configuration for 20ms TTI case.

Conversational / speech / UL:(12.2 7.95 5.9 4.75) DL(12.2 7.95 5.9 4.75) kbps / CS RAB + Conversational / unknown / UL:64 DL:64 kbps / CS RAB+ UL:3.4 DL: 3.4 kbps SRBs for DCCH (40ms TTI). Test to verify establishment and data transfer of reference radio bearer configuration for 40 ms TTI case .

Conversational / unknown / UL:64 DL:64 kbps / CS RAB + Conversational / unknown / UL:64 DL:64 kbps / CS RAB + UL:3.4 DL:3.4 kbps SRBs for DCCH. Test to verify establishment and data transfer of reference radio bearer configuration for the 20 ms TTI case.

Conversational / unknown / UL:64 DL:64 kbps / CS RAB + Conversational / unknown / UL:64 DL:64 kbps / CS RAB + UL:3.4 DL:3.4 kbps SRBs for DCCH / 40 ms TTI. Test to verify establishment and data transfer of reference radio bearer configuration for the 40 ms TTI case.

Conversational / unknown / UL:64 DL:64 kbps / CS RAB + Interactive or background / UL:64 DL:64 kbps / PS RAB + UL:3.4 DL:3.4 kbps SRBs for DCCH. Test to verify establishment and data transfer of reference radio bearer configuration for the 20 ms TTI case.

Conversational / unknown / UL:64 DL:64 kbps / CS RAB / 40 ms TTI + Interactive or background / UL:64 DL:64 kbps / PS RAB. Test to verify establishment and data transfer of reference radio bearer configuration for the 40 ms TTI case.

Conversational / unknown / UL:64 DL:64 kbps / CS RAB + Interactive or Background / UL:8 DL:8 kbps / PS RAB + UL:3.4 DL:3.4 kbps SRBs for DCCH. Test to verify establishment and data transfer of reference radio bearer configuration for 20 ms TTI case.

Conversational / unknown / UL:64 DL:64 kbps / CS RAB / 40 ms TTI + Interactive or background / UL:8 DL:8 kbps / PS RAB. Test to verify establishment and data transfer of reference radio bearer configuration for the 40 ms TTI case.

Conversational / unknown / UL:64 DL:64 kbps / CS RAB / 20 ms TTI + Interactive or background / UL:16 DL:64 kbps / PS RAB. Test to verify establishment and data transfer of reference radio bearer configuration for the 20 ms TTI case.

Conversational / unknown / UL:64 DL:64 kbps / CS RAB / 40 ms TTI + Interactive or background / UL:16 DL:64 kbps / PS RAB. Test to verify establishment and data transfer of reference radio bearer configuration for the 40 ms TTI case.

Conversational / unknown / UL:64 DL:64 kbps / CS RAB + Interactive or background / UL:64 DL:128 kbps / PS RAB + UL:3.4 DL:3.4 kbps SRBs for DCCH. Test to verify establishment and data transfer of reference radio bearer configuration for the 20 ms TTI case.

Conversational / unknown / UL:64 DL:64 kbps / CS RAB / 40 ms TTI + Interactive or background / UL:64 DL:128 kbps / PS RAB. Test to verify establishment and data transfer of reference radio bearer configuration for the 40 ms TTI case.

Conversational / unknown / UL:64 DL:64 kbps / CS RAB + Interactive or background / UL:128 DL:128 kbps / PS RAB + UL:3.4 DL:3.4 kbps SRBs for DCCH. Test to verify establishment and data transfer of reference radio bearer configuration for the 20 ms TTI case.

Conversational / unknown / UL:64 DL:64 kbps / CS RAB / 40 ms TTI + Interactive or background / UL:128 DL:128 kbps / PS RAB. Test to verify establishment and data transfer of reference radio bearer configuration for the 40 ms TTI case.

Interactive or background / UL:8 DL:8 kbps / PS RAB + Interactive or background / UL:8 DL:8 kbps / PS RAB + UL:3.4 DL:3.4 kbps SRBs for DCCH. Test to verify establishment and data transfer of reference radio bearer configuration.

Interactive or background / UL:64 DL:64 kbps / PS RAB + Interactive or background / UL:64 DL:64 kbps / PS RAB + UL:3.4 DL:3.4 kbps SRBs for DCCH. Test to verify establishment and data transfer of reference radio bearer configuration.

Streaming / unknown / UL:16 DL:64 kbps / PS RAB + Interactive or background / UL:8 DL:8 kbps / PS RAB + UL:3.4 DL:3.4 kbps SRBs for DCCH. Test to verify establishment and data transfer of reference radio bearer configuration.

Streaming / unknown / UL:16 DL:128 kbps / PS RAB + Interactive or background / UL:8 DL:8 kbps / PS RAB + UL:3.4 DL:3.4 kbps SRBs for DCCH. Test to verify establishment and data transfer of reference radio bearer configuration.

Conversational / speech / UL:(12.65 8.85 6.6) DL:(12.65 8.85 6.6) kbps / CS RAB + UL:3.4 DL:3.4 kbps SRBs for DCCH + DL:0.15 kbps SRBs for DCCH. Test to verify establishment and data transfer of reference radio bearer configuration.

Interactive or background / UL:64 DL:768 kbps / PS RAB + UL:3.4 DL:3.4 kbps SRBs for DCCH. Test to verify establishment and data transfer of reference radio bearer configuration for the downlink 10 ms TTI case.

Interactive or background / UL:64 DL:768 kbps / PS RAB + UL:3.4 DL: 3.4 kbps SRBs for DCCH / 20 ms TTI. Test to verify establishment and data transfer of reference radio bearer configuration for the downlink 20 ms TTI case.

Combinations on PDSCH and DPCH

Interactive or background / UL:64 DL:384 kbps / PS RAB + UL:3.4 DL: 3.4 kbps SRBs for DCCH. Test to verify establishment and data transfer of reference radio bearer configuration for the downlink 10 ms TTI case.

Interactive or background / UL:64 DL:384 kbps / PS RAB / 20 ms TTI. Test to verify establishment and data transfer of reference radio bearer configuration for the downlink 20 ms TTI case.

Interactive or background / UL:64 DL:2048 kbps / PS RAB + UL:3.4 DL: 3.4 kbps SRBs for DCCH. Test to verify establishment and data transfer of reference radio bearer configuration for the downlink 10 ms TTI case.

Interactive or background / UL:64 DL:2048 kbps / PS RAB / 20 ms TTI. Test to verify establishment and data transfer of reference radio bearer configuration for the downlink 20 ms TTI case.

Conversational / speech / UL:12.2 DL:12.2 kbps / CS RAB + Interactive or background / UL:64 DL:384 kbps / PS RAB + UL:3.4 DL:3.4 kbps SRBs for DCCH. Test to verify establishment and data transfer of reference radio bearer configuration for the downlink 10 ms TTI case.

Conversational / speech / UL:12.2 DL:12.2 kbps / CS RAB + Interactive or background / UL:64 DL:384 kbps / PS RAB + UL:3.4 DL:3.4 kbps SRBs for DCCH / 20 ms TTI. Test to verify establishment and data transfer of reference radio bearer configuration for the downlink 20 ms TTI case.

Conversational / speech / UL:12.2 DL:12.2 kbps / CS RAB + Interactive or background / UL:64 DL:2048 kbps / PS RAB + UL:3.4 DL:3.4 kbps SRBs for DCCH. Test to verify establishment and data transfer of reference radio bearer configuration for the downlink 10 ms TTI case.

Conversational / speech / UL:12.2 DL:12.2 kbps / CS RAB + Interactive or background / UL:64 DL:2048 kbps / PS RAB + UL:3.4 DL:3.4 kbps SRBs for DCCH / 10 ms TTI. Test to verify establishment and data transfer of reference radio bearer configuration for the downlink 20 ms TTI case.

Combinations on SCCPCH

One SCCPCH: Interactive/Background 32 kbps PS RAB + SRBs for CCCH + SRB for DCCH + SRB for BCCH. Test to confirm establishment and data transfer of reference radio bearer configuration for the case when two SCCPCHs are used in this SYSTEM INFORMATION configuration. The first SCCPCH carries the PCH and the second SCCPCH carries the FACH for Interactive/Background 32 kbps PS RAB and the FACH for SRBs on CCCH/ DCCH/ BCCH. To be able to test the downlink radio bearer using the UE loopback function, the reference radio bearer configuration (Interactive/Background 32 kbps PS RAB + SRB for CCCH + SRB for DCCH on PRACH) is used in uplink.

Two SCCPCHs: Interactive/Background 32 kbps PS RAB + SRBs for CCCH + SRB for DCCH + SRB for BCCH. Test to confirm establishment and data transfer of reference radio bearer configuration for the case when three

SCCPCHs are used in this SYSTEM INFORMATION configuration. The first SCCPCH carries the PCH and both the second and third SCCPCHs carry the FACH for Interactive/Background 32 kbps PS RAB and the FACH for SRBs on CCCH/ DCCH/ BCCH. To be able to test the downlink radio bearer using the UE loopback function, the reference radio bearer configuration (Interactive/ Background 32 kbps PS RAB + SRB for CCCH + SRB for DCCH on PRACH) is used in uplink.

One SCCPCH/connected mode: Interactive/Background 32 kbps PS RAB + SRBs for CCCH + SRB for DCCH + SRB for BCCH. Test to confirm establishment and data transfer of reference radio bearer configuration for the case when three SCCPCHs are used in this SYSTEM INFORMATION configuration. The first SCCPCH carries the PCH. The second SCCPCH carries the FACH for CTCH (Cell Broadcast Service) and the FACH for SRBs on CCCH/ BCCH for idle mode UEs. The third SCCPCH carries the FACH for Interactive/Background 32 kbps PS RAB and the FACH for SRBs on CCCH/ DCCH/ BCCH for connected mode UEs. To be able to test the downlink radio bearer using the UE loopback function, the reference radio bearer configuration (Interactive/Background 32 kbps PS RAB + SRB for CCCH + SRB for DCCH on PRACH) is used in uplink.

Interactive/Background 32 kbps PS RAB + Interactive/Background 32 kbps PS RAB + SRBs for CCCH + SRB for DCCH + SRB for BCCH. Test to verify establishment and data transfer of reference radio bearer configuration. Test to confirm establishment and data transfer of reference radio bearer configuration for the case when two SCCPCHs are used in this SYSTEM INFORMATION configuration. The first SCCPCH carries the PCH and the second SCCPCH carries the FACH for two Interactive/Background 32 kbps PS RABs and the FACH for SRBs on CCCH/ DCCH/ BCCH. To be able to test the downlink radio bearer using the UE loopback function, the reference radio bearer configuration (Interactive/Background 32 kbps PS RAB + Interactive/Background 32 kbps PS RAB + SRB for CCCH + SRB for DCCH on PRACH) is used in uplink.

Two SCCPCHs: Interactive/Background 32 kbps PS RAB + Interactive/Background 32 kbps PS RAB + SRBs for CCCH + SRB for DCCH + SRB for BCCH. Test to confirm establishment and data transfer of reference radio bearer configuration for the case when three SCCPCHs are used in this SYSTEM INFORMATION configuration. The first SCCPCH carries the PCH and both the second and third SCCPCHs carry the FACH for two Interactive/Background 32 kbps PS RABs and the FACH for SRBs on CCCH/ DCCH/ BCCH. To be able to test the downlink radio bearer using the UE loopback function, the reference radio bearer configuration (Interactive/Background 32 kbps PS RAB + Interactive/Background 32 kbps PS RAB + SRB for CCCH + SRB for DCCH on PRACH) is used in uplink.

One SCCPCH/connected mode: Interactive/Background 32 kbps PS RAB + Interactive/Background 32 kbps PS RAB + SRBs for CCCH + SRB

for DCCH + SRB for BCCH. Test to confirm establishment and data transfer of reference radio bearer configuration for the case when three SCCPCHs are used in this SYSTEM INFORMATION configuration. The first SCCPCH carries the PCH. The second SCCPCH carries the FACH for CTCH (Cell Broadcast Service) and the FACH for SRBs on CCCH/ BCCH for idle mode UEs. The third SCCPCH carries the FACH for two Interactive/Background 32 kbps PS RABs and the FACH for SRBs on CCCH/ DCCH/ BCCH for connected mode UEs. To be able to test the downlink radio bearer using the UE loopback function, the reference radio bearer configuration (Interactive/Background 32 kbps PS RAB + Interactive/Background 32 kbps PS RAB + SRB for CCCH + SRB for DCCH on PRACH) is used in uplink.

Interactive/Background 32 kbps RAB + SRBs for PCCH + SRB for CCCH + SRB for DCCH + SRB for BCCH. Test to confirm establishment and data transfer of reference radio bearer configuration for the case when one SCCPCH is used in this SYSTEM INFORMATION (BCCH) configuration. The SCCPCH carries the PCH, the FACH for Interactive/Background 32 kbps PS RAB and the FACH for SRBs on CCCH/ DCCH/ BCCH. To be able to test the downlink radio bearer using the UE loopback function, the reference radio bearer configuration (Interactive/Background 32 kbps PS RAB + SRB for CCCH + SRB for DCCH on PRACH) is used in uplink.

RB for CTCH + SRB for CCCH +SRB for BCCH. Test to verify establishment and data transfer of reference radio bearer configuration for the case when three SCCPCHs are used in this SYSTEM INFORMATION configuration. The first SCCPCH carries the PCH. The second SCCPCH carries the FACH for CTCH (Cell Broadcast Service) and the FACH for SRBs on CCCH/ BCCH for idle mode UEs. The third SCCPCH carries the FACH for Interactive/Background 32 kbps PS RAB and the FACH for SRBs on CCCH/ DCCH/ BCCH for connected mode UEs. Test to confirm establishment and data transfer of reference radio bearer configuration. Data transfer on CTCH is tested similar to testing BMC for a UE in idle mode, data transfer on CCCH is tested by establishing a RRC connection.

Combinations on DPCH and HS-PDSCH

Interactive or background / UL:64 DL: [max bit rate depending on UE category] / PS RAB + UL:3.4 DL:3.4 kbps SRBs for DCCH. Test to confirm radio bearer establishment and correct data transfer for reference radio bearer configuration.

Interactive or background / UL:128 DL: [max bit rate depending on UE category] / PS RAB + UL:3.4 DL:3.4 kbps SRBs for DCCH. Test to confirm radio bearer establishment and correct data transfer for reference radio bearer configuration.

Interactive or background / UL:384 DL: [max bit rate depending on UE category] / PS RAB + UL:3.4 DL:3.4 kbps SRBs for DCCH. Test to confirm radio bearer establishment and correct data transfer for reference radio bearer configuration.

Conversational / speech / UL:12.2 DL:12.2 kbps / CS RAB + Interactive or background / UL:384 DL:[Bit rate depending on the UE category] / PS RAB + UL:3.4 DL:3.4 kbps SRBs for DCCH. Test to confirm radio bearer establishment and correct data transfer for reference radio bearer configuration.

Conversational / speech / UL:12.2 DL:12.2 kbps / CS RAB + Interactive or background / UL: 64 DL:[Bit rate depending on the UE category] / PS RAB + UL:3.4 DL:3.4 kbps SRBs for DCCH. Test to confirm radio bearer establishment and correct data transfer for reference radio bearer configuration for the uplink 64 kbps case.

Conversational / unknown / UL:64 DL:64 kbps / CS RAB + Interactive or background / UL:384 DL:[Bit rate depending on the UE category] / PS RAB + UL:3.4 DL:3.4 kbps SRBs for DCCH. Test to confirm radio bearer establishment and correct data transfer for reference radio bearer configuration.

Conversational / unknown / UL:64 DL:64 kbps / CS RAB + Interactive or background / UL:64 DL:[Bit rate depending on the UE category] / PS RAB + UL:3.4 DL:3.4 kbps SRBs for DCCH. Test to confirm radio bearer establishment and correct data transfer for reference radio bearer configuration.

Interactive or background / UL:384 DL:[Bit rate depending on the UE category] / PS RAB + Interactive or background / UL:384 DL:[Bit rate depending on the UE category] / PS RAB + UL:3.4 DL:3.4 kbps SRBs for DCCH. Test to confirm radio bearer establishment and correct data transfer for reference radio bearer configuration.

Interactive or background / UL:64 DL:[Bit rate depending on the UE category] / PS RAB + Interactive or background / UL:64 DL:[Bit rate depending on the UE category] / PS RAB + UL:3.4 DL:3.4 kbps SRBs for DCCH. Test to confirm radio bearer establishment and correct data transfer for reference radio bearer configuration

Streaming / unknown / UL:128 DL: [guaranteed 128, max bit rate depending on UE category] kbps / PS RAB + Interactive or background / UL:128 DL: [max bit rate depending on UE category] / PS RAB + UL:3.4 DL:3.4 kbps SRBs for DCCH. Test to confirm radio bearer establishment and correct data transfer for reference radio bearer configuration.

Conversational / speech / UL:12.2 DL:12.2 kbps / CS RAB + Streaming / unknown / UL:128 DL: [guaranteed 128, max bit rate depending on UE category] kbps / PS RAB + Interactive or background / UL:128 DL: [max bit rate depending on UE category] / PS RAB + UL:3.4 DL:3.4 kbps SRBs for DCCH. Test to confirm radio bearer establishment and correct data transfer for reference radio bearer configuration.

Conversational / speech / UL:(12.65 8.85 6.6) DL:(12.65 8.85 6.6) kbps / CS RAB + Interactive or Background / UL:384 DL:[Bit rate depending on the UE category] / PS RAB+ UL:3.4 DL:3.4 kbps SRBs for DCCH + DL:0.15 kbps SRB#5 for DCCH. Test to confirm radio bearer establishment and correct data transfer for reference radio bearer configuration.

Combinations on DPCH, HS-PDSCH and E-DPDCH

The testing of radio bearer combinations on DPCH, HSPDSCH and E-DPDCH is focusing on verifying that the UE is capable to establish the radio bearer combination and to verify correct data transfer. This is using all the possible uplink TFS for transport channels mapped on DCH and all the configured MAC-d PDU sizes for the transport channel mapped to E-DCH. The verification of MAC-e/es transport block size selection and that the UE is capable of transmitting all the possible transport block sizes within the UE capability is tested in the MAC-e/es.

Streaming or interactive or background / UL: [max bit rate depending on UE category and TTI] DL: [max bit rate depending on UE category] / PS RAB + UL:3.4 DL:3.4 kbps SRBs for DCCH on DCH. Test to confirm that the UE is able to establish the radio bearer combination. Test to confirm correct data transfer using all the possible MAC-d PDU sizes of the transport channel mapped to E-DCH.

Streaming or interactive or background / UL: [max bit rate depending on UE category and TTI] DL: [max bit rate depending on UE category] / PS RAB + UL:[max bit rate depending on UE category and TTI] DL:3.4 kbps SRBs for DCCH on E-DCH and DL DCH. Test to confirm that the UE is able to establish the radio bearer combination. Test to confirm correct data transfer using all the possible MAC-d PDU sizes of the transport channel mapped to E-DCH.

Streaming or interactive or background / UL: [max bit rate depending on UE category and TTI] DL: [max bit rate depending on UE category] / PS RAB + UL: [max bit rate depending on UE category and TTI] DL: [max bit rate depending on UE category] SRBs for DCCH on E-DCH and HS-DSCH. Test to confirm that the UE is able to establish the radio bearer combination. Test to confirm correct data transfer using all the possible MAC-d PDU sizes of the transport channel mapped to E-DCH.

Conversational / speech / UL:12.2 DL:12.2 kbps / CS RAB + Streaming or interactive or background / UL: [max bit rate depending on UE category and TTI] DL: [max bit rate depending on UE category] / PS RAB + UL:3.4 DL:3.4 kbps SRBs for DCCH. Test to confirm that the UE is able to establish the radio bearer combination. Test to confirm correct data transfer using all the possible MAC-d PDU sizes of the transport channel mapped to E-DCH in combination with the possible TFCI of the conversional speech radio bearer.

Streaming or interative or background / UL:[max bit rate depending on UE category and TTI] DL: [max bit rate depending on UE category] kbps / PS RAB + Streaming or interactive or background / UL: [max bit rate depending on UE category and TTI] DL: [max bit rate depending on UE category] / PS RAB + UL:[max bit rate depending on UE category and TTI] DL:3.4 kbps SRBs for DCCH on E-DCH and DL DCH. Test to confirm that the UE is able to establish the radio bearer combination. Test to confirm correct data transfer using all the possible MAC-d PDU sizes of the transport channel mapped to E-DCH.

Conversational / unknown or speech / UL:[max bit rate depending on UE category and TTI] DL: [max bit rate depending on UE category] kbps / PS RAB + Streaming or Interactive or background / UL: [max bit rate depending on UE category and TTI] DL: [max bit rate depending on UE category] / PS RAB + UL:[max bit rate depending on UE category and TTI] DL: :[max bit rate depending on UE category] SRBs for DCCH on E-DCH and HS-DSCH. Test to confirm that the UE is able to establish the radio bearer combination. Test to confirm correct data transfer using all the possible MAC-d PDU sizes of the transport channel mapped to E-DCH.

Conversational / unknown or speech / UL:[max bit rate depending on UE category and TTI] DL: [max bit rate depending on UE category] kbps / PS RAB + Streaming or Interactive or background / UL: [max bit rate depending on UE category and TTI] DL: [max bit rate depending on UE category] / PS RAB + Streaming or Interactive or background / UL: [max bit rate depending on UE category and TTI] DL: [max bit rate depending on UE category] / PS RAB + UL:[max bit rate depending on UE category and TTI] DL: :[max   bit rate depending on UE category] SRBs for DCCH on E-DCH and HS-DSCH. Test to confirm that the UE is able to establish the radio bearer combination. Test to confirm correct data transfer using all the possible MAC-d PDU sizes of the transport channel mapped to E-DCH.

Conversational / speech / UL:(12.65 8.85 6.6) DL:(12.65 8.85 6.6) kbps / CS RAB + Streaming or interactive or background / UL: [max bit rate depending on UE category and TTI] DL: [max bit rate depending on UE category] / PS RAB + UL:3.4 DL:3.4 kbps SRBs for DCCH + DL:0.15 kbps SRB#5 for DCCH. Test to confirm that the UE is able to establish the radio bearer combination. Test to confirm correct data transfer using all the possible MAC-d PDU sizes of the transport channel mapped to E-DCH.

Short message service (SMS)

These tests are to verify that the UE can handle Iu mode system functions when submitting or receiving Short Messages (SM) between UE and a short message service centre. The procedures are based upon services provided by the Mobility Management (MM) sublayer and GPRS Mobility Management (GMM) sublayer which are not tested in this case.

This service comprises three basic services; SMS point to point services on CS mode, on PS mode and SMS cell broadcast service. The SMS point to point services on CS mode shall work in an active UE at any time independent of whether or not there is a speech or data call in progress. The SMS point to point services on PS mode shall work in an active UE at any time independent of whether or not there is a PDP context in progress. The SMS cell broadcast service only works when the UE is in idle mode.

SMS point to point on CS mode

SMS UE terminated. Test to confirm the ability of a UE to receive and decode the SMS where provided for the point to point service.

SMS UE originated. Test to confirm that the UE is able to correctly send a short message where the SMS is provided for the point to point service.

Memory available. Test of memory full condition and memory available notification: The Memory Available Notification provides a means for the UE to notify the network that it has memory available to receive one or more short messages. The SMS status field in the USIM contains status information on the "memory available" notification flag. 1) Test to verify that the UE sends the correct acknowledgement when its memory in the USIM becomes full. 2) Test to verify that the UE sends the correct acknowledgement when its memory in the ME and the USIM becomes full, and sets the "memory exceeded" notification flag in the USIM. 3) Test to verify that the UE performs the "memory available" procedure when its message store becomes available for receiving short messages, and only at this moment.

Test of the status report capabilities and of SMS-COMMAND: This test applies to UEs which support the status report capabilities. 1) Test to verify that the UE is able to accept a SMS-STATUS-REPORT TPDU. 2) Test to verify that the UE is able to use the SMS-COMMAND functionality correctly and sends an SMS-COMMAND TPDU with the correct TP-Message-Reference.

Short message class 0. Test to confirm that the UE will accept and indicate but not store a class 0 message, and that it will accept and indicate a class 0 message if its message store is full.

Test of class 1 short messages. This test shall apply to UEs which support: storing of received Class 1 Short Messages; and indicating of stored Short Messages. This procedure verifies that the UE acts correctly on receiving a class 1 message, i.e. that it stores the message in the ME or USIM and sends an acknowledgement (at RP and CP-Layer).

Test of class 2 short messages. This procedure verifies that the UE acts correctly on receiving a class 2 message, i.e. that it stores the message correctly in the USIM, and if this is not possible, returns a protocol error message, with the correct error cause, to the network.. There are 2 cases: 1) if the UE supports storing of short messages in the USIM and in the ME, and storage in the ME is not full, and the short message cannot be stored in the USIM, the error cause shall be "protocol error, unspecified"; 2)   if the UE supports storing of short messages in the USIM and not in the ME, and storage in the ME is not full, and the short message cannot be stored in the USIM, the error cause shall be "memory capacity exceeded".

Test of short message type 0. Test to confirm that the UE will acknowledge receipt of the short message to the SC. The UE shall discard its contents. This means that 1) the UE shall be able to receive the type 0 short message irrespective of whether there is memory available in the (U)SIM or ME or not, 2) the UE shall not indicate the receipt of the type 0 short message to the user, 3) the short message shall neither be stored in the (U)SIM nor ME.

Test of the reply path scheme. This procedure verifies that the UE is able to send a Reply Short Message back to the correct originating SME even if in the meantime it receives another Short Message.

Multiple SMS UE originated, UE in idle mode. This test applies to UE supporting the ability of sending multiple short messages on the same RRC connection when there is no call in progress. Test to confirm that the UE is able to correctly concatenate multiple short messages on the same RRC connection when using a DCCH.

Multiple SMS UE originated, UE in active mode. This test applies to UE supporting the ability of sending concatenated multiple short messages when there is a call in progress. Test to confirm that the UE is able to correctly concatenate multiple short messages on the same RRC connection when sent parallel to a call. Test of capabilities of simultaneously receiving a short message whilst sending a UE originated short message. The test verifies that the UE is capable of simultaneously receiving a network originated SM whilst sending a UE originated SM.

SMS point to point on PS mode

SMS UE terminated. Test to confirm the ability of a UE to receive and decode the SMS where provided for the point to point service.

SMS UE originated. Test to confirm that the UE is able to correctly send a short message where the SMS is provided for the point to point service.

Test of memory full condition and memory available notification: The Memory Available Notification provides a means for the UE to notify the network that it has memory available to receive one or more short messages. The SMS status field in the USIM contains status information on the "memory available" notification flag. 1) Test to verify that the UE sends the correct acknowledgement when its memory in the USIM becomes full. 2) Test to verify that the UE sends the correct acknowledgement when its memory in the ME and the USIM becomes full, and sets the "memory exceeded" notification flag in the USIM. 3) Test to verify that the UE performs the "memory available" procedure when its message store becomes available for receiving short messages, and only at this moment.

Test of the status report capabilities and of SMS-COMMAND: This test applies to UEs which support the status report capabilities. 1) Test to verify that the UE is able to accept a SMS-STATUS-REPORT TPDU. 2) Test to verify that the UE is able to use the SMS-COMMAND functionality correctly and sends an SMS-COMMAND TPDU with the correct TP-Message-Reference.

Short message class 0. Test to confirm that the UE will accept and indicate but not store a class 0 message and that it will accept and indicate a class 0 message if its message store is full.

Test of class 1 short messages. This test shall apply to UEs which support: storing of received Class 1 Short Messages; and indicating of stored Short Messages. This procedure verifies that the UE acts correctly on receiving a class 1 message, i.e. that it stores the message in the ME or USIM and sends an acknowledgement (at RP and CP-Layer).

Test of class 2 short messages. This procedure verifies that the UE acts correctly on receiving a class 2 message, i.e. that it stores the message correctly in the USIM, and if this is not possible, returns a protocol error message, with the correct error cause, to the network. There are 2 cases: 1) if the UE supports storing of short messages in the USIM and in the ME, and storage in the ME is not full, and the short message cannot be stored in the USIM, the error cause shall be "protocol error, unspecified"; 2)   if the UE supports storing of short messages in the USIM and not in the ME, and storage in the ME is not full, and the short message cannot be stored in the USIM, the error cause shall be "memory capacity exceeded".

Test of short message type 0. Test to confirm that the UE will acknowledge receipt of the short message to the SC. The UE shall discard its contents. This means that: the UE shall be able to receive the type 0 short message irrespective of whether there is memory available in the (U)SIM or ME or not, the UE shall not indicate the receipt of the type 0 short message to the user, the short message shall neither be stored in the (U)SIM nor ME.

Test of the replace mechanism for SM type 1-7. This procedure verifies the correct implementation of the replace mechanism for Replace Short Messages.

Test of the reply path scheme. This procedure verifies that the UE is able to send a Reply Short Message back to the correct originating SME even if in the meantime it receives another Short Message.

Multiple SMS UE originated. Test of capabilities of simultaneously receiving a short message whilst sending a UE originated short message. The test verifies that the UE is capable of simultaneously receiving a network originated SM whilst sending a UE originated SM.

Short message service cell broadcast. This test verifies that an UE supporting SMS-CB is able to receive SMS-CB messages and is able to ignore repeated broadcasts of CBS messages.

A-GPS

Signalling tests

LCS Network Induced location request/ UE-Based GPS/ Emergency Call / with USIM. Test to confirm when an emergency call is initiated by a UE with a USIM, and the network performs a location request using the RRC measurement control procedure by sending Measurement Control message , then the UE respond with a Measurement Report containing UE location.

LCS Network Induced location request/ UE-Based GPS/ Emergency Call / without USIM. Test to confirm when an emergency call is initiated by a UE in the "MM idle, no IMSI" state (no USIM inserted) and the network performs a location request using the RRC measurement control procedure by sending Measurement Control message , then the UE respond with a Measurement Report containing UE location.

LCS Network induced location request/ UE-Assisted GPS/ Emergency call/ With USIM. Test to confirm when an emergency call is initiated by a UE with a USIM, and the network performs a location request using the RRC measurement control procedure by sending Measurement Control message , then the UE respond with a Measurement Report containing "UE positioning GPS measured results".

LCS Network induced location request/ UE-Assisted GPS/ Emergency call/ Without USIM. Test to confirm that when an emergency call is initiated by a UE with no USIM, and the network performs a network-induced location request using UE-assisted A-GPS, the UE responds with a Measurement Report containing the IE "UE positioning GPS measured results".

LCS UE originated location request/ UE-Based GPS/ Position estimate request/ Success. Test to confirm the UE behaviour at a UE originated location request procedure using network-assisted UE-based GPS.

LCS UE originated location request/ UE-Based or UE-Assisted GPS/ Assistance data request/ Success. Test to confirm the UE behaviour at a UE originated location request procedure using network-assisted network assisted GPS.

LCS UE originated location request/ UE-Assisted GPS/ Position Estimate/ Success. Test to confirm the UE behaviour in the UE-originated location request procedure using network-assisted UE-assisted GPS to request a position estimate from the network.

LCS UE originated location request/ UE-Based GPS/ Transfer to third party/ Success. Test to confirm the UE behaviour in the UE-originated location request procedure using network-assisted UE-based GPS to request a position estimate from the network for transfer to a third-party LCS client.

LCS UE originated location request/ UE-Assisted GPS/ Transfer to third party/ Success. Test to confirm the UE behaviour in the UE-originated location request procedure using network-assisted UE-assisted GPS to request a position estimate from the network for transfer to a third-party LCS client.

LCS UE originated location request/ UE-Based or UE-Assisted GPS/ Assistance data request/ Failure. Test to confirm the UE behaviour at a UE originated location request for GPS assistance data where the network is unable to provide the requested GPS assistance data.

LCS UE originated location request/ UE-Based GPS/ Position estimate request/ Failure. Test to confirm the UE behaviour at a UE originated location request procedure using network-assisted UE-based GPS when the MO-LR procedure fails due to failure of positioning method.

LCS UE terminated location request/ UE-Based GPS. Test to confirm that when the UE receives a REGISTER message during an established CS call, containing a LCS Location Notification Invoke component set to NotifyLocation-Allowed, the UE displays information about the LCS client correctly and sends a RELEASE COMPLETE message containing a LocationNotification return result with verificationResponse set to permissionGranted. Test to confirm that the UE responds with a Measurement Report message containing UE location when the assistance data is divided between several Measurement Control messages using Measurement Command "Modify".

LCS UE-terminated location request/UE-Based GPS/ Request for additional assistance data/ Success. Test to confirm the UE's behavior in a UE-terminated location request procedure using UE-based A-GPS with assistance data from the network. Test to confirm that the UE in CELL_DCH state accepts assistance data received in multiple MEASUREMENT CONTROL messages. Test to confirm that the UE includes the IE "GPS Additional Assistance Data Request" to request assistance data when it does not have enough assistance data to compute a position.

LCS UE-terminated location request/UE-Based GPS/ Failure – Not Enough Satellites. Test to confirm the UE's behaviour in a UE-terminated location request procedure using UE-based A-GPS with assistance data from the network. Test to confirm that the UE in CELL_DCH state accepts assistance data received in multiple MEASUREMENT CONTROL messages. Test to confirm that the UE sets the IE Error Reason in 'UE Postioning Error' to 'Not Enough GPS Satellites' when it does not receive enough satellite signals to compute a position.

LCS UE terminated location request/ UE-Assisted GPS/ Success. Test to confirm the UE behaviour in the UE-terminated location request procedure using network-assisted UE-assisted GPS to deliver UE positioning measurements to the network.

LCS UE terminated location request/ UE-Assisted GPS/ Request for additional assistance data/ Success. Test to confirm the UE behaviour in the UE-terminated location request procedure using network-assisted UE-assisted GPS to deliver UE positioning measurements to the network. Test to confirm that the UE includes the IE "GPS Additional Assistance Data Request" to request additional assistance data when it does not have enough assistance data to perform the requested measurements.

LCS UE terminated location request/ UE-Based GPS/ Privacy Verification/ Location allowed if no response. Test to confirm that when the UE receives a REGISTER message, containing a LCS Location Notification Invoke component set to notifyAndVerify-LocationAllowedIfNoResponse, the UE notifies the user of the request and indicates that the default response is location allowed if no response and providing the opportunity to accept or deny the request and sends a RELEASE COMPLETE message containing a LocationNotification return result with verificationResponse set to permissionDenied or permissionGranted as appropriate.

LCS UE terminated location request/ UE-Based GPS/ Privacy Verification/ Location Not Allowed if No Response. Test to confirm that when the UE receives a REGISTER message, containing a LCS Location Notification Invoke component set to notifyAndVerify-LocationNotAllowedIfNoResponse, the UE notifies the user of the request and indicates that the default response is location not allowed if no response and providing the opportunity to accept or deny the request and sends a RELEASE COMPLETE message containing a LocationNotification return result with verificationResponse set to permissionDenied or permissionGranted as appropriate.

LCS UE terminated location request/ UE-Assisted GPS/ Privacy Verification/ Location Allowed if No Response. Test to confirm that when the UE receives a REGISTER message, containing a LCS Location Notification Invoke component set to notifyAndVerify-LocationAllowedIfNoResponse, the UE notifies the user of the request and indicates that the default response is location allowed if no response and providing the opportunity to accept or deny the request and sends a RELEASE COMPLETE message containing a Location Notification return result with verificationResponse set to permissionDenied or permissionGranted as appropriate.

LCS UE terminated location request/ UE-Assisted GPS/ Privacy Verification/ Location Not Allowed if No Response. Test to confirm that when the UE receives a REGISTER message, containing a LCS Location Notification Invoke component set to notifyAndVerify-LocationNotAllowedIfNoResponse, the UE notifies the user of the request and indicates that the default response is location not allowed if no response and providing the opportunity to accept or deny the request and sends a RELEASE COMPLETE message containing a LocationNotification return result with verificationResponse set to permission Denied or permissionGranted as appropriate.

LCS UE terminated location request/ UE-Based or UE-Assisted GPS/ Configuration Incomplete. Test to confirm that the UE sends a MEASUREMENT CONTROL FAILURE message, after receiving a MEASUREMENT CONTROL message with IE "Method Type" set a value which is inconsistent with the UE positioning capabilities. Test to confirm that the UE set the "failure cause" IE to value "configuration incomplete" in the uplink MEASUREMENT CONTROL FAILURE message.

Performance Tests

This section identifies the minimum performance tests for both UE based and UE assisted FDD A GPS terminals. If a terminal supports both modes then it shall be tested in both modes. Tests are defined for CELL_DCH and CELL_FACH states. All tests should be performed in CELL_DCH state and the Nominal Accuracy Performance test case should be also performed in CELL_FACH state.

Sensitivity Coarse Time Assistance. Sensitivity with coarse time assistance is the minimum level of GPS satellite signals required for the UE to make an A-GPS position estimate to a specific accuracy and within a specific response time when the network only provides coarse time assistance. Test to confirm the UE's first position estimate meets the minimum requirements under GPS satellite signal conditions that represent weak signal conditions and with only Coarse Time Assistance provided by the SS.

Sensitivity Fine Time Assistance. Sensitivity with fine time assistance is the minimum level of GPS satellite signals required for the UE to make an A-GPS position estimate to a specific accuracy and within a specific response time when the network provides fine time assistance in addition to coarse time assistance. Test to confirm the UE's first position estimate meets the minimum requirements under GPS satellite signal conditions that represent weak signal conditions and with Fine Time Assistance provided by the SS.

Nominal Accuracy. Nominal accuracy is the accuracy of the UE's A-GPS position estimate under ideal GPS signal conditions. The requirements and this test apply to all types of UTRA for the FDD UE that supports A-GPS. Test to confirm the UE's first position estimate meets the minimum requirements under GPS satellite signal conditions that represent ideal conditions.

Dynamic Range. Test to confirm the UE's first position estimate meets the minimum requirements under GPS satellite signal conditions that have a wide dynamic range. Strong satellites are likely to degrade the acquisition of weaker satellites due to their cross correlation products.

Multi-path Performance. Multi-path performance measures the accuracy and response time of the UE's A-GPS position estimate in a specific GPS signal multi-path environment. Test to confirm the UE's first position estimate meets the minimum requirements under GPS satellite signal conditions that represent simple multi-path conditions.

Moving Scenario and Periodic Update Performance. Moving scenario and periodic update performance measures the accuracy of the UE's A-GPS position estimates and the periodic update capability of the UE in a moving scenario. Test to confirm the UE's position estimates, after the first reported position estimate, meet the minimum requirements under GPS satellite signal conditions that simulate a moving scenario. A good tracking performance, with regular position estimate reporting is essential for certain location services.

Acoustic Testing

Connections with handset UE, Sending Loudness Rating (SLR)
　　Connections with handset UE, Receiving Loudness Rating (RLR)
　　Idle channel noise (handset and headset UE), Sending.
　　Idle channel noise (handset and headset UE), Receiving.
　　Sensitivity/frequency characteristics, Handset UE sending
　　Sensitivity/frequency characteristics, Handset UE receiving

Stability loss. A gain equivalent to the minimum stability margin is inserted in the loop between the go and return paths of the reference speech coder in the SS and any acoustic echo control is enabled.

Acoustic echo control in a handset UE. The handset is suspended in free air in such a way that the inherent mechanical coupling of the handset is not affected. The testing shall be made under real use environmental conditions. The ambient noise level shall be less than -64 dBPa(A). The attenuation from reference point input to reference point output shall be measured using the speech like test signal.

USIM Testing

Subscription related tests

IMSI / TMSI handling

UE identification by short IMSI. Test to confirm that the Terminal uses the IMSI of the USIM. Test to confirm that the Terminal can handle an IMSI of less than the maximum length. Test to confirm that the READ EFIMSI command is performed correctly by the terminal

UE identification by short IMSI using a 2 digit MNC. Test to confirm that the Terminal can handle an IMSI consistence of a 2 digit MNC.

UE identification by "short" TMSI. Test to confirm that the Terminal uses the TMSI stored in the USIM. Test to confirm that the Terminal can handle a TMSI of less than maximum length.

UE identification by "long" TMSI. Test to confirm that the Terminal uses the TMSI stored in the USIM. Test to confirm that the Terminal can handle a TMSI of maximum length. Test to confirm that the Terminal does not respond to page requests containing a previous TMSI.

UE identification by long IMSI, TMSI updating and key set identifier assignment. Test to confirm that the Terminal uses the IMSI stored in the USIM. Test to confirm that the Terminal does not respond to page requests containing a previous IMSI. Test to confirm that the Terminal can handle an IMSI of maximum length. Test to confirm that the Terminal correctly updates the key set identifier respectively the ciphering key sequence number at call termination. Test to confirm that the Terminal correctly updates the TMSI at call termination. Test to confirm that the UPDATE EFLOCI command is performed correctly by the terminal

Access Control handling

Access Control information handling. Test to confirm that the Terminal reads the access control value as part of the USIM-Terminal initialisation procedure, and subsequently adopts this value. Test to confirm that the UE controls its network access in accordance with its access control class and the conditions imposed by the serving network.

Security related Tests

PIN handling

Entry of PIN. Test to confirm that the PIN verification procedure is performed by the Terminal correctly. Test to confirm that the basic public MMI string is supported.

Change of PIN. Test to confirm that the PIN substitution procedure is performed correctly by the Terminal. Test to confirm that the basic public MMI string is supported.

Unblock PIN. Test to confirm that the PIN unblocking procedure is performed correctly. Test to confirm that the basic public MMI string is supported.

Entry of PIN2. Test to confirm that the PIN2 verification procedure is performed by the Terminal correctly. Test to confirm that the basic public MMI string is supported.

Change of PIN2. Test to confirm that the PIN2 substitution procedure is performed correctly by the Terminal. Test to confirm that the basic public MMI string is supported.

Unblock PIN2. Test to confirm that the PIN2 unblocking procedure is performed correctly. Test to confirm that the basic public MMI string is supported.

Replacement of PIN. Test to confirm that the PIN replacement is supported by the Terminal correctly. Test to confirm that the PIN replacement procedure is performed by the Terminal correctly. Test to confirm that the procedure to disable the PIN replacement is performed by the Terminal correctly.

Change of Universal PIN. Test to confirm that the PIN substitution procedure is performed correctly by the Terminal.

Unblock Universal PIN. Test to confirm that the PIN unblocking procedure is performed correctly.

Entry of PIN on multi-verification capable UICCs. Test to confirm that the PIN verification procedure is performed by the Terminal correctly. Test to confirm that the basic public MMI string is supported. Test to confirm that the Terminal supports key references in the range of "01" to "08" as PIN.

Change of PIN on multi-verification capable UICCs. Test to confirm that the PIN substitution procedure is performed correctly by the Terminal. Test to confirm that the basic public MMI string is supported. Test to confirm that the Terminal supports key references in the range of "01" to "08" as PIN.

Unblock PIN on multi-verification capable UICCs. Test to confirm that the PIN unblocking procedure is performed correctly. Test to confirm that the basic public MMI string is supported. Test to confirm that the Terminal supports key references in the range of "01" to "08" as PIN.

Entry of PIN2 on multi-verification capable UICCs. Test to confirm that the PIN2 verification procedure is performed by the Terminal correctly. Test to confirm that the basic public MMI string is supported. Test to confirm that the Terminal supports key references in the range of "81" to "88" as PIN2.

Change of PIN2 on multi-verification capable UICCs. Test to confirm that the PIN2 substitution procedure is performed correctly by the Terminal. Test to confirm that the basic public MMI string is supported. Test to confirm that the Terminal supports key references in the range of "81" to "88" as PIN2.

Unblock PIN2 on multi-verification capable UICCs. Test to confirm that the PIN2 unblocking procedure is performed correctly. Test to confirm that the basic public MMI string is supported. Test to confirm that the Terminal supports key references in the range of "81" to "88" as PIN2.

Replacement of PIN with key reference "07". Test to confirm that the PIN replacement is supported by the Terminal correctly. Test to confirm that the PIN replacement procedure is performed by the Terminal correctly. Test to confirm that the procedure to disable the PIN replacement is performed by the Terminal correctly. Test to confirm that the Terminal supports key references in the range of "01" to "08" as PIN.

Fixed Dialling Numbers (FDN) handling

Terminal and USIM with FDN enabled, EFADN readable and updateable. Test to confirm that the Terminal allows call set-up to a FDN number. Test to confirm that the Terminal allows call set-up to a FDN number extended by some digits in the end. Test to confirm that the Terminal rejects call set-up to number having no reference in EFFDN. Test to confirm that the Terminal rejects call set-up to a FDN number not completely corresponding to an entry in EFFDN. Test to confirm that the Terminal does not allow emergency call set-up using the emergency number stored in the Terminal except "112", "911", the emergency numbers stored on the SIM/USIM and emergency numbers downloaded from the serving network (if any). Test to confirm that the Terminal allows emergency call set-up using the emergency number stored in the UISM.

Terminal and USIM with FDN disabled. Test to confirm that the Terminal as a result of the state of the USIM correctly performs the UICC-Terminal initialisation procedure. Test to confirm that the Terminal allows call set-up to a FDN number. Test to confirm that the Terminal allows call set-up to a ADN number. Test to confirm that the Terminal allows call set-up to manually given number.

Enabling, disabling and updating of FDN. Test to confirm that the Terminal correctly performs the update of a number in EF_{FDN}. Test to confirm that the Terminal correctly disables FDN service. Test to confirm that the Terminal recognises disabling of FDN and allows access to EF_{ADN}.

Terminal and USIM with FDN enabled, EF_{ADN} readable and updateable. Test to confirm that the Terminal allows call set-up to a FDN number. Test to confirm that the Terminal allows call set-up to a FDN number extended by some digits in the end. Test to confirm that the Terminal rejects call set-up to number having no reference in EF_{FDN}. Test to confirm that the Terminal rejects call set-up to a FDN number not completely corresponding to an entry in EF_{FDN}. Test to confirm that the Terminal does not allow emergency call set-up using the emergency number stored in the Terminal except "112", "911", the emergency numbers stored on the SIM/USIM and emergency numbers downloaded from the serving network (if any). Test to confirm that the Terminal allows emergency call set-up using the emergency number stored in the UISM. Test to confirm that the Terminal reads correctly the emergency service category.

Advice of charge (AoC) handling

AoC not supported by USIM. Test to confirm that an UE not supporting AoCC (where the Terminal does support AoCC but the USIM does not) and in the outgoing call / U4 call delivered state, on receipt of a CONNECT message containing AoCC information shall acknowledge the CONNECT message but ignore and not acknowledge the AoCC information sent within the CONNECT. Test to confirm that an UE not supporting AoCC (where the Terminal does support AoCC but the USIM does not) and in the outgoing call / U4 call delivered state, on receipt of a FACILITY message containing AoCC information shall ignore and not acknowledge the AoCC information sent within the FACILITY. Test to confirm that an UE not supporting AoCC (where the Terminal does support AoCC but the USIM does not) and in the incoming call / U9 call confirmed state, on receipt of a FACILITY message containing AoCC information shall ignore and not acknowledge the AoCC information sent within the FACILITY. Test to confirm that an UE not supporting AoCC (where the Terminal does support AoCC but the USIM does not) and in the U10 call active state, on receipt of a FACILITY message containing AoCC information, shall ignore and not acknowledge the AoCC information sent within the FACILITY.

Maximum frequency of ACM updating. Test to confirm that the Terminal, during a call, increments the ACM every 5 s when e2 is less or equal to 5 s. Test to confirm that the Terminal is able to handle other values than '1C' as SFI of EF_{ACM}.

Call terminated when ACM greater than ACMmax. Test to confirm that the Terminal increments the ACM by the correct number of units, even though this may take ACM above ACMmax. Test to confirm that the Terminal terminates the call. Test to confirm that the INCREMENT EF_{ACM} command is performed correctly by the terminal. Test to confirm that the Terminal is able to handle other values than '1C' as SFI of EF_{ACM}.

Response codes of increase command of ACM. Test to confirm that the Terminal clears a charged call if the USIM indicates that the ACM cannot be increased. Test to confirm that the Terminal is able to handle other values than "1C" as SFI of EF_{ACM}.

PLMN related tests

FPLMN handling
Adding FPLMN to the Forbidden PLMN list. Test to confirm that in automatic PLMN selection mode the UE does not attempt to access PLMNs stored in EFFPLMN on the USIM. Test to confirm that the EFFPLMN is correctly updated by the Terminal after receipt of a (I) LOCATION UPDATING REJECT message with cause "PLMN not allowed" during registration on CS or (II) ATTACH REJECT message with cause "PLMN not allowed" during registration on PS or (III) LOCATION UPDATING REJECT and/or ATTACH REJECT message with cause "PLMN not allowed" during registration on CS/PS. Test to confirm that (I) the EFLOCI has been correctly updated by the Terminal during registration on CS or (II) the EFPSLOCI has been correctly updated by the Terminal during registration on PS or (III) the the EFLOCI and EFPSLOCI have been correctly updated by the Terminal during registration on CS/PS.

UE updating forbidden PLMNs. Test to confirm that the UE correctly updates the EFFPLMN, i.e. fill up existing gaps in the elementary file before overwriting any existing entries.

UE deleting forbidden PLMNs. Test to confirm that the 2G UE is able to perform a LOCATION UPDATING on a forbidden PLMN in manual PLMN selection mode or to verify that the 3G UE is able to perform (I) a LOCATION UPDATING REQUEST during registration on CS on a forbidden PLMN in manual PLMN selection mode or (II) a ATTACH REQUEST during registration on PS on a forbidden PLMN in manual PLMN selection mode or (III) a LOCA-TION UPDATING REQUEST and/or ATTACH REQUEST during registration on CS/PS on a forbidden PLMN in manual PLMN selection mode. Test to confirm that the UE after a successful registration attempt deletes the PLMN in the EFFPLMN on the USIM.

User controlled PLMN selector handling

UE updating the User controlled PLMN selector list. Test to confirm that the UE correctly updates the EF$_{PLMNwACT}$.

UE recognising the priority order of the User controlled PLMN selector list with the same access technology. Test to confirm that the UPLMN with the higher priority (defined by its position in EF$_{PLMNwACT}$) takes precedence over the UPLMN with the lower priority when the UE performs a network selection.

UE recognising the priority order of the User controlled PLMN selector list using an ACT preference. Test to confirm that the ACT with the higher priority (defined by its position in EF$_{PLMNwACT}$) takes precedence over the UPLMN with the lower priority when the UE performs a network selection.

Operator controlled PLMN selector handling

UE recognising the priority order of the Operator controlled PLMN selector list. Test to confirm that the OPLMN with the higher priority (defined by its position in EF$_{OPLMNwACT}$) takes precedence over the OPLMN with the lower priority when the UE performs a network selection.

UE recognising the priority order of the User controlled PLMN selector over the Operator controlled PLMN selector list. Test to confirm that the User controlled PLMN with a lower priority (defined by its position in EF$_{PLMNwACT}$) takes precedence over the OPLMN with a higher priority when the UE performs a network selection.

Higher priority PLMN search handling

UE recognising the search period of the Higher priority PLMN. Test to confirm that the higher priority PLMN timer is read and the Higher priority PLMN takes precedence over the VPLMN in which the UE is currently registered in.

GSM/UMTS dual mode UEs recognising the search period of the Higher priority PLMN. Test to confirm that the higher priority PLMN timer is read and the higher priority PLMN with the higher priority (defined by its position in

$EF_{HPLMNwACT}$) takes precedence over the VPLMN in which the UE is currently registered in.

Subscription independent tests

Phone book procedures

Recognition of a previously changed phonebook. Test to confirm that the Terminal has recognised that the phonebook has been altered by a GSM Terminal. Test to confirm that the Terminal does the synchronising of the changed phonebook entries. Test to confirm that the Terminal updates the EF_{PBC} and EF_{CC}.

Update of the Phonebook Synchronisation Counter (PSC). Test to confirm that the Terminal has recognised that the values of UID and CC have changed. Test to confirm that the Terminal resets the value of EF_{UID} and EF_{CC}. Test to confirm that the Terminal updates EF_{PSC}.

Phonebook content handling

Handling of BCD number/ SSC content extension. Test to confirm that the terminal is able to read and update BCD numbers/ SSC content with and without extension correctly in EF_{ADN} and EF_{EXT1}.

Phonebook selection. Test to confirm that the terminal offers a possibility to select which phonebook the user would like to select if both, the global and the local phonebook, co-exist. Test to confirm that the data contained in the local phonebook can be read and updated correctly. Test to confirm that the data contained in the global phonebook can be read and updated correctly.

Local Phonebook handling. Test to confirm that the terminal supports the local phonebook without existence of the global phonebook. Test to confirm that the data contained in the local phonebook can be read and updated correctly.

Short message handling report

Correct storage of a SM on the USIM. Test to confirm that the Terminal stored correctly the class 2 SMS on the USIM. Test to confirm that the Terminal sets the status of a received, and not yet read SMS to "3" (SMS to be read). Test to confirm that the Terminal sets the memory full flag in EF_{SMSS} if the terminal notifies the network that the terminal has been unable to accept a short message because its memory capacity has been exceeded.

Correct reading of a SM on the USIM. Test to confirm that the Terminal read correctly the SMS on the USIM. Test to confirm that the Terminal changes the status of a read SMS to "1" (SMS read).

MMS related tests

UE recognising the priority order of MMS Issuer Connectivity Parameters. Test to confirm that the Terminal's MMS User Agent uses the MMS connectivity parameter stored on the USIM to connect to the network for MMS purposes. Test to confirm that the Terminal's MMS User Agent uses the first stored set of supported parameters in EF_{MMSICP} as default. Test to confirm that the Terminal's

MMS User Agent uses the MMS user preference information stored on the USIM for user assistance in preparation of terminal-originated MMS.

UE recognising the priority order of MMS User Connectivity Parameters. Test to confirm that the Terminal's MMS User Agent uses the MMS connectivity parameter stored on the USIM to connect to the network for MMS purposes. Test to confirm that when using the MMS User Connectivity Parameters to connect to the network for MMS purposes the Terminal's MMS User Agent uses the set of supported parameters in EF_{MMSUCP} with the highest priority (as defined by its position in EF_{MMSUCP}). Test to confirm that the Terminal's MMS User Agent uses the MMS user preference information stored on the USIM for user assistance in preparation of terminal-originated MMS.

UE recognising the priority order of MMS Issuer Connectivity Parameters over the MMS User Connectivity Parameters. Test to confirm that the Terminal's MMS User Agent uses the MMS connectivity parameter stored on the USIM to connect to the network for MMS purposes. Test to confirm that a MMS Issuer Connectivity Parameter set with lower priority (as defined by its position in EF_{MMSICP}) takes precedence over a MMS User Connectivity Parameter set with a higher priority.

Usage of MMS notification. Test to confirm that the Terminal stores and updates MMS notifications with the associated status on the USIM correctly.

UICC presence detection

Test to verify that there are no periods of inactivity on the UICC Terminal interface greater than 30 seconds during a call. Test to confirm that the terminal terminates a call within 5s at the latest after having received an invalid response to the STATUS command.

USIM service handling

Access Point Name Control List handling

Access Point Name Control List handling for terminals supporting ACL. Test to confirm that the terminal takes into account the status of the APN Control List service as indicated in EF_{UST} and EF_{EST}. Test to confirm that the terminal checks that the entire APN of any PDP context is listed in EF_{ACL} before requesting this PDP context activation from the network if the ACL service is enabled. Test to confirm that the terminal does not request the corresponding PDP context activation from the network if the ACL service is enabled and the APN is not present in EF_{ACL}.

Network provided APN handling for terminals supporting ACL. Test to confirm that if ACL is enabled and if no APN is indicated in the PDP context the terminal request the PDP context activation only if "network provided APN" is contained within EF_{ACL}. Test to confirm that the user is able to set an APN in EF_{ACL} entry to the value "network provided APN". Test to confirm that the minimum set of APN entries in EF_{ACL} is ensured when the user deletes APN entries.

Access Point Name Control List handling for terminals not supporting ACL. Test to confirm that if ACL is enabled, an ME which does not support ACL, does not send any APN to the network to request a PDP context activation.

Service Dialling Numbers handling

This is to ensure that the terminal takes into account the status of the Service Dialling Numbers service as indicated in EF_{UST}. Also to verify that the user can use the Service Dialling Numbers to make outgoing calls and to verify that the terminal is able to handle SDNs with an extended dialling number string. Test to verify that the terminal is able to handle an empty alpha identifier in EF_{SDN}. Test to confirm that the terminal is able to handle an alpha identifier of maximum length in EF_{SDN}.

USAT Testing

Initialization of USIM Application Toolkit Enabled UICC by USIM Application Toolkit Enabled ME (Profile Download). Test to confirm that the ME sends a TERMINAL PROFILE command in accordance with the above requirements.

Contents of the TERMINAL PROFILE command. Verify that the TERMINAL PROFILE indicates that Profile Download facility is supported.

Servicing of proactive UICC commands. Test to confirm that the ME uses the FE_{TCH} command to obtain the proactive UICC command, after detection of a pending proactive UICC command. The pending proactive UICC command is indicated by the response parameters '91 xx' from the UICC. Test to confirm that the ME transmits the result of execution of the proactive UICC command to the UICC in the TERMINAL RESPONSE command.

Proactive UICC commands

Display Text

(Normal). Test to confirm that the ME displays the text contained in the DISPLAY TEXT proactive UICC command, and returns a successful result in the TERMINAL RESPONSE command send to the UICC.

(Support of "No response from user"). Test to confirm that the ME displays the text contained in the DISPLAY TEXT proactive UICC command, and returns a "No response from user" result value in the TERMINAL RESPONSE command send to the UICC.

(Display of extension text). Test to confirm that the ME displays the extension text contained in the DISPLAY TEXT proactive UICC command, and returns a successful result in the TERMINAL RESPONSE command send to the UICC.

(Sustained text). Test to confirm that the ME displays the text contained in the DISPLAY TEXT proactive UICC command, returns a successful result in the TERMINAL RESPONSE command send to the UICC and sustain the display beyond sending the TERMINAL response.

(Display of icons). Test to confirm that the ME displays the icons which are referred to in the contents of the DISPLAY TEXT proactive UICC command, and returns a successful result in the TERMINAL RESPONSE command send to the UICC.

(UCS2 display in Cyrillic). Test to confirm that the ME displays the text contained in the DISPLAY TEXT proactive UICC command, and returns a successful result in the TERMINAL RESPONSE command send to the UICC.

(Variable Time out). Test to confirm that the ME displays the text contained in the DISPLAY TEXT proactive UICC command, and returns a successful result in the TERMINAL RESPONSE command send to the UICC.

DISPLAY TEXT (Support of Text Attribute – Left Alignment). Test to confirm that the ME displays the text formatted according to the left alignment text attribute configuration contained in the DISPLAY TEXT proactive UICC command, and returns a successful result in the TERMINAL RESPONSE command send to the UICC.

(Support of Text Attribute – Center Alignment). Test to confirm that the ME displays the text formatted according to the center alignment text attribute configuration contained in the DISPLAY TEXT proactive UICC command, and returns a successful result in the TERMINAL RESPONSE command send to the UICC.

(Support of Text Attribute – Right Alignment). Test to confirm that the ME displays the text formatted according to the right alignment text attribute configuration contained in the DISPLAY TEXT proactive UICC command, and returns a successful result in the TERMINAL RESPONSE command send to the UICC.

(Support of Text Attribute – Large Font Size). Test to confirm that the ME displays the text formatted according to the large size font text attribute configuration contained in the DISPLAY TEXT proactive UICC command, and returns a successful result in the TERMINAL RESPONSE command send to the UICC.

(Support of Text Attribute – Small Font Size). Test to confirm that the ME displays the text formatted according to the small size font text attribute configuration contained in the DISPLAY TEXT proactive UICC command, and returns a successful result in the TERMINAL RESPONSE command send to the UICC.

(Support of Text Attribute – Bold On). Test to confirm that the ME displays the text formatted according to the bold text attribute configuration contained in the DISPLAY TEXT proactive UICC command, and returns a successful result in the TERMINAL RESPONSE command send to the UICC.

(Support of Text Attribute – Italic On). Test to confirm that the ME displays the text formatted according to the italic text attribute configuration contained in the DISPLAY TEXT proactive UICC command, and returns a successful result in the TERMINAL RESPONSE command send to the UICC.

(Support of Text Attribute – Underline On). Test to confirm that the ME displays the text formatted according to the underline text attribute configuration contained in the DISPLAY TEXT proactive UICC command, and returns a successful result in the TERMINAL RESPONSE command send to the UICC.

(Support of Text Attribute – Strikethrough On). Test to confirm that the ME displays the text formatted according to the strikethrough text attribute configuration contained in the DISPLAY TEXT proactive UICC command, and returns a successful result in the TERMINAL RESPONSE command send to the UICC.

(Attribute with Strikethrough On). Test to confirm that the ME displays the text formatted according to the foreground and background colour text attribute configuration contained in the DISPLAY TEXT proactive UICC command, and returns a successful result in the TERMINAL RESPONSE command send to the UICC.

(UCS2 display in Chinese). Test to confirm that the ME displays the text contained in the DISPLAY TEXT proactive UICC command, and returns a successful result in the TERMINAL RESPONSE command send to the UICC.

(UCS2 display in Katakana). Test to confirm that the ME displays the text contained in the DISPLAY TEXT proactive UICC command, and returns a successful result in the TERMINAL RESPONSE command send to the UICC.

GET INKEY

(normal). Test to confirm that the ME displays the text contained in the GET INKEY proactive UICC command, and returns the single character entered in the TERMINAL RESPONSE command sent to the UICC.

(No response from User). Test to confirm that the ME displays the text contained in the GET INKEY proactive UICC command, and returns a "No response from user" result value in the TERMINAL RESPONSE command send to the UICC. Test to confirm that the ME displays the text contained in the GET INKEY proactive UICC command, and returns the text string entered in the TERMINAL RESPONSE command sent to the UICC.

(UCS2 entry in Cyrillic). Test to confirm that the ME displays the text contained in the GET INKEY proactive UICC command, and returns the text string entered in the TERMINAL RESPONSE command sent to the UICC.

("Yes/No" Response). Test to confirm that the ME displays the text contained in the GET INKEY proactive UICC command, and returns the text string entered in the TERMINAL RESPONSE command sent to the UICC.

(display of Icon). Test to confirm that the ME displays the Icon contained in the GET INKEY proactive UICC command, and returns the text string entered in the TERMINAL RESPONSE command sent to the UICC.

(Help Information). Test to confirm that the ME displays the text contained in the GET INKEY proactive UICC command, and returns the text string entered in the TERMINAL RESPONSE command sent to the UICC.

(Variable Time out). Test to confirm that the ME displays the text contained in the GET INKEY proactive UICC command, and returns the text string entered in the TERMINAL RESPONSE command sent to the UICC.

(Support of Text Attribute – Left Alignment). Test to confirm that the ME displays the text formatted according to the left alignment text attribute configuration contained in the GET INKEY proactive UICC command, and returns the text string entered in the TERMINAL RESPONSE command sent to the UICC.

(Support of Text Attribute – Centre Alignment). Test to confirm that the ME displays the text formatted according to the centre alignment text attribute configuration contained in the GET INKEY proactive UICC command, and returns the text string entered in the TERMINAL RESPONSE command sent to the UICC.

(Support of Text Attribute – Right Alignment). Test to confirm that the ME displays the text formatted according to the right alignment text attribute configuration contained in the GET INKEY proactive UICC command, and returns the text string entered in the TERMINAL RESPONSE command sent to the UICC.

(Support of Text Attribute – Large Font Size). Test to confirm that the ME displays the text formatted according to the large font size text attribute configuration contained in the GET INKEY proactive UICC command, and returns the text string entered in the TERMINAL RESPONSE command sent to the UICC.

(Support of Text Attribute – Small Font Size). Test to confirm that the ME displays the text formatted according to the small font size text attribute configuration contained in the GET INKEY proactive UICC command, and returns the text string entered in the TERMINAL RESPONSE command sent to the UICC.

(Support of Text Attribute – Bold On). Test to confirm that the ME displays the text formatted according to the bold text attribute configuration contained in the GET INKEY proactive UICC command, and returns the text string entered in the TERMINAL RESPONSE command sent to the UICC.

(Support of Text Attribute – Italic On). Test to confirm that the ME displays the text formatted according to the italic text attribute configuration contained in the GET INKEY proactive UICC command, and returns the text string entered in the TERMINAL RESPONSE command sent to the UICC.

(Support of Text Attribute – Underline On). Test to confirm that the ME displays the text formatted according to the underline text attribute configuration contained in the GET INKEY proactive UICC command, and returns the text string entered in the TERMINAL RESPONSE command sent to the UICC.

(Support of Text Attribute – Strikethrough On). Test to confirm that the ME displays the text formatted according to the strikethrough text attribute configuration contained in the GET INKEY proactive UICC command, and returns the text string entered in the TERMINAL RESPONSE command sent to the UICC.

(Support of Text Attribute – Foreground and Background Colour). Test to confirm that the ME displays the text formatted according to the foreground and background colour text attribute configuration contained in the GET INKEY proactive UICC command, and returns the text string entered in the TERMINAL RESPONSE command sent to the UICC.

(UCS2 display in Chinese). Test to confirm that the ME displays the text contained in the GET INKEY proactive UICC command, and returns the text string entered in the TERMINAL RESPONSE command sent to the UICC.

(UCS2 entry in Chinese). Test to confirm that the ME displays the text contained in the GET INKEY proactive UICC command, and returns the text string entered in the TERMINAL RESPONSE command sent to the UICC.

(UCS2 display in Katakana). Test to confirm that the ME displays the text contained in the GET INKEY proactive UICC command, and returns the text string entered in the TERMINAL RESPONSE command sent to the UICC.

(UCS2 entry in Katakana). Test to confirm that the ME displays the text contained in the GET INKEY proactive UICC command, and returns the text string entered in the TERMINAL RESPONSE command sent to the UICC.

GET INPUT

(normal). Test to confirm that the ME displays the text contained in the GET INPUT proactive UICC command, and returns the text string entered in the TERMINAL RESPONSE command sent to the UICC.

(No response from User). Test to confirm that the ME displays the text contained in the GET INPUT proactive UICC command, and returns a "No response from user" result value in the TERMINAL RESPONSE command send to the UICC.

(UCS2 display in Cyrillic). Test to confirm that the ME displays the text contained in the GET INPUT proactive UICC command, and returns the text string entered in the TERMINAL RESPONSE command sent to the UICC.

(UCS2 entry in Cyrillic). Test to confirm that the ME displays the text contained in the GET INPUT proactive UICC command, and returns the text string entered in the TERMINAL RESPONSE command sent to the UICC.

(default text). Test to confirm that the ME displays the text contained in the GET INPUT proactive UICC command, and returns the text string entered in the TERMINAL RESPONSE command sent to the UICC.

(display of Icon). Test to confirm that the ME displays the Icon contained in the GET INPUT proactive UICC command, and returns the text string entered in the TERMINAL RESPONSE command sent to the UICC.

(Help Information). Test to confirm that the ME displays the text contained in the GET INPUT proactive UICC command, and returns a 'help information required by the user' result value in the TERMINAL RESPONSE command sent to the UICC if the user has indicated the need to get help information.

(Support of Text Attribute – Left Alignment). Test to confirm that the ME displays the text formatted according to the left alignment text attribute configuration contained in the GET INPUT proactive UICC command, and returns a successful response in the TERMINAL RESPONSE command sent to the UICC.

(Support of Text Attribute – Centre Alignment). Test to confirm that the ME displays the text formatted according to the centre alignment text attribute configuration contained in the GET INPUT proactive UICC command, and returns a successful response in the TERMINAL RESPONSE command sent to the UICC.

(Support of Text Attribute – Right Alignment). Test to confirm that the ME displays the text formatted according to the right alignment text attribute configuration contained in the GET INPUT proactive UICC command, and returns a successful response in the TERMINAL RESPONSE command sent to the UICC.

(Support of Text Attribute – Large Font Size). Test to confirm that the ME displays the text formatted according to the large font size text attribute configuration contained in the GET INPUT proactive UICC command, and returns a successful response in the TERMINAL RESPONSE command sent to the UICC.

(Support of Text Attribute – Small Font Size). Test to confirm that the ME displays the text formatted according to the small font size text attribute configuration contained in the GET INPUT proactive UICC command, and returns a successful response in the TERMINAL RESPONSE command sent to the UICC.

(Support of Text Attribute – Bold On). Test to confirm that the ME displays the text formatted according to the bold text attribute configuration contained in the GET INPUT proactive UICC command, and returns a successful response in the TERMINAL RESPONSE command sent to the UICC.

(Support of Text Attribute – Italic On). Test to confirm that the ME displays the text formatted according to the italic text attribute configuration contained in the GET INPUT proactive UICC command, and returns a successful response in the TERMINAL RESPONSE command sent to the UICC.

(Support of Text Attribute – Underline On). Test to confirm that the ME displays the text formatted according to the underline text attribute configuration contained in the GET INPUT proactive UICC command, and returns a successful response in the TERMINAL RESPONSE command sent to the UICC.

(Support of Text Attribute – Strikethrough On). Test to confirm that the ME displays the text formatted according to the strikethrough text attribute configuration contained in the GET INPUT proactive UICC command, and returns a successful response in the TERMINAL RESPONSE command sent to the UICC.

(Support of Text Attribute – Foreground and Background Colour). Test to confirm that the ME displays the text formatted according to the fore- and background colour text attribute configuration contained in the GET INPUT proactive UICC command, and returns a successful response in the TERMINAL RESPONSE command sent to the UICC.

(UCS2 display in Chinese). Test to confirm that the ME displays the text contained in the GET INPUT proactive UICC command, and returns the text string entered in the TERMINAL RESPONSE command sent to the UICC.

(UCS2 entry in Chinese). Test to confirm that the ME displays the text contained in the GET INPUT proactive UICC command, and returns the text string entered in the TERMINAL RESPONSE command sent to the UICC.

(UCS2 display in Katakana). Test to confirm that the ME displays the text contained in the GET INPUT proactive UICC command, and returns the text string entered in the TERMINAL RESPONSE command sent to the UICC.

(UCS2 entry in Katakana). Test to confirm that the ME displays the text contained in the GET INPUT proactive UICC command, and returns the text string entered in the TERMINAL RESPONSE command sent to the UICC.

MORE TIME

MORE TIME. Test to confirm that the ME shall send a TERMINAL RESPONSE (OK) to the UICC after the ME receives the MORE TIME proactive UICC command.

PLAY TONE

(Normal). Test to confirm that the ME plays an audio tone of a type and duration contained in the PLAY TONE proactive UICC command, and returns a successful response in the TERMINAL RESPONSE command sent to the UICC. Test to confirm that the ME plays the requested audio tone through the earpiece whilst not in call and shall superimpose the tone on top of the downlink audio whilst in call. Test to confirm that the ME displays the text contained in the PLAY TONE proactive UICC command.

(UCS2 display in Cyrillic). Test to confirm that the ME displays the text contained in the PLAY TONE proactive UICC command, and returns a successful response in the TERMINAL RESPONSE command sent to the UICC. Test to confirm that the ME plays the requested audio tone through the earpiece.

(display of Icon). Test to confirm that the ME plays an audio tone of a type and duration contained in the PLAY TONE proactive UICC command, and returns a successful response in the TERMINAL RESPONSE command sent to the UICC. Test to confirm that the ME plays the requested audio tone through the earpiece. Test to confirm that the ME displays the icon contained in the PLAY TONE proactive UICC command.

(Support of Text Attribute – Left Alignment). Test to confirm that the ME displays the text formatted according to the left alignment text attribute configuration contained in the PLAY TONE proactive UICC command, and returns a successful response in the TERMINAL RESPONSE command sent to the UICC.

(Support of Text Attribute – Centre Alignment). Test to confirm that the ME displays the text formatted according to the centre alignment text attribute configuration contained in the PLAY TONE proactive UICC command, and returns a successful response in the TERMINAL RESPONSE command sent to the UICC.

(Support of Text Attribute – Right Alignment). Test to confirm that the ME displays the text formatted according to the right alignment text attribute configuration contained in the PLAY TONE proactive UICC command, and returns a successful response in the TERMINAL RESPONSE command sent to the UICC.

(Support of Text Attribute – Large Font Size). Test to confirm that the ME displays the text formatted according to the large font size text attribute configuration contained in the PLAY TONE proactive UICC command, and returns a successful response in the TERMINAL RESPONSE command sent to the UICC.

(Support of Text Attribute – Small Font Size). Test to confirm that the ME displays the text formatted according to the small font size text attribute configuration contained in the PLAY TONE proactive UICC command, and returns a successful response in the TERMINAL RESPONSE command sent to the UICC.

(Support of Text Attribute – Bold On). Test to confirm that the ME displays the text formatted according to the bold text attribute configuration contained in the PLAY TONE proactive UICC command, and returns a successful response in the TERMINAL RESPONSE command sent to the UICC.

(Support of Text Attribute – Italic On). Test to confirm that the ME displays the text formatted according to the italic text attribute configuration contained in the PLAY TONE proactive UICC command, and returns a successful response in the TERMINAL RESPONSE command sent to the UICC.

(Support of Text Attribute – Underline On). Test to confirm that the ME displays the text formatted according to the underline text attribute configuration contained in the PLAY TONE proactive UICC command, and returns a successful response in the TERMINAL RESPONSE command sent to the UICC.

(Support of Text Attribute – Strikethrough On). Test to confirm that the ME displays the text formatted according to the strikethrough text attribute configuration contained in the PLAY TONE proactive UICC command, and returns a successful response in the TERMINAL RESPONSE command sent to the UICC.

(Support of Text Attribute – Foreground and Background Colour). Test to confirm that the ME displays the text formatted according to the foreground and background colour text attribute configuration contained in the PLAY TONE proactive UICC command, and returns a successful response in the TERMINAL RESPONSE command sent to the UICC.

(UCS2 display in Chinese). Test to confirm that the ME displays the text contained in the PLAY TONE proactive UICC command, and returns a successful response in the TERMINAL RESPONSE command sent to the UICC. Test to confirm that the ME plays the requested audio tone through the earpiece.

(UCS2 display in Katakana). Test to confirm that the ME displays the text contained in the PLAY TONE proactive UICC command, and returns a successful response in the TERMINAL RESPONSE command sent to the UICC. Test to confirm that the ME plays the requested audio tone through the earpiece.

POLL INTERVAL

POLL INTERVAL. Test to confirm that the ME shall send a TERMINAL RE-SPONSE (OK) to the UICC after the ME receives the POLL INTERVAL proactive UICC command. Test to confirm that the ME gives a valid response to the polling interval requested by the UICC. Test to confirm that the ME sends STATUS commands to the UICC at an interval no longer than the interval negotiated by the UICC.

REFRESH

(normal). Test to confirm that the ME performs the Proactive Command – REFRESH in accordance with the Command Qualifier. This shall require the ME to perform: UICC and USIM initialization, re-read of the contents and structure of the EFs on the UICC that have been notified as changed and are either part of initialization or used during the test, restart of the card session, and successful return of the result of the execution of the command in the TERMINAL RESPONSE command send to the UICC.

(IMSI changing procedure). Test to confirm that the ME performs the Proactive Command – REFRESH in accordance with the Command Qualifier and the IMSI changing procedure. This may require the ME to perform: USIM initialization, re-read of the contents and structure of the IMSI on the USIM, restart of the card session, and successful return of the result of the execution of the command in the TERMINAL RESPONSE command sent to the UICC.

SET UP MENU and ENVELOPE MENU SELECTION

SET UP MENU (Normal). Test to confirm that the ME correctly integrates the menu items contained in the SET UP MENU proactive UICC command, and returns a successful response in the TERMINAL RESPONSE command sent to the UICC. Test to confirm that the ME replaces the current list of menu items with the list of menu items contained in the SET UP MENU command. Test to confirm that the ME removes the current list of menu items following receipt of a SET UP MENU command with no items. Test to confirm that the ME correctly passes the identifier of the selected menu item to the UICC using the ENVELOPE (MENU SELECTION) command. Test to confirm that when the help is available for the command and the user gas indicated the need to get help information on one of the items, the ME informs properly the UICC about an HELP REQUEST, using the MENU SELECTION mechanism.

SET UP MENU (help request support) and ENVELOPE MENU SELECTION. Test to confirm that the ME correctly integrates the menu items contained in the SET UP MENU proactive UICC command, and returns a successful response in the TERMINAL RESPONSE command sent to the UICC. Test to confirm that when the help is available for the command and the user has indicated the need to get help information on one of the items, the ME informs properly the UICC about an HELP REQUEST, using the MENU SELECTION mechanism. Test to confirm that the ME correctly passes the identifier of the selected menu item to the UICC using the ENVELOPE (MENU SELECTION) command.

SET UP MENU (next action support) and ENVELOPE MENU SELECTION. Test to confirm that the ME correctly integrates the menu items contained in the SET UP MENU proactive UICC command, and returns a successful response in the TERMINAL RESPONSE command sent to the UICC. Test to confirm that the next action indicator is supported. Test to confirm that the ME correctly passes the identifier of the selected menu item to the UICC using the ENVELOPE (MENU SELECTION) command.

SET UP MENU (display of icons) and ENVELOPE MENU SELECTION. Test to confirm that the ME correctly integrates the menu items contained in the SET UP MENU proactive UICC command, and returns a successful response in the TERMINAL RESPONSE command sent to the UICC. Test to confirm that icons are displayed with the command Set Up Menu in the Alpha Identifier and Items Data Objects. Test to confirm that the ME correctly passes the identifier of the selected menu item to the UICC using the ENVELOPE (MENU SELECTION) command.

SET UP MENU (soft keys support) and ENVELOPE MENU SELECTION. Test to confirm that the ME correctly integrates the menu items contained in the SET UP MENU proactive UICC command, and returns a successful response in the TERMINAL RESPONSE command sent to the UICC. Test to confirm that if soft key preferred is indicated in the command details and soft key for SET UP MENU is supported by the ME and the number of icon items does not exceed the number of soft keys available, then the ME displays those icons as soft key. Test to confirm that the ME correctly passes the identifier of the selected menu item to the UICC using the ENVELOPE (MENU SELECTION) command.

SET UP MENU (support of Text Attribute – Left Alignment) and ENVELOPE MENU SELECTION. Test to confirm that the ME correctly integrates the menu items contained in the SET UP MENU proactive UICC command, and returns a successful response in the TERMINAL RESPONSE command sent to the UICC. Test to confirm that text is displayed according to the left alignment text attribute configuration within the command Set Up Menu and the ME correctly passes the identifier of the selected menu item to the UICC using the ENVELOPE (MENU SELECTION) command.

SET UP MENU (support of Text Attribute – Centre Alignment) and ENVELOPE MENU SELECTION. Test to confirm that the ME correctly integrates the menu items contained in the SET UP MENU proactive UICC command, and returns a successful response in the TERMINAL RESPONSE command sent to the UICC. Test to confirm that text is displayed according to the centre alignment text attribute configuration within the command Set Up Menu and the ME correctly passes the identifier of the selected menu item to the UICC using the ENVELOPE (MENU SELECTION) command.

SET UP MENU (support of Text Attribute – Right Alignment) and ENVELOPE MENU SELECTION. Test to confirm that the ME correctly integrates the menu items contained in the SET UP MENU proactive UICC command, and returns a successful response in the TERMINAL RESPONSE command sent to the UICC. Test to confirm that text is displayed according to the right alignment text attribute configuration within the command Set Up Menu and the ME correctly passes the identifier of the selected menu item to the UICC using the ENVELOPE (MENU SELECTION) command.

SET UP MENU (support of Text Attribute – Large Font Size) and ENVELOPE MENU SELECTION. Test to confirm that the ME correctly integrates the menu items contained in the SET UP MENU proactive UICC command, and returns a successful response in the TERMINAL RESPONSE command sent to the UICC. Test to confirm that text is displayed according to the large font size text attribute configuration within the command Set Up Menu and the ME correctly passes the identifier of the selected menu item to the UICC using the ENVELOPE (MENU SELECTION) command.

SET UP MENU (support of Text Attribute – Small Font Size) and ENVELOPE MENU SELECTION. Test to confirm that the ME correctly integrates the menu items contained in the SET UP MENU proactive UICC command, and returns a successful response in the TERMINAL RESPONSE command sent to the UICC. Test to verify that text is displayed according to the with small font size text attribute configuration within the command Set Up Menu and the ME correctly passes the identifier of the selected menu item to the UICC using the ENVELOPE (MENU SELECTION) command.

SET UP MENU (support of Text Attribute – Bold On) and ENVELOPE MENU SELECTION. Test to confirm that the ME correctly integrates the menu items contained in the SET UP MENU proactive UICC command, and returns a successful response in the TERMINAL RESPONSE command sent to the UICC. Test to confirm that text is displayed according to the text attribute configuration within the command Set Up Menu and the ME correctly passes the identifier of the selected menu item to the UICC using the ENVELOPE (MENU SELECTION) command.

SET UP MENU (support of Text Attribute – Italic On) and ENVELOPE MENU SELECTION. Test to confirm that the ME correctly integrates the menu items contained in the SET UP MENU proactive UICC command, and returns a successful response in the TERMINAL RESPONSE command sent to the UICC. Test to confirm that text is displayed according to the text attribute configuration within the command Set Up Menu and the ME correctly passes the identifier of the selected menu item to the UICC using the ENVELOPE (MENU SELECTION) command.

SET UP MENU (support of Text Attribute – Underline On) and ENVELOPE MENU SELECTION. Test to confirm that the ME correctly integrates the menu items contained in the SET UP MENU proactive UICC command, and returns a successful response in the TERMINAL RESPONSE command sent to the UICC. Test to confirm that text is displayed according to the text attribute configuration within the command Set Up Menu and the ME correctly passes the identifier of the selected menu item to the UICC using the ENVELOPE (MENU SELECTION) command.

SET UP MENU (support of Text Attribute – Strikethrough On) and ENVELOPE MENU SELECTION. Test to confirm that the ME correctly integrates the menu items contained in the SET UP MENU proactive UICC command,

and returns a successful response in the TERMINAL RESPONSE command sent to the UICC. Test to confirm that text is displayed according to the text attribute configuration within the command Set Up Menu and the ME correctly passes the identifier of the selected menu item to the UICC using the ENVELOPE (MENU SELECTION) command.

SET UP MENU (support of Text Attribute – Foreground and Background Colour) and ENVELOPE MENU SELECTION. Test to confirm that the ME correctly integrates the menu items contained in the SET UP MENU proactive UICC command, and returns a successful response in the TERMINAL RESPONSE command sent to the UICC. Test to confirm that text is displayed according to the text attribute configuration within the command Set Up Menu and the ME correctly passes the identifier of the selected menu item to the UICC using the ENVELOPE (MENU SELECTION) command.

SET UP MENU (UCS2 display in Cyrillic) and ENVELOPE MENU SELECTION. Test to confirm that the ME correctly integrates the menu items in UCS2 coding contained in the SET UP MENU proactive UICC command, and returns a successful response in the TERMINAL RESPONSE command sent to the UICC. Test to confirm that the ME replaces the current list of menu items with the list of menu items contained in the SET UP MENU command. Test to confirm that the ME removes the current list of menu items following receipt of a SET UP MENU command with no items. Test to confirm that the ME correctly passes the identifier of the selected menu item to the UICC using the ENVELOPE (MENU SELECTION) command. Test to confirm that when the help is available for the command and the user gas indicated the need to get help information on one of the items, the ME informs properly the UICC about an HELP REQUEST, using the MENU SELECTION mechanism.

SET UP MENU (UCS2 display in Chinese) and ENVELOPE MENU SELECTION. Test to confirm that the ME correctly integrates the menu items in UCS2 coding contained in the SET UP MENU proactive UICC command, and returns a successful response in the TERMINAL RESPONSE command sent to the UICC. Test to confirm that the ME replaces the current list of menu items with the list of menu items contained in the SET UP MENU command. Test to confirm that the ME removes the current list of menu items following receipt of a SET UP MENU command with no items. Test to confirm that the ME correctly passes the identifier of the selected menu item to the UICC using the ENVELOPE (MENU SELECTION) command. Test to confirm that when the help is available for the command and the user gas indicated the need to get help information on one of the items, the ME informs properly the UICC about an HELP REQUEST, using the MENU SELECTION mechanism.

SET UP MENU (UCS2 display in Katakana) and ENVELOPE MENU SELECTION. Test to confirm that the ME correctly integrates the menu items in UCS2 coding contained in the SET UP MENU proactive UICC command, and returns a successful response in the TERMINAL RESPONSE command sent to the UICC. Test to confirm that the ME replaces the current list of menu items with the list of menu items contained in the SET UP MENU command. Test to confirm that the ME removes the current list of menu items following receipt of a SET UP MENU command with no items. Test to confirm that the ME correctly passes the identifier of the selected menu item to the UICC using the ENVELOPE (MENU SELECTION) command. Test to confirm that when the help is available for the command and the user gas indicated the need to get help information on one of the items, the ME informs properly the UICC about an HELP REQUEST, using the MENU SELECTION mechanism.

SELECT ITEM

(mandatory features for ME supporting SELECT ITEM). Test to confirm that the ME correctly presents the set of items contained in the SELECT ITEM proactive UICC command, and returns a TERMINAL RESPONSE command to the UICC with the identifier of the item chosen. Test to confirm that the ME allows a SELECT ITEM proactive UICC command within the maximum 255 byte BER-TLV boundary. Test to confirm that the ME returns a TERMINAL RESPONSE with "Proactive UICC application session terminated by the user", if the user has indicated the need to end the proactive UICC session. Test to confirm that the ME returns a TERMINAL RESPONSE with "Backwards move in the proactive UICC application session requested by the user", if the user has indicated the need to go backwards in the proactive UICC application session.

(next action support). Test to confirm that the UE supports next action indicator mode.

(default item support). Test to confirm that the UE supports "default item" mode.

(help request support). Test to confirm that the UE supports "help request" for the command Select Item.

(icons support). Test to confirm that the UE displays icons with the command Select Item.

(presentation style). Test to confirm that the UE supports the "presentation style" with the command Select Item.

(soft keys support). Test to confirm that the UE supports the "soft keys" with the command Select Item.

(Support of "No response from user"). Test to confirm that after a period of user inactivity the ME returns a "No response from user" result value in the TERMINAL RESPONSE command sent to the UICC.

(Support of Text Attribute – Left Alignment). Test to confirm that the ME displays text formatted according to the left alignment text attribute configuration within the command Select Item.

(Support of Text Attribute – Centre Alignment). Test to confirm that the ME displays text formatted according to the centre alignment text attribute configuration within the command Select Item.

(Support of Text Attribute – Right Alignment). Test to confirm that the ME displays text formatted according to the right alignment text attribute configuration within the command Select Item.

(Support of Text Attribute – Large Font Size). Test to confirm that the ME displays text formatted according to the large font size text attribute configuration within the command Select Item.

(Support of Text Attribute – Small Font Size). Test to confirm that the ME displays text formatted according to the small font size text attribute configuration within the command Select Item.

(Support of Text Attribute – Bold On). Test to confirm that the ME displays text formatted according to the bold text attribute configuration within the command Select Item.

(Support of Text Attribute – Italic On). Test to confirm that the ME displays text formatted according to the italic text attribute configuration within the command Select Item.

(Support of Text Attribute – Underline On). Test to confirm that the ME displays text formatted according to the underline text attribute configuration within the command Select Item.

(Support of Text Attribute – Strikethrough On). Test to confirm that the ME displays text formatted according to the strikethrough text attribute configuration within the command Select Item.

(Support of Text Attribute – Foreground and Background Colour). Test to confirm that the ME displays text formatted according to the foreground and background colour text attribute configuration within the command Select Item.

(UCS2 display in Cyrillic). Test to confirm that the ME correctly presents the set of items in UCS2 coding contained in the SELECT ITEM proactive UICC command, and returns a TERMINAL RESPONSE command to the UICC with the identifier of the item chosen. Test to confirm that the ME allows a SELECT ITEM proactive UICC command within the maximum 255 byte BER-TLV boundary. Test to confirm that the ME returns a TERMINAL RESPONSE with "Proactive UICC application session terminated by the user", if the user has indicated the need to end the proactive UICC session. Test to confirm that the ME returns a TERMINAL RESPONSE with "Backwards move in the proactive UICC application session requested by the user", if the user has indicated the need to go backwards in the proactive UICC application session.

(UCS2 display in Chinese). Test to confirm that the ME correctly presents the set of items in UCS2 coding contained in the SELECT ITEM proactive UICC command, and returns a TERMINAL RESPONSE command to the UICC with the identifier of the item chosen. Test to confirm that the ME allows a SELECT ITEM proactive UICC command within the maximum 255 byte BER-TLV boundary. Test to confirm that the ME returns a TERMINAL RESPONSE with "Proactive UICC application session terminated by the user", if the user has indicated the need to end the proactive UICC session. Test to confirm that the ME returns a TERMINAL RESPONSE with "Backwards move in the proactive UICC application session requested by the user", if the user has indicated the need to go backwards in the proactive UICC application session.

(UCS2 display in Katakana). Test to confirm that the ME correctly presents the set of items in UCS2 coding contained in the SELECT ITEM proactive UICC command, and returns a TERMINAL RESPONSE command to the UICC with the identifier of the item chosen. Test to confirm that the ME allows a SELECT ITEM proactive UICC command within the maximum 255 byte BER-TLV boundary. Test to confirm that the ME returns a TERMINAL RESPONSE with "Proactive UICC application session terminated by the user", if the user has indicated the need to end the proactive UICC session. Test to confirm that the ME returns a TERMINAL RESPONSE with "Backwards move in the proactive UICC application session requested by the user", if the user has indicated the need to go backwards in the proactive UICC application session.

SEND SHORT MESSAGE

(normal). Test to confirm that the ME correctly formats and sends a short message to the network (USS) as indicated in the SEND SHORT MESSAGE proactive UICC command, and returns a TERMINAL RESPONSE command to the UICC indicating the status of the transmission of the Short Message.

(UCS2 display in Cyrillic). Test to confirm that the ME correctly formats and sends a short message to the network (USS) as indicated in the SEND SHORT MESSAGE proactive UICC command, and returns a TERMINAL RESPONSE command to the UICC indicating the status of the transmission of the Short Message.

(icon support). Test to confirm that the ME correctly formats and sends a short message to the network (USS) as indicated in the SEND SHORT MESSAGE proactive UICC command, and returns a TERMINAL RESPONSE command to the UICC indicating the status of the transmission of the Short Message.

(Support of Text Attribute – Left Alignment). Test to confirm that the ME correctly formats and sends a short message to the network (USS) and display the alpha identifier according to the left alignment text attribute configuration as indicated in the SEND SHORT MESSAGE proactive UICC command, and returns a TERMINAL RESPONSE command to the UICC indicating the status of the transmission of the Short Message.

(Support of Text Attribute – Centre Alignment). Test to confirm that the ME correctly formats and sends a short message to the network (USS) and display the alpha identifier according to the centre alignment text attribute configuration as indicated in the SEND SHORT MESSAGE proactive UICC command, and returns a TERMINAL RESPONSE command to the UICC indicating the status of the transmission of the Short Message.

(Support of Text Attribute – Right Alignment). Test to confirm that the ME correctly formats and sends a short message to the network (USS) and display the alpha identifier according to the right alignment text attribute configuration as indicated in the SEND SHORT MESSAGE proactive UICC command, and returns a TERMINAL RESPONSE command to the UICC indicating the status of the transmission of the Short Message.

(Support of Text Attribute – Large Font Size). Test to confirm that the ME correctly formats and sends a short message to the network (USS) and display the alpha identifier according to the large font size text attribute configuration as indicated in the SEND SHORT MESSAGE proactive UICC command, and returns a TERMINAL RESPONSE command to the UICC indicating the status of the transmission of the Short Message.

(Support of Text Attribute – Small Font Size). Test to confirm that the ME correctly formats and sends a short message to the network (USS) and display the alpha identifier according to the small font size text attribute configuration as indicated in the SEND SHORT MESSAGE proactive UICC command, and returns a TERMINAL RESPONSE command to the UICC indicating the status of the transmission of the Short Message.

(Support of Text Attribute – Bold On). Test to confirm that the ME correctly formats and sends a short message to the network (USS) and display the alpha identifier according to the bold text attribute configuration as indicated in the SEND SHORT MESSAGE proactive UICC command, and returns a TERMINAL RESPONSE command to the UICC indicating the status of the transmission of the Short Message.

(Support of Text Attribute – Italic On). Test to confirm that the ME correctly formats and sends a short message to the network (USS) and display the alpha identifier according to the italic text attribute configuration as indicated in the SEND SHORT MESSAGE proactive UICC command, and returns a TERMINAL RESPONSE command to the UICC indicating the status of the transmission of the Short Message.

(Support of Text Attribute – Underline On). Test to confirm that the ME correctly formats and sends a short message to the network (USS) and display the alpha identifier according to the underline text attribute configuration as indicated in the SEND SHORT MESSAGE proactive UICC command, and returns a TERMINAL RESPONSE command to the UICC indicating the status of the transmission of the Short Message.

(Support of Text Attribute – Strikethrough On). Test to confirm that the ME correctly formats and sends a short message to the network (USS) and display the alpha identifier according to the strikethrough text attribute configuration as indicated in the SEND SHORT MESSAGE proactive UICC command, and returns a TERMINAL RESPONSE command to the UICC indicating the status of the transmission of the Short Message.

(Support of Text Attribute – Foreground and Background Colour). Test to confirm that the ME correctly formats and sends a short message to the network (USS) and display the alpha identifier according to the foreground and background colour text attribute configuration as indicated in the SEND SHORT MESSAGE proactive UICC command, and returns a TERMINAL RESPONSE command to the UICC indicating the status of the transmission of the Short Message.

(UCS2 display in Chinese). Test to confirm that the ME correctly formats and sends a short message to the network (USS) as indicated in the SEND SHORT MESSAGE proactive UICC command, and returns a TERMINAL RESPONSE command to the UICC indicating the status of the transmission of the Short Message.

(UCS2 display in Katakana). Test to confirm that the ME correctly formats and sends a short message to the network (USS) as indicated in the SEND SHORT MESSAGE proactive UICC command, and returns a TERMINAL RESPONSE command to the UICC indicating the status of the transmission of the Short Message.

SEND SS

(normal). Test to confirm that the ME correctly translates and sends the supplementary service request indicated in the SEND SS proactive UICC command to the USS. Test to confirm that the ME returns a TERMINAL RESPONSE command to the UICC indicating the status of the transmission of the SS and any contents of the SS result as additional data.

(Icon support). Test to confirm that the ME displays the text contained in the SEND SS proactive UICC command, and returns a successful result in the TERMINAL RESPONSE command send to the UICC. In addition to verify that if an icon is provided by the UICC, the icon indicated in the command may be used by the ME to inform the user, in addition to, or instead of the alpha identifier, as indicated with the icon qualifier.

(UCS2 display in Cyrillic). Test to confirm that the ME displays the UCS2 text contained in the SEND SS proactive UICC command, and returns a successful result in the TERMINAL RESPONSE command send to the UICC.

(support of Text Attribute – Left Alignment). Test to confirm that the ME displays the alpha identifier according to the left alignment text attribute configuration in the SEND SS proactive UICC command, and returns a successful result in the TERMINAL RESPONSE command send to the UICC.

(support of Text Attribute – Centre Alignment). Test to confirm that the ME displays the alpha identifier according to the centre alignment text attribute configuration in the SEND SS proactive UICC command, and returns a successful result in the TERMINAL RESPONSE command send to the UICC.

(support of Text Attribute – Right Alignment). Test to confirm that the ME displays the alpha identifier according to the right alignment text attribute configuration in the SEND SS proactive UICC command, and returns a successful result in the TERMINAL RESPONSE command send to the UICC.

(support of Text Attribute – Large Font Size). Test to confirm that the ME displays the alpha identifier according to the large font size text attribute configuration in the SEND SS proactive UICC command, and returns a successful result in the TERMINAL RESPONSE command send to the UICC.

SEND SS (support of Text Attribute – Small Font Size). Test to confirm that the ME displays the alpha identifier according to the small font size text attribute configuration in the SEND SS proactive UICC command, and returns a successful result in the TERMINAL RESPONSE command send to the UICC.

(support of Text Attribute – Bold On). Test to confirm that the ME displays the alpha identifier according to the bold text attribute configuration in the SEND SS proactive UICC command, and returns a successful result in the TERMINAL RESPONSE command send to the UICC.

(support of Text Attribute – Italic On). Test to confirm that the ME displays the alpha identifier according to the italic text attribute configuration in the SEND SS proactive UICC command, and returns a successful result in the TERMINAL RESPONSE command send to the UICC.

(support of Text Attribute – Underline On). Test to confirm that the ME displays the alpha identifier according to the underline text attribute configuration in the SEND SS proactive UICC command, and returns a successful result in the TERMINAL RESPONSE command send to the UICC.

(support of Text Attribute – Strikethrough On). Test to confirm that the ME displays the alpha identifier according to the strikethrough text attribute configuration in the SEND SS proactive UICC command, and returns a successful result in the TERMINAL RESPONSE command send to the UICC.

(support of Text Attribute – Foreground and Background Colour). Test to confirm that the ME displays the alpha identifier according to the foreground and background colour text attribute configuration in the SEND SS proactive UICC command, and returns a successful result in the TERMINAL RESPONSE command send to the UICC.

(UCS2 display in Chinese). Test to confirm that the ME displays the UCS2 text contained in the SEND SS proactive UICC command, and returns a successful result in the TERMINAL RESPONSE command send to the UICC.

(UCS2 display in Katakana). Test to confirm that the ME displays the UCS2 text contained in the SEND SS proactive UICC command, and returns a successful result in the TERMINAL RESPONSE command send to the UICC.

SEND USSD

(normal). Test to confirm that the ME correctly translates and sends the unstructured supplementary service request indicated in the SEND USSD proactive UICC command to the USS. Test to confirm that the ME returns a TERMINAL RESPONSE command to the UICC indicating the status of the transmission of the USSD request and including a USSD result as a text string in the TERMINAL RESPONSE.

(Icon support). Test to confirm that the ME displays the text contained in the SEND USSD proactive UICC command, and returns a successful result in the

TERMINAL RESPONSE command send to the UICC. In addition to verify that if an icon is provided by the UICC, the icon indicated in the command may be used by the ME to inform the user, in addition to, or instead of the alpha identifier, as indicated with the icon qualifier.

(UCS2 display in Cyrillic). Test to confirm that the ME displays the UCS2 text contained in the SEND USSD proactive UICC command, and returns a successful result in the TERMINAL RESPONSE command send to the UICC.

(support of Text Attribute – Left Alignment). Test to confirm that the ME displays the alpha identifier according to the left alignment text attribute configuration in the SEND USSD proactive UICC command, and returns a successful result in the TERMINAL RESPONSE command send to the UICC.

(support of Text Attribute – Center Alignment). Test to confirm that the ME displays the alpha identifier according to the center alignment text attribute configuration in the SEND USSD proactive UICC command, and returns a successful result in the TERMINAL RESPONSE command send to the UICC.

(support of Text Attribute – Right Alignment). Test to confirm that the ME displays the alpha identifier according to the right alignment text attribute configuration in the SEND USSD proactive UICC command, and returns a successful result in the TERMINAL RESPONSE command send to the UICC.

(support of Text Attribute – Large Font Size). Test to confirm that the ME displays the alpha identifier according to the large font size text attribute configuration in the SEND USSD proactive UICC command, and returns a successful result in the TERMINAL RESPONSE command send to the UICC.

(support of Text Attribute – Small Font Size). Test to confirm that the ME displays the alpha identifier according to the small font size text attribute configuration in the SEND USSD proactive UICC command, and returns a successful result in the TERMINAL RESPONSE command send to the UICC.

(support of Text Attribute – Bold On). Test to confirm that the ME displays the alpha identifier according to the bold text attribute configuration in the SEND USSD proactive UICC command, and returns a successful result in the TERMINAL RESPONSE command send to the UICC.

(support of Text Attribute – Italic On). Test to confirm that the ME displays the alpha identifier according to the italic text attribute configuration in the SEND USSD proactive UICC command, and returns a successful result in the TERMINAL RESPONSE command send to the UICC.

(support of Text Attribute – Underline On). Test to confirm that the ME displays the alpha identifier according to the underline text attribute configuration in the SEND USSD proactive UICC command, and returns a successful result in the TERMINAL RESPONSE command send to the UICC.

(support of Text Attribute – Strikethrough On). Test to confirm that the ME displays the alpha identifier according to the strikethrough text attribute configuration in the SEND USSD proactive UICC command, and returns a successful result in the TERMINAL RESPONSE command send to the UICC.

(support of Text Attribute – Foreground and Background Colour). Test to confirm that the ME displays the alpha identifier according to the foreground and background colour text attribute configuration in the SEND USSD proactive UICC command, and returns a successful result in the TERMINAL RESPONSE command send to the UICC.

(UCS2 display in Chinese). Test to confirm that the ME displays the UCS2 text contained in the SEND USSD proactive UICC command, and returns a successful result in the TERMINAL RESPONSE command send to the UICC.

(UCS2 display in Katakana). Test to confirm that the ME displays the UCS2 text contained in the SEND USSD proactive UICC command, and returns a successful result in the TERMINAL RESPONSE command send to the UICC.

SET UP CALL

(normal). Test to confirm that the ME accepts the Proactive Command - Set Up Call, displays the alpha identifier to the user, attempts to set up a call to the address and returns the result in the TERMINAL RESPONSE.

(second alpha identifier). Test to confirm that the ME accepts a Proactive Command - Set Up Call, displays the alpha identifiers to the user, attempts to set up a call to the address and returns the result in the TERMINAL RESPONSE.

(display of icons). Test to confirm that the ME accepts a Proactive Set Up Call , displays the message or icon to the user ,attempts to set up a call to the address, returns the result in the TERMINAL response.

(support of Text Attribute – Left Alignment). Test to confirm that the ME accepts the Proactive Command - Set Up Call, displays the alpha identifier according to the left alignment text attribute configuration to the user, attempts to set up a call to the address and returns the result in the TERMINAL RESPONSE.

(support of Text Attribute – Centre Alignment). Test to confirm that the ME accepts the Proactive Command - Set Up Call, displays the alpha identifier according to the centre alignment text attribute configuration to the user, attempts to set up a call to the address and returns the result in the TERMINAL RESPONSE.

(support of Text Attribute – Right Alignment). Test to confirm that the ME accepts the Proactive Command - Set Up Call, displays the alpha identifier according to the right alignment text attribute configuration to the user, attempts to set up a call to the address and returns the result in the TERMINAL RESPONSE.

(support of Text Attribute – Large Font Size). Test to confirm that the ME accepts the Proactive Command - Set Up Call, displays the alpha identifier according to the large font size text attribute configuration to the user, attempts to set up a call to the address and returns the result in the TERMINAL RESPONSE.

(support of Text Attribute – Small Font Size). Test to confirm that the ME accepts the Proactive Command - Set Up Call, displays the alpha identifier according to the small font size text attribute configuration to the user, attempts to set up a call to the address and returns the result in the TERMINAL RESPONSE.

(support of Text Attribute – Bold On). Test to confirm that the ME accepts the Proactive Command - Set Up Call, displays the alpha identifier according to the bold text attribute configuration to the user, attempts to set up a call to the address and returns the result in the TERMINAL RESPONSE.

(support of Text Attribute – Italic On). Test to confirm that the ME accepts the Proactive Command - Set Up Call, displays the alpha identifier according to the italic text attribute configuration to the user, attempts to set up a call to the address and returns the result in the TERMINAL RESPONSE.

(support of Text Attribute – Underline On). Test to verify that the ME accepts the Proactive Command - Set Up Call, displays the alpha identifier according to the underline text attribute configuration to the user, attempts to set up a call to the address and returns the result in the TERMINAL RESPONSE.

(support of Text Attribute – Strikethrough On). Test to confirm that the ME accepts the Proactive Command - Set Up Call, displays the alpha identifier according to the strikethrough text attribute configuration to the user, attempts to set up a call to the address and returns the result in the TERMINAL RESPONSE.

(support of Text Attribute – Foreground and Background Colour). Test to confirm that the ME accepts the Proactive Command - Set Up Call, displays the alpha identifier according to the foreground and background colour text attribute configuration to the user, attempts to set up a call to the address and returns the result in the TERMINAL RESPONSE.

(UCS2 Display in Cyrillic). Test to confirm that the ME accepts the Proactive Command - Set Up Call, displays the alpha identifier with UCS2 coding to the user, attempts to set up a call to the address and returns the result in the TERMINAL RESPONSE.

SET UP EVENT LIST

SET UP EVENT LIST (normal). Test to confirm that the ME accepts a list of events that it shall monitor the current list of events supplied by the UICC, is able to have this current list of events replaced and is able to have the list of events removed. Test to confirm that when the ME has successfully accepted or removed the list of events, it shall send TERMINAL RESPONSE (OK) to the UICC and when the ME is not able to successfully accept or remove the list of events, it shall send TERMINAL RESPONSE (Command beyond ME's capabilities).

PERFORM CARD APDU

PERFORM CARD APDU (normal). Test to confirm that the ME sends an APDU command to the additional card identified in the PERFORM CARD APDU proactive UICC command, and successfully returns the result of the execution of the command in the TERMINAL RESPONSE command send to the UICC. The ME Manufacturer can assign the card reader identifier from 0 to 7. This test applies for MEs with only one additional card reader. In this particular case the card reader identifier 1 is chosen. In this particular case a special Test-SIM (TestSIM) with T=0 protocol is chosen as additional card for the additional ME card reader.

SET UP IDLE MODE TEXT

SET UP IDLE MODE TEXT (Icon support). Test to confirm that the ME text and / or icon passed to the ME is displayed by the ME as an idle mode text. Test to confirm that the icon identifier provided with the text string can replace the text string or accompany it. Test to confirm that if both an alpha identifier or text string, and an icon are provided with a proactive command, and both are requested to be displayed, but the ME is not able to display both together on the screen, then the alpha identifier or text string takes precedence over the icon. Test to confirm that if the UICC provides an icon identifier with a proactive command, then the ME shall inform the UICC if the icon could not be displayed by sending the general result "Command performed successfully, but requested icon could not be displayed". Test to confirm that if the ME receives an icon identifier with a proactive command, and either an empty, or no alpha identifier / text string is given by the UICC, than the ME shall reject the command with general result "Command data not understood by ME".

SET UP IDLE MODE TEXT (UCS2 support). Test to confirm that the UCS2 coded text string is displayed by the ME as an idle mode text.

SET UP IDLE MODE TEXT (support of Text Attribute – Left Alignment). Test to confirm that the text passed to the ME is displayed as idle mode text according to the left alignment text attribute configuration.

SET UP IDLE MODE TEXT (support of Text Attribute – Centre Alignment). Test to confirm that the text passed to the ME is displayed as idle mode text according to the centre alignment text attribute configuration.

SET UP IDLE MODE TEXT (support of Text Attribute – Right Alignment). Test to confirm that the text passed to the ME is displayed as idle mode text according to the right alignment text attribute configuration.

SET UP IDLE MODE TEXT (support of Text Attribute – Large Font Size). Test to confirm that the text passed to the ME is displayed as idle mode text according to the large font size text attribute configuration.

SET UP IDLE MODE TEXT (support of Text Attribute – Small Font Size). Test to confirm that the text passed to the ME is displayed as idle mode text according to the small font size text attribute configuration.

SET UP IDLE MODE TEXT (support of Text Attribute – Bold On). Test to confirm that the text passed to the ME is displayed as idle mode text according to the bold text attribute configuration.

SET UP IDLE MODE TEXT (support of Text Attribute – Italic On). Test to confirm that the text passed to the ME is displayed as idle mode text according to the italic text attribute configuration.

SET UP IDLE MODE TEXT (support of Text Attribute – Underline On). Test to confirm that the text passed to the ME is displayed as idle mode text according to the underline text attribute configuration.

SET UP IDLE MODE TEXT (support of Text Attribute – Strikethrough On). Test to confirm that the text passed to the ME is displayed as idle mode text according to the strikethrough text attribute configuration.

SET UP IDLE MODE TEXT (support of Text Attribute – Foreground and Background Colour). Test to confirm that the text passed to the ME is displayed as idle mode text according to the foreground and background colour text attribute configuration.

SET UP IDLE MODE TEXT (UCS2 display in Chinese). Test to confirm that the UCS2 coded text string is displayed by the ME as an idle mode text.

SET UP IDLE MODE TEXT (UCS2 display in Katakana). Test to confirm that the UCS2 coded text string is displayed by the ME as an idle mode text.

RUN AT COMMAND

(normal). Test to confirm that the ME responds to an AT Command contained within a RUN AT COMMAND as though it were initiated by an attached TE, and returns an AT Response within a TERMINAL RESPONSE to the UICC. Test to confirm that the ME responds to an AT Command contained within a RUN AT COMMAND as though it were initiated by an attached TE, and returns an AT Response within a TERMINAL RESPONSE to the UICC. In addition to verify that if an icon is provided by the UICC, the icon indicated in the command may be used by the ME to inform the user, in addition to, or instead of the alpha identifier, as indicated with the icon qualifier.

(support of Text Attribute – Left Alignment). Test to confirm that the ME responds to an AT Command contained within a RUN AT COMMAND with left alignment text attribute as though it were initiated by an attached TE, and returns an AT Response within a TERMINAL RESPONSE to the UICC.

(support of Text Attribute – Centre Alignment). Test to confirm that the ME responds to an AT Command contained within a RUN AT COMMAND with centre alignment text attribute as though it were initiated by an attached TE, and returns an AT Response within a TERMINAL RESPONSE to the UICC.

(support of Text Attribute – Right Alignment). Test to confirm that the ME responds to an AT Command contained within a RUN AT COMMAND with right alignment text attribute as though it were initiated by an attached TE, and returns an AT Response within a TERMINAL RESPONSE to the UICC.

(support of Text Attribute – Large Font Size). Test to confirm that the ME responds to an AT Command contained within a RUN AT COMMAND with large font size as though it were initiated by an attached TE, and returns an AT Response within a TERMINAL RESPONSE to the UICC.

(support of Text Attribute – Small Font Size). Test to confirm that the ME responds to an AT Command contained within a RUN AT COMMAND with small font size as though it were initiated by an attached TE, and returns an AT Response within a TERMINAL RESPONSE to the UICC.

(support of Text Attribute – Bold On). Test to confirm that the ME responds to an AT Command contained within a RUN AT COMMAND with bold text attribute as though it were initiated by an attached TE, and returns an AT Response within a TERMINAL RESPONSE to the UICC.

(support of Text Attribute – Italic On). Test to confirm that the ME responds to an AT Command contained within a RUN AT COMMAND with italic text attribute as though it were initiated by an attached TE, and returns an AT Response within a TERMINAL RESPONSE to the UICC.

(support of Text Attribute – Underline On). Test to confirm that the ME responds to an AT Command contained within a RUN AT COMMAND with underline text attribute as though it were initiated by an attached TE, and returns an AT Response within a TERMINAL RESPONSE to the UICC.

(support of Text Attribute – Strikethrough On). Test to confirm that the ME responds to an AT Command contained within a RUN AT COMMAND with strikethrough text attribute as though it were initiated by an attached TE, and returns an AT Response within a TERMINAL RESPONSE to the UICC.

(support of Text Attribute – Foreground and Background Colour). Test to confirm that the ME responds to an AT Command contained within a RUN AT COMMAND with foreground and background colour text attribute as though it were initiated by an attached TE, and returns an AT Response within a TERMINAL RESPONSE to the UICC.

(UCS2 display in Cyrillic). Test to confirm that the ME responds to an AT Command contained within a RUN AT COMMAND with UCS2 alpha identifier as though it were initiated by an attached TE, and returns an AT Response within a TERMINAL RESPONSE to the UICC.

(UCS2 display in Chinese). Test to confirm that the ME responds to an AT Command contained within a RUN AT COMMAND with UCS2 alpha identifier as though it were initiated by an attached TE, and returns an AT Response within a TERMINAL RESPONSE to the UICC.

(UCS2 display in Katakana). Test to confirm that the ME responds to an AT Command contained within a RUN AT COMMAND with UCS2 alpha identifier as though it were initiated by an attached TE, and returns an AT Response within a TERMINAL RESPONSE to the UICC.

SEND DTMF

(Normal). Test to confirm that after a call has been successfully established the ME sends the DTMF string contained in the SEND DTMF proactive UICC command to the network, and returns a successful response in the TERMINAL RESPONSE command sent to the UICC.. Test to confirm that the ME does not locally generate audible DTMF tones and play them to the user. Test to confirm that if the ME is in idle mode it informs the UICC using TERMINAL RESPONSE '20' with the additional information "Not in speech call". Test to confirm that the ME displays the text contained in the SEND DTMF proactive UICC command. Test to confirm that if an alpha identifier is provided by the UICC and is a null data object the ME does not give any information to the user on the fact that the ME is performing a SEND DTMF command.

(Display of icons). Test to confirm that after a call has been successfully established the ME send the DTMF string contained in the SEND DTMF proactive UICC command to the network, and returns a successful response in the TERMINAL RESPONSE command sent to the UICC. Test to confirm that the ME do not locally generate audible DTMF tones and play them to the user. Test to confirm that the ME displays the text contained in the SEND DTMF proactive UICC command. Test to confirm that the ME displays the icons which are referred to in the contents of the SEND DTMF proactive UICC command.

(UCS2 display in Cyrillic). Test to confirm that the ME displays the UCS2 text contained in the SEND DTMF proactive UICC command, and returns a successful result in the TERMINAL RESPONSE command send to the UICC.

(support of Text Attribute – Left Alignment). Test to confirm that after a call has been successfully established the ME sends the DTMF string contained in the SEND DTMF proactive UICC command to the network, and returns a successful response in the TERMINAL RESPONSE command sent to the UICC. Test to confirm that the ME does not locally generate audible DTMF tones and play them to the user. Test to confirm that if the ME is in idle mode it informs the UICC using TERMINAL RESPONSE '20' with the additional information "Not in speech call". Test to confirm that the ME displays the text contained in the SEND DTMF proactive UICC command. Test to confirm that the ME displays the alpha identifier according to the left alignment text attribute configuration which are referred to in the contents of the SEND DTMF proactive UICC command.

(support of Text Attribute – Center Alignment). Test to confirm that after a call has been successfully established the ME sends the DTMF string contained in the SEND DTMF proactive UICC command to the network, and returns a successful response in the TERMINAL RESPONSE command sent to the UICC. Test to confirm that the ME does not locally generate audible DTMF tones and play them to the user. Test to confirm that if the ME is in idle mode it informs the UICC using TERMINAL RESPONSE '20' with the additional information "Not in speech call". Test to confirm that the ME displays the text contained in the SEND DTMF proactive UICC command. Test to confirm that the ME displays the alpha identifier according to the centre alignment text attribute configuration which are referred to in the contents of the SEND DTMF proactive UICC command.

(support of Text Attribute – Right Alignment). Test to confirm that after a call has been successfully established the ME sends the DTMF string contained in the SEND DTMF proactive UICC command to the network, and returns a successful response in the TERMINAL RESPONSE command sent to the UICC. Test to confirm that the ME does not locally generate audible DTMF tones and play them to the user. Test to confirm that if the ME is in idle mode it informs the UICC using TERMINAL RESPONSE '20' with the additional information "Not in speech call". Test to confirm that the ME displays the text contained in the SEND DTMF proactive UICC command. Test to confirm that the ME displays the alpha identifier according to the right alignment text attribute configuration which are referred to in the contents of the SEND DTMF proactive UICC command.

(support of Text Attribute – Large Font Size). Test to confirm that after a call has been successfully established the ME sends the DTMF string contained in the SEND DTMF proactive UICC command to the network, and returns a successful response in the TERMINAL RESPONSE command sent to the UICC. Test to confirm that the ME does not locally generate audible DTMF tones and play them to the user. Test to confirm that if the ME is in idle mode it informs the UICC using TERMINAL RESPONSE '20' with the additional information "Not in speech call". Test to confirm that the ME displays the text contained in the SEND DTMF proactive UICC command. Test to confirm that the ME displays the alpha identifier according to the large font size text attribute configuration which are referred to in the contents of the SEND DTMF proactive UICC command.

(support of Text Attribute – Small Font Size). Test to confirm that after a call has been successfully established the ME sends the DTMF string contained in the SEND DTMF proactive UICC command to the network, and returns a successful response in the TERMINAL RESPONSE command sent to the UICC. Test to confirm that the ME does not locally generate audible DTMF tones and play them to the user. Test to confirm that if the ME is in idle mode it informs the UICC using TERMINAL RESPONSE '20' with the additional information "Not in speech call". Test to confirm that the ME displays the text contained in the SEND DTMF proactive UICC command. Test to confirm that the ME displays the alpha identifier according to the small font size text attribute configuration which are referred to in the contents of the SEND DTMF proactive UICC command.

SEND DTMF (support of Text Attribute – Bold On). Test to confirm that after a call has been successfully established the ME sends the DTMF string contained in the SEND DTMF proactive UICC command to the network, and returns a successful response in the TERMINAL RESPONSE command sent to the UICC. Test to confirm that the ME does not locally generate audible DTMF tones and play them to the user. Test to confirm that if the ME is in idle mode it informs the UICC using TERMINAL RESPONSE '20' with the additional information "Not in speech call". Test to confirm that the ME displays the text contained in the SEND DTMF proactive UICC command. Test to confirm that the ME displays the alpha identifier according to the bold text attribute configuration which are referred to in the contents of the SEND DTMF proactive UICC command.

(support of Text Attribute – Italic On). Test to confirm that after a call has been successfully established the ME sends the DTMF string contained in the SEND DTMF proactive UICC command to the network, and returns a successful response in the TERMINAL RESPONSE command sent to the UICC. Test to confirm that the ME does not locally generate audible DTMF tones and play them to the user. Test to confirm that if the ME is in idle mode it informs the UICC using TERMINAL RESPONSE '20' with the additional information "Not in speech call". Test to confirm that the ME displays the text contained in the SEND DTMF proactive UICC command. Test to confirm that the ME displays the alpha identifier according to the italic text attribute configuration which are referred to in the contents of the SEND DTMF proactive UICC command.

(support of Text Attribute – Underline On). Test to confirm that after a call has been successfully established the ME sends the DTMF string contained in the SEND DTMF proactive UICC command to the network, and returns a successful response in the TERMINAL RESPONSE command sent to the UICC. Test to confirm that the ME does not locally generate audible DTMF tones and play them to the user. Test to confirm that if the ME is in idle mode it informs the UICC using TERMINAL RESPONSE '20' with the additional information "Not in speech call". Test to confirm that the ME displays the text contained in the SEND DTMF proactive UICC command. Test to confirm that the ME displays the alpha identifier according to the underline text attribute configuration which are referred to in the contents of the SEND DTMF proactive UICC command.

(support of Text Attribute – Strikethrough On). Test to confirm that after a call has been successfully established the ME sends the DTMF string contained in the SEND DTMF proactive UICC command to the network, and returns a successful response in the TERMINAL RESPONSE command sent to the UICC. Test to confirm that the ME does not locally generate audible DTMF tones and play them to the user. Test to confirm that if the ME is in idle mode it informs the UICC using TERMINAL RESPONSE '20' with the additional information "Not in speech call". Test to confirm that the ME displays the text contained in the SEND DTMF proactive UICC command. Test to confirm that the ME displays the alpha identifier according to the strikethrough text attribute configuration which are referred to in the contents of the SEND DTMF proactive UICC command.

(support of Text Attribute – Foreground and Background Colour). Test to confirm that after a call has been successfully established the ME sends the DTMF string contained in the SEND DTMF proactive UICC command to the network, and returns a successful response in the TERMINAL RESPONSE command sent to the UICC. Test to confirm that the ME does not locally generate audible DTMF tones and play them to the user. Test to confirm that if the ME is in idle mode it informs the UICC using TERMINAL RESPONSE '20' with the additional information "Not in speech call". Test to confirm that the ME displays the text contained in the SEND DTMF proactive UICC command. Test to confirm that the ME displays the alpha identifier according to the foreground and background colour text attribute configuration which are referred to in the contents of the SEND DTMF proactive UICC command.

(UCS2 Display in Chinese). Test to confirm that the ME displays the UCS2 text contained in the SEND DTMF proactive UICC command, and returns a successful result in the TERMINAL RESPONSE command send to the UICC.

(UCS2 Display in Katakana). Test to confirm that the ME displays the UCS2 text contained in the SEND DTMF proactive UICC command, and returns a successful result in the TERMINAL RESPONSE command send to the UICC.

LANGUAGE NOTIFICATION

This test is to verify that the ME shall send a TERMINAL RESPONSE (OK) to the UICC after the ME receives the LANGUAGE NOTIFICATION proactive UICC command.

LAUNCH BROWSER

(No session already launched). Test to confirm that when the ME is in idle state, it launches properly the browser session required in LAUNCH BROWSER, and returns a successful result in the TERMINAL RESPONSE command.

(Interaction with current session). Test to confirm that when the ME is already busy in a browser session, it launches properly the browser session required in LAUNCH BROWSER, and returns a successful result in the TERMINAL RESPONSE.

(UCS2 display in Cyrillic). Test to confirm that the ME performs a proper user confirmation with an USC2 alpha identifier, launches the WAP session required in LAUNCH BROWSER and returns a successful result in the TERMINAL RESPONSE command send to the UICC.

(icons support). Test to confirm that the ME performs a proper user confirmation with an icon identifier, launches the browser session required in LAUNCH BROWSER and returns a successful result in the TERMINAL RESPONSE command send to the UICC.

(support of Text Attribute – Left Alignment). Test to confirm that the ME performs a proper user confirmation with an alpha identifier according to the left alignment text attribute configuration, launches the WAP session required in LAUNCH BROWSER and returns a successful result in the TERMINAL RESPONSE command send to the UICC.

(support of Text Attribute – Centre Alignment). Test to confirm that the ME performs a proper user confirmation with an alpha identifier according to the centre alignment text attribute configuration, launches the WAP session required in LAUNCH BROWSER and returns a successful result in the TERMINAL RESPONSE command send to the UICC.

(support of Text Attribute – Right Alignment). Test to confirm that the ME performs a proper user confirmation with an alpha identifier according to the right alignment text attribute configuration, launches the WAP session required in LAUNCH BROWSER and returns a successful result in the TERMINAL RESPONSE command send to the UICC.

(support of Text Attribute – Large Font Size). Test to confirm that the ME performs a proper user confirmation with an alpha identifier according to the large font size text attribute configuration, launches the WAP session required in LAUNCH BROWSER and returns a successful result in the TERMINAL RESPONSE command send to the UICC.

(support of Text Attribute – Small Font Size). Test to confirm that the ME performs a proper user confirmation with an alpha identifier according to the small font size text attribute configuration, launches the WAP session required in LAUNCH BROWSER and returns a successful result in the TERMINAL RESPONSE command send to the UICC.

(support of Text Attribute – Bold on). Test to confirm that the ME performs a proper user confirmation with an alpha identifier according to the bold text attribute configuration, launches the WAP session required in LAUNCH BROWSER and returns a successful result in the TERMINAL RESPONSE command send to the UICC.

(support of Text Attribute – Italic On). Test to confirm that the ME performs a proper user confirmation with an alpha identifier according to the italic text attribute configuration, launches the WAP session required in LAUNCH BROWSER and returns a successful result in the TERMINAL RESPONSE command send to the UICC.

(support of Text Attribute – Underline On). Test to confirm that the ME performs a proper user confirmation with an alpha identifier according to the underline text attribute configuration, launches the WAP session required in LAUNCH BROWSER and returns a successful result in the TERMINAL RESPONSE command send to the UICC.

(support of Text Attribute – Strikethrough On). Test to confirm that the ME performs a proper user confirmation with an alpha identifier according to the strikethrough text attribute configuration, launches the WAP session required in LAUNCH BROWSER and returns a successful result in the TERMINAL RESPONSE command send to the UICC.

(support of Text Attribute – Foreground and Background Colour). Test to confirm that the ME performs a proper user confirmation with an alpha identifier according to the foreground and background colour text attribute configuration, launches the WAP session required in LAUNCH BROWSER and returns a successful result in the TERMINAL RESPONSE command send to the UICC.

(UCS2 Display in Chinese). Test to confirm that the ME performs a proper user confirmation with an USC2 alpha identifier, launches the WAP session required in LAUNCH BROWSER and returns a successful result in the TERMINAL RESPONSE command send to the UICC.

(UCS2 Display in Katakana). Test to confirm that the ME performs a proper user confirmation with an USC2 alpha identifier, launches the WAP session required in LAUNCH BROWSER and returns a successful result in the TERMINAL RESPONSE command send to the UICC.

OPEN CHANNEL

Open Channel (related to GPRS). Test to confirm that the ME shall send:

TERMINAL RESPONSE (OK); or

TERMINAL RESPONSE (Command performed with modification); or

TERMINAL RESPONSE (User did not accept the proactive command);

TERMINAL RESPONSE (ME currently unable to process command); to the UICC after the ME receives the OPEN CHANNEL proactive command. The TERMINAL RESPONSE sent back to the UICC is the result of the ME and the network capabilities against requested parameters by the UICC.

(GPRS, support of Text Attribute – Left Alignment). Test to confirm that the ME displays an alpha identifier according to the left alignment text attribute configuration in OPEN CHANNEL and returns a successful result in the TERMINAL RESPONSE command send to the UICC.

(GPRS, support of Text Attribute – Centre Alignment). Test to confirm that the ME displays an alpha identifier according to the centre alignment text attribute configuration in OPEN CHANNEL and returns a successful result in the TERMINAL RESPONSE command send to the UICC.

(GPRS, support of Text Attribute – Right Alignment). Test to confirm that the ME displays an alpha identifier according to the right alignment text attribute configuration in OPEN CHANNEL and returns a successful result in the TERMINAL RESPONSE command send to the UICC.

(GPRS, support of Text Attribute – Large Font Size). Test to confirm that the ME displays an alpha identifier according to the large font size text attribute configuration in OPEN CHANNEL and returns a successful result in the TERMINAL RESPONSE command send to the UICC.

(GPRS, support of Text Attribute – Small Font Size). Test to confirm that the ME displays an alpha identifier according to the small font size text attribute configuration in OPEN CHANNEL and returns a successful result in the TERMINAL RESPONSE command send to the UICC.

(GPRS, support of Text Attribute – Bold On). Test to confirm that the ME displays an alpha identifier according to the bold text attribute configuration in OPEN CHANNEL and returns a successful result in the TERMINAL RESPONSE command send to the UICC.

(GPRS, support of Text Attribute – Italic On). Test to confirm that the ME displays an alpha identifier according to the italic text attribute configuration in OPEN CHANNEL and returns a successful result in the TERMINAL RESPONSE command send to the UICC.

(GPRS, support of Text Attribute – Underline On). Test to confirm that the ME displays an alpha identifier according to the underline text attribute configuration in OPEN CHANNEL and returns a successful result in the TERMINAL RESPONSE command send to the UICC.

(GPRS, support of Text Attribute – Strikethrough On). Test to confirm that the ME displays an alpha identifier according to the strikethrough text attribute configuration in OPEN CHANNEL and returns a successful result in the TERMINAL RESPONSE command send to the UICC.

(GPRS, support of Text Attribute – Foreground and Background Colour). Test to confirm that the ME displays an alpha identifier according to the foreground and background colour text attribute configuration in OPEN CHAN-NEL and returns a successful result in the TERMINAL RESPONSE command send to the UICC.

CLOSE CHANNEL

(normal). Test to confirm that the ME shall send a: TERMINAL RESPONSE (Command Performed Successfully); or TERMINAL RESPONSE (Bearer Independent Protocol Error); to the UICC after the ME receives the CLOSE CHANNEL proactive command. The TERMINAL RESPONSE sent back to the UICC is function of the ME and the network capabilities against asked parameters by the UICC.

(support of Text Attribute – Left Alignment). Test to confirm that the ME shall display the alpha identifier according to the left alignment text attribute configuration in the CLOSE CHANNEL proactive command and send a TERMINAL RESPONSE (Command Performed Successfully) to the UICC.

(support of Text Attribute – Centre Alignment). Test to confirm that the ME shall display the alpha identifier according to the centre alignment text attribute configuration in the CLOSE CHANNEL proactive command and send a TERMINAL RESPONSE (Command Performed Successfully) to the UICC.

(support of Text Attribute – Right Alignment). Test to confirm that the ME shall display the alpha identifier according to the right alignment text attribute configuration in the CLOSE CHANNEL proactive command and send a TERMINAL RESPONSE (Command Performed Successfully) to the UICC.

(support of Text Attribute – Large Font Size). Test to confirm that the ME shall display the alpha identifier according to the large font size text attribute configuration in the CLOSE CHANNEL proactive command and send a TERMINAL RESPONSE (Command Performed Successfully) to the UICC.

(support of Text Attribute – Small Font Size). Test to confirm that the ME shall display the alpha identifier according to the small font size text attribute configuration in the CLOSE CHANNEL proactive command and send a TERMINAL RESPONSE (Command Performed Successfully) to the UICC.

(support of Text Attribute – Bold On). Test to confirm that the ME shall display the alpha identifier according to the bold text attribute configuration in the CLOSE CHANNEL proactive command and send a TERMINAL RESPONSE (Command Performed Successfully) to the UICC.

(support of Text Attribute – Italic On). Test to confirm that the ME shall display the alpha identifier according to the italic text attribute configuration in the CLOSE CHANNEL proactive command and send a TERMINAL RESPONSE (Command Performed Successfully) to the UICC.

(support of Text Attribute – Underline On). Test to confirm that the ME shall display the alpha identifier according to the underline text attribute configuration in the CLOSE CHANNEL proactive command and send a TERMINAL RESPONSE (Command Performed Successfully) to the UICC.

(support of Text Attribute – Strikethrough On). Test to confirm that the ME shall display the alpha identifier according to the strikethrough text attribute configuration in the CLOSE CHANNEL proactive command and send a TER-MINAL RESPONSE (Command Performed Successfully) to the UICC.

(support of Text Attribute – Foreground and Background Colour). Test to confirm that the ME shall display the alpha identifier according to the foreground and background colour text attribute configuration in the CLOSE CHANNEL proactive command and send a TERMINAL RESPONSE (Command Performed Successfully) to the UICC.

RECEIVE DATA

(normal). Test to confirm that the ME shall send a: TERMINAL RESPONSE (Command Performed Successfully); or TERMINAL RESPONSE (ME currently unable to process command); or TERMINAL RESPONSE (Bearer Independent Protocol Error); to the UICC after the ME receives the RECEIVE DATA proactive command. The TERMINAL RESPONSE sent back to the UICC is function of the ME and the network capabilities against asked parameters by the UICC.

(support of Text Attribute – Left Alignment). Test to confirm that the ME shall display the alpha identifier according to the left alignment text attribute configuration in the RECEIVE DATA proactive command and send a TERMINAL RESPONSE (Command Performed Successfully) to the UICC.

(support of Text Attribute – Centre Alignment). Test to confirm that the ME shall display the alpha identifier according to the centre alignment text attribute configuration in the RECEIVE DATA proactive command and send a TERMINAL RESPONSE (Command Performed Successfully) to the UICC.

(support of Text Attribute – Right Alignment). Test to confirm that the ME shall display the alpha identifier according to the right alignment text attribute configuration in the RECEIVE DATA proactive command and send a TERMI-NAL RESPONSE (Command Performed Successfully) to the UICC.

(support of Text Attribute – Large Font Size). Test to confirm that the ME shall display the alpha identifier according to the large font size text attribute configuration in the RECEIVE DATA proactive command and send a TERMINAL RESPONSE (Command Performed Successfully) to the UICC.

(support of Text Attribute – Small Font Size). Test to confirm that the ME shall display the alpha identifier according to small font size the text attribute configuration in the RECEIVE DATA proactive command and send a TERMINAL RESPONSE (Command Performed Successfully) to the UICC.

(support of Text Attribute – Bold On). Test to confirm that the ME shall display the alpha identifier according to the bold text attribute configuration in the RECEIVE DATA proactive command and send a TERMINAL RESPONSE (Command Performed Successfully) to the UICC.

(support of Text Attribute – Italic On). Test to confirm that the ME shall display the alpha identifier according to the italic text attribute configuration in the RECEIVE DATA proactive command and send a TERMINAL RESPONSE (Command Performed Successfully) to the UICC.

(support of Text Attribute – Underline On). Test to confirm that the ME shall display the alpha identifier according to the underline text attribute configuration in the RECEIVE DATA proactive command and send a TERMINAL RESPONSE (Command Performed Successfully) to the UICC.

(support of Text Attribute – Strikethrough On). Test to confirm that the ME shall display the alpha identifier according to the strikethrough text attribute configuration in the RECEIVE DATA proactive command and send a TERMINAL RESPONSE (Command Performed Successfully) to the UICC.

(support of Text Attribute – Foreground and Background Colour). Test to confirm that the ME shall display the alpha identifier according to the foreground and background colour text attribute configuration in the RECEIVE DATA proactive command and send a TERMINAL RESPONSE (Command Performed Successfully) to the UICC.

SEND DATA

(normal). Test to confirm that the ME shall send a: TERMINAL RESPONSE (Command Performed Successfully); or TERMINAL RESPONSE (ME currently unable to process command); or TERMINAL RESPONSE (Bearer Independent Protocol Error); TERMINAL RESPONSE (Proactive USIM session terminated by the user); to the UICC after the ME receives the SEND DATA proactive command. The TERMINAL RESPONSE sent back to the UICC is the result of the ME and the network capabilities against requested parameters by the UICC.

(support of Text Attribute – Left Alignment). Test to confirm that the ME shall display the alpha identifier according to the left alignment text attribute configuration in the SEND DATA proactive command and send a TERMINAL RESPONSE (Command Performed Successfully) to the UICC.

(support of Text Attribute – Centre Alignment). Test to confirm that the ME shall display the alpha identifier according to the centre alignment text attribute configuration in the SEND DATA proactive command and send a TERMINAL RESPONSE (Command Performed Successfully) to the UICC.

(support of Text Attribute – Right Alignment). Test to confirm that the ME shall display the alpha identifier according to the right alignment text attribute configuration in the SEND DATA proactive command and send a TERMINAL RESPONSE (Command Performed Successfully) to the UICC.

(support of Text Attribute – Small Font Size). Test to confirm that the ME shall display the alpha identifier according to the small font size text attribute configuration in the SEND DATA proactive command and send a TERMINAL RESPONSE (Command Performed Successfully) to the UICC.

(support of Text Attribute – Bold On). Test to confirm that the ME shall display the alpha identifier according to the bold text attribute configuration in the

SEND DATA proactive command and send a TERMINAL RESPONSE (Command Performed Successfully) to the UICC.

(support of Text Attribute – Italic On). Test to confirm that the ME shall display the alpha identifier according to the italic text attribute configuration in the SEND DATA proactive command and send a TERMINAL RESPONSE (Command Performed Successfully) to the UICC.

(support of Text Attribute – Underline On). Test to confirm that the ME shall display the alpha identifier according to the underline text attribute configuration in the SEND DATA proactive command and send a TERMINAL RESPONSE (Command Performed Successfully) to the UICC.

(support of Text Attribute – Strikethrough On). Test to confirm that the ME shall display the alpha identifier according to the strikethrough text attribute configuration in the SEND DATA proactive command and send a TERMINAL RESPONSE (Command Performed Successfully) to the UICC.

(support of Text Attribute – Foreground and Background Colour). Test to confirm that the ME shall display the alpha identifier according to the foreground and background colour text attribute configuration in the SEND DATA proactive command and send a TERMINAL RESPONSE (Command Performed Successfully) to the UICC.

GET CHANNEL STATUS

Test to verify that the ME shall send a TERMINAL RESPONSE (Command Performed Successfully) to the UICC after the ME receives the GET STATUS proactive command. The TERMINAL RESPONSE sent back to the UICC is function of the ME and the network capabilities against asked parameters by the UICC.

Data Download to UICC

SMS-PP Data Download. Test to confirm that the ME transparently passes the "data download via SMS Point-to-point" messages to the UICC. Test to confirm that the ME returns the RP-ACK message back to the USS, if the UICC responds with '90 00' or '91 XX'. Test to verify that the ME returns the RP-ERROR message back to the system Simulator, if the UICC responds with '62 XX' or '63 XX'. Test to confirm that the ME returns the response data from the UICC back to the USS in the TP-User-Data element of the RP-ACK message, if the UICC returns response data'.

SMS-CB Data Download. Test to confirm that the ME transparently passes the "data download via SMS Cell Broadcast" messages to the UICC, which contain a message identifier found in EFCBMID.

CALL CONTROL BY USIM

Procedure for UE Originated calls. Test to confirm that for all call set-up attempts , even those resulting from a SET UP CALL proactive UICC command, the ME shall first pass the call set-up details (dialled digits and associated parameters) to the UICC, using the ENVELOPE (CALL CONTROL). Test to confirm that if the UICC responds with '90 00', the ME shall set up the call with the dialled digits and other parameters as sent to the UICC. Test to confirm that if the UICC returns response data, the ME shall use the response data appropriately to set up the call as proposed, not set up the call, or set up a call using the data supplied by the UICC. Test to confirm that, in the case where the initial call set-up request results from a proactive SET UP CALL, if the call control result is "not allowed" or "allowed with modifications", the ME shall inform the UICC using TERMINAL RESPONSE "interaction with call control by UICC or MO short message control by UICC, action not allowed". Test to confirm that it is possible for the UICC to request the ME to set up an emergency call by supplying the number "112" as the response data.

Procedure for Supplementary (SS) Services. Test to confirm that the ME first pass the supplementary service control string corresponding to the supplementary service operation to the USIM, using the ENVELOPE (CALL CONTROL) command. Test to confirm that, if the UICC responds with '90 00', the ME shall send the supplementary service operation with the information as sent to the UICC. Test to confirm that, if the UICC returns response data, the ME shall use the response data appropriately to send the supplementary service operation as proposed, not send the SS operation, or instead send the USS operation using the data supplied by the UICC.

Interaction with Fixed Dialling Number (FDN). Test to confirm that the ME checks that the number entered through the MMI is on the FDN list. Test to confirm that, if the MMI input does not pass the FDN check, the call shall not be set up. Test to confirm that, if the MMI input does pass the FDN check, the ME shall pass the dialled digits and other parameters to the UICC, using the ENVELOPE (CALL CONTROL) command. Test to confirm that, if the UICC responds with "allowed, no modification", the ME shall set up the call as proposed. Test to confirm that, if the UICC responds with "not allowed", the ME shall not set up the call. Test to confirm that, if the UICC responds with "allowed with modifications", the ME shall set up the call in accordance with the response from the UICC. If the modifications involve changing the dialled digits, the ME shall not re-check this modified number against the FDN list.

Support of Barred Dialling Number (BDN) service. Test to confirm that the Terminal rejects call set-up to any number that has an entry in EF_{BDN} if BDN service is enabled. Test to confirm that the Terminal allows call set-up to any number not stored in EF_{BDN}. Test to confirm that the Terminal allows emergency call set-up even if the number is stored in EF_{BDN}. Test to confirm that, if the UICC responds with "not allowed", the ME does not set up the call. Test to confirm that, if the UICC responds with "allowed, no modification", the ME shall set up the call (or the supplementary service operation) as proposed. Test to confirm that, if the UICC responds with "allowed with modifications", the ME sets up the call in accordance with the response from the UICC. If the modifications involve changing the dialled number the ME does not re-check this modified number against the

FDN list when FDN is enabled. Test to confirm that updating EF_{BDN} or changing the status of BDN service shall be performed by the use of second application PIN only. Test to confirm that the ME allows call set up to a BDN number if BDN service is disabled.

Barred Dialling Number (BDN) service handling for terminals not supporting BDN. Test to confirm that the Terminal rejects MO-CS call set-up to any number except to emergency call numbers if BDN service is enabled. Test to confirm that the Terminal allows emergency call set-up even if the BDN service is enabled.

EVENT DOWNLOAD

MT Call Event (normal). Test to confirm that the ME informs the UICC that an Event: MT Call has occurred using the ENVELOPE (EVENT DOWNLOAD - MT Call) command.

Call Connected Event (MT and MO call). Test to confirm that the ME informs the UICC that an Event: Call Connected has occurred using the ENVELOPE (EVENT DOWNLOAD -Call Connected) command.

Call Connected Event (ME supporting SET UP CALL). Test to confirm that the ME informs the UICC that an Event: Call Connected has occurred using the ENVELOPE (EVENT DOWNLOAD -Call Connected) command.

Call Disconnected Event. Test to confirm that the ME informs the UICC that an Event: Call Disconnected has occurred using the ENVELOPE (EVENT DOWNLOAD -Call Disconnected) command.

Location Status Event (normal). Test to confirm that the ME informs the UICC that an Event: MM_IDLE state has occurred using the ENVELOPE (EVENT DOWNLOAD - Location Status) command.

User Activity Event (normal). Test to confirm that the ME performed correctly the procedure of USER ACTIVITY EVENT.

Idle Screen Available (normal). Test to confirm that the ME informs the UICC that an Event: Idle Screen Available has occurred using the ENVELOPE (EVENT DOWNLOAD - IDLE SCREEN AVAILABLE) command.

Card Reader Status (normal). Test to confirm that the ME informs the UICC that an Event: Card Reader Status has changed using the ENVELOPE (EVENT DOWNLOAD - Card Reader Status) command.

Card Reader Status(detachable card reader). Test to confirm that the ME informs the UICC that an Event: Card Reader Status has changed using the ENVELOPE (EVENT DOWNLOAD - Card Reader Status) command.

Language selection event (normal). Test to confirm that the ME informs the UICC that an Event: Language selection has occurred using the ENVELOPE (EVENT DOWNLOAD - LANGUAGE SELECTION) command.

Browser termination (normal). Test to confirm that the ME informs the UICC of an Event: Browser termination using the ENVELOPE (EVENT DOWNLOAD - Browser Termination) command.

Data available event. Test to confirm that the ME shall send an ENVELOPE (EVENT DOWNLOAD - Data available) to the UICC after the ME receives a packet of data from the server by the BIP channel previously opened.

Channel Status event. Test to confirm that the ME shall send an ENVELOPE (EVENT DOWNLOAD - Channel Status) to the UICC after the link dropped between the NETWORK and the ME.

MO SHORT MESSAGE CONTROL BY USIM

Test to verify that for all SMS sending attempts, even those resulting from a SEND SHORT MESSAGE proactive UICC command, the ME shall first pass the RP_destination_address of the service centre and the TP_Destination_Address to the UICC, using the ENVELOPE (MO Short Message CONTROL). Test to confirm that if the UICC responds with '90 00', the ME shall send the SMS with the address unchanged. Test to confirm that if the UICC responds with '93 00', the ME shall not send the SMS and may retry the command. Test to confirm that if the UICC returns response data, the ME shall use the response data appropriately to send the SM as proposed, not send the SM, or send the SM using the data supplied by the UICC. Test to confirm that, in the case where the initial SM request results from a proactive SEND SHORT MESSAGE, if the MO SMS CONTROL result is "not allowed" or "allowed with modifications", the ME shall inform the UICC using TERMINAL RESPONSE "interaction with call control by UICC or MO short message control by USIM, action not allowed".

Handling of command number

Test to verify that the ME sends a Terminal Response with the Command number equivalent to the value in the corresponding proactive command.

IMS

General Purpose PDP Context Establishment (UE Requests for a Dedicated PDP Context). Test to confirm that the UE sends a correctly composed Activate PDP context request by setting the IM CN Subsystem Signalling Flag to the GGSN within the Protocol Configuration Options IE.

Dedicated PDP Context Establishment. Test to confirm that on receiving Activate PDP Context accept with IM CN Subsystem Signalling Flag included within the Protocol Configuration Options IE, UE shall consider the PDP context as a Dedicated PDP context for SIP signalling.

P-CSCF Discovery

P-CSCF Discovery via PDP Context. Test to confirm that the UE sends a correctly composed Activate PDP context request message requesting for P-CSCF address(es) to the GGSN within the Protocol Configuration Options IE. On receiving Activate PDP Context accept with IM CN Subsystem Signalling Flag not included within the Protocol Configuration Options IE and list of P-CSCF

IPv6/IPv4 addresses included, UE shall consider the PDP context as a general purpose PDP context for SIP signalling and P-CSCF discovery procedure to be successful.

P-CSCF Discovery via DHCP – IPv4. Test to confirm UE shall initiate and successfully complete a P-CSCF discovery procedure via DHCP when P-CSCF address is not provided as part of PDP Context Activation procedure.

P-CSCF Discovery via DHCP – IPv4 (UE Requests P-CSCF discovery via PCO). Test to confirm that the UE sends a correctly composed Activate PDP context request message requesting for P-CSCF address(es) to the GGSN within the Protocol Configuration Options IE. On receiving Activate PDP Context accept not including P-CSCF address(es) in PCO, UE will initiate a P-CSCF discovery procedure employing DHCP/DNS.

P-CSCF Discovery by DHCP-IPv6 (UE Requests P-CSCF discovery by PCO). UE will initiate P-CSCF discovery procedure employing DHCP. Test to confirm that the UE sends a correctly composed Activate PDP context requesting for P-CSCF address(es) to the GGSN within the Protocol Configuration Options IE. On receiving Activate PDP Context accept not including P-CSCF address(es) in PCO IE, will initiate a P-CSCF discovery procedure employing DHCP.

P-CSCF Discovery by DHCP – IPv6 (UE does not Request P-CSCF discovery by PCO, SS includes P-CSCF Address(es) in PCO). Test to confirm that a UE, which has not requested for P-CSCF address in PDP context activate message, receives P-CSCF address, may accept the P-CSCF address or ignore it and hence initiate P-CSCF discovery by DHCP.

P-CSCF Discovery (UE Receives list of FQDNs / IPv6 addresses). Test to verify that on receiving a list of P-CSCF FQDNs / IPv6 addresses, UE shall assume the list to be prioritised and try them in the order of their presence. The test is applicable for IMS support or early IMS security. Test to confirm that on receiving a list of P-CSCF FQDNs / IPv6 addresses UE shall assume the list to be prioritised and try them in the order of their presence.

P-CSCF Discovery (UE Receives list of FQDNs/IPv4 addresses). Test to confirm that on receiving a list of P-CSCF FQDNs / IPv4 addresses UE shall assume the list to be prioritised and try them in the order of their presence.

Registration
Initial Registration. Test to confirm that UE correctly derives a private user identity, a temporary public user identity and a home network domain name from the IMSI parameter in the USIM

User Initiated Re-Registration. Test to confirm that the UE can re-register a previously registered public user identity at either 600 seconds before the expiration time if the initial registration was for greater than 1200 seconds, or when half of the time has expired if the initial registration was for 1200 seconds or less; and Extract or derive from the UICC a public user identity, the private user identity, and the domain name to be used in the Request-URI in the registration; and Test to verify that the UE populates the header field in the REGISTER request with From, To, Via, Contact, Authorization, Expires, Security-Client, Security-verify, Supported, and P-Access-Network-Info headers; and Upon receiving 200 OK for REGISTER, the UE shall store the new expiration time of the registration for this public user identity, the list of URIs contained in the P-Associated-URI header value and use these values in the next re-register request.

UE Initiated Deregistration. Test to confirm that the UE sends a correctly composed initial REGISTER request with an Expires header or expires parameter set to 0 to S-CSCF via the discovered P-CSCF.

Invalid behaviour- 423 Interval too brief. verify that after receiving a valid 423 (Interval Too Brief) response to the REGISTER request, the UE sends another REGISTER request populating the Expires header or the expires parameter in the Contact header with an expiration timer of at least the value received in the Min-Expires header of the 423 (Interval Too Brief) response.

Initial registration for early IMS security. verify that UE correctly derives a temporary public user identity from the IMSI parameter according to the procedures and Test to verify that UE correctly derives a home network domain name from the IMSI parameter or alternatively uses the values retrieved from ISIM; and Test to verify that the UE sends a correctly composed initial REGISTER request to S-CSCF via the discovered P-CSCF; and Test to verify that after receiving a valid 200 OK response from S-CSCF for the REGISTER, the UE stores the default public user identity and information about barred user identities; and Test to verify that after receiving a valid 200 OK response from S-CSCF for the REGISTER, the UE subscribes to the reg event package for the public user identity registered at the users registrar (S-CSCF) and Test to verify that after receiving a valid 200 OK response from S-CSCF to the SUBSCRIBE sent for registration event package, the UE maintains the generated dialog; and Test to verify that the UE responds the received valid NOTIFY with 200 OK.

Initial registration for combined IMS support and early IMS security against a network with early IMS support only. Test to confirm that UE correctly derives a private user identity, a temporary public user identity and a home network domain name from the IMSI parameter in the USIM, or alternatively uses the values retrieved from ISIM; and Test to verify that UE correctly derives a home network domain name from the IMSI parameter in the USIM, according to the procedures or alternatively uses the values retrieved from ISIM; and Test to verify that after receiving a 420 (Bad Extension) response from S-CSCF for the initial REGISTER sent, the UE sends a correctly composed initial REGISTER request to S-CSCF via the discovered P-CSCF, and Test to verify that after receiving a valid 200 OK response from S-CSCF for the REGISTER, the UE subscribes to the reg event package for the public user identity registered at the users registrar (S-CSCF) and Test to verify that after receiving a valid 200 OK response from S-CSCF to the SUBSCRIBE sent for registration event package, the UE maintains the generated dialog; and Test to verify that the UE responds the received valid NOTIFY with 200 OK.

Initial registration for combined IMS support and early IMS security with SIM application. Test to confirm that the UE initiate the early IMS security registration procedure when a SIM application is in use, even if the UE has IMS support; and Test to verify that UE correctly derives a temporary public user identity from the IMSI parameter in the USIM according to the procedures; and Test to verify that UE correctly derives a home network domain name from the IMSI parameter in the USIM, according to the procedures or alternatively uses the values retrieved from ISIM; and Test to verify that the UE sends a correctly composed initial REGISTER request to S-CSCF via the discovered P-CSCF; and Test to verify that after receiving a valid 200 OK response from S-CSCF for the REGISTER, the UE subscribes to the reg event package for the public user identity registered at the users registrar (S-CSCF) and Test to verify that after receiving a valid 200 OK response from S-CSCF to the SUBSCRIBE sent for registration event package, the UE maintains the generated dialog; and Test to verify that the UE responds the received valid NOTIFY with 200 OK.

Authentication

Invalid Behaviour – MAC Parameter Invalid. Test to confirm that after receiving a 401 (Unauthorized) response from S-CSCF for the initial REGISTER sent, the UE checks the validity of the received authentication challenge, i.e. the locally calculated XMAC must match the MAC parameter derived from the AUTN part of the challenge. If, the value of MAC derived from the AUTN part of the 401 (Unauthorized) received by the UE does not match the value of locally calculated XMAC: the UE responds with a further REGISTER indicating to the S-CSCF that the challenge has been deemed invalid and: this subsequent REGISTER request contains no AUTS directive and an empty response directive, i.e. no authentication challenge response-populates a new Security-Client header within the REGISTER request, set to specify the security mechanism it supports, the IPsec layer algorithms it supports and the parameters needed for the new security association setup; and does not create a temporary set of security associations.

Invalid Behaviour – SQN out of range. Test to confirm that after receiving a 401 (Unauthorized) response for the initial REGISTER sent, the UE checks that the SQN parameter derived from the AUTN part of the authentication challenge is within the correct range. If, the value of SQN derived from the AUTN part of the 401 (Unauthorized) received by the UE is out of range the UE reacts correctly: Test to verify after a failed authentication attempt if the UE on receives a valid 401 (Unauthorized) message from the network in response to the Register request sent, the UE is able to perform the authentication and registration successfully

Subscription

Invalid Behaviour – 503 Service Unavailable. Test to confirm that after receiving a 503 (Service Unavailable) response to a SUBSCRIBE request, containing a Retry-After header, the UE shall not automatically reattempt the request until after the period indicated by the Retry-After header contents. This can happen when the server is temporarily unable to process the request due to a temporary over-loading or maintenance of the server.

Notification

Network-initiated deregistration. Test to confirm that UE will not try registration after getting a NOTIFY with all <registration> element(s) set to "terminated" and "rejected".

Network initiated re-authentication. Test to confirm that UE adjusts the expiration time for a public user identity as indicated within the received NOTIFY related to reg event package; and Test to verify that the UE will start the re-authentication procedures at the appropriate time before the registration expires.

Call Control

MO Call Successful with preconditions. Test to confirm that when initiating MO call the UE performs correct exchange of SIP protocol signalling messages for setting up the session; and Test to verify that within SIP signalling the UE performs the correct exchange of SDP messages for negotiating media and indicating preconditions for resource reservation. Test to confirm that the UE is able to release the call.

MO Call – 503 Service Unavailable. Test to confirm that when the UE receives a 503 (Service Unavailable) response to an initial INVITE request containing a Retry-After header, then the UE shall not automatically reattempt the request until after the period indicated by the Retry-After header contents.

Call initiation – UE termination. Test to confirm that after receiving a valid INVITE for call initiation, the UE correctly generates and sends the first 183 Session Progress response; and Test to verify that the UE includes the proper SDP answer to the SDP offer in the INVITE; and Test to verify that the UE inserts a P-Access-Network-Info header into any response to a request for a dialog, any subsequent request (except CANCEL requests) or response (except CANCEL responses) within a dialog or any response to a standalone method. This header shall contain information concerning the access network technology and, if applicable, the cell ID; and Test to verify that the UE includes the protected server port in any Contact header; and Test to verify that the UE does not encrypt the SDP payload; and Test to verify that the UE supports and handles the precondition extension properly Test to verify that the UE can release the call on receiving BYE from the SS

Signalling Compression (SIGComp)

SigComp in the Initial registration. Test to confirm that the UE performs initial registration, subscription and notification. The UE can send messages compressed or not compressed. The UE can announce to support SIP Compression "comp=sigcomp"; and Test to verify that the UE uses the SIP/SDP dictionary at least in the first message sent;; and Test to verify that the UE decompresses all the SIP messages sent by the SS. This is tested implicitly by checking the messages sent by the UE verifying the correct exchange of SIP protocol signalling messages.

SigComp in the MO Call. Test to confirm that, when initiating MO call, the UE performs the session setup. The UE can send messages compressed or not compressed The UE can announce to support SIP Compression "comp=sigcomp"; and Test to verify that the UE decompresses all the SIP messages sent by the SS. This is tested implicitly by verifying the correct exchange of SIP protocol signalling messages..

SigComp in the MT Call. Test to confirm that, when initiating MT call, the UE performs the session setup with compression set to on. The UE can announce to support SIP Compression "comp=sigcomp"; and Test to verify that the UE decompresses all the SIP messages sent by the SS. This is tested implicitly by verifying the correct exchange of SIP protocol signalling messages..

Invalid Behaviour - State creation before authentication. Test to confirm that, when the P-CSCF try to create a state with the first REGISTER request in th UE state handler the UE does not create it.

Emergency Service
Emergency Call Initiation – Using CS domain. Test to confirm that when calling an emergency number the UE attempts an emergency call setup.

REFERENCES

[1] 3GPP TS 34.123-1 V6.5.0, Technical Specification, 3rd Generation Partnership Project; Technical Specification Group Radio Access Network; User Equipment (UE) conformance specification; Part 1: **Protocol conformance specification**

[2] 3GPP TS 34.123-2 V6.5.0, Technical Specification, 3rd Generation Partnership Project; Technical Specification Group Radio Access Network; User Equipment (UE) conformance specification; Part 2: **Implementation Conformance Statement (ICS) proforma specification**

[3] 3GPP TS 26.132 V6.0.0, Technical Specification, 3rd Generation Partnership Project; Technical Specification Group Services and System Aspects; **Speech and video telephony terminal acoustic test specification**

[4] 3GPP TS 31.121 V6.6.0, Technical Specification, 3rd Generation Partnership Project; Technical Specification Group Core Network and Terminals; **UICC-terminal interface; Universal Subscriber Identity Module (USIM) application test specification**

[5] 3GPP TS 31.124 V6.7.0, Technical Specification, 3rd Generation Partnership Project; Technical Specification Group Core Network and Terminals; **Mobile Equipment (ME) conformance test specification; Universal Subscriber Identity Module Application Toolkit (USAT) conformance test specification**

[6] 3GPP TS 34.131 V6.0.1, Technical Specification, 3rd Generation Partnership Project; Technical Specification Group Terminals; **Test Specification for C-language binding to (Universal) Subscriber Interface Module ((U)SIM) Application Programming Interface (API)**

[7] 3GPP TS 34.171 V6.5.0, Technical Specification, 3rd Generation Partnership Project; Technical Specification Group Radio Access Network; **Terminal conformance specification; Assisted Global Positioning System (A-GPS); Frequency Division Duplex (FDD)**

[8] 3GPP TS 34.229-1 V6.0.0, Technical Specification, 3rd Generation Partnership Project; Technical Specification Group Radio Access Network; Internet Protocol (IP) multimedia call control protocol based on Session Initiation Protocol (SIP) and Session Description Protocol (SDP); User Equipment (UE) conformance specification; Part 1: **Protocol conformance specification**

[9] 3GPP TS 51.010-1 V7.4.0 (2006-12), Technical Specification, 3rd Generation Partnership Project; Technical Specification Group, GSM/EDGE Radio Access Network, Digital cellular telecommunications system (Phase 2+); **Mobile Station (MS) conformance specification; Part 1: Conformance specification**

Chapter Three

UMTS Conformance RF Testing

RF PERFORMANCE TESTING

In addition to the protocol conformance, the RF performance of the UE must also be verified. Many measurements of the transmitter and receiver performance are performed in a number of areas, e.g. out-of-band emissions. Measurements of the radio resource management (RRM) are performed to ensure that the control capability of the UE is operating according to the standards. The RRM is the component used to control the physical or RF layers in accordance with the requirements of the protocols from the upper layers. There are, for instance, very tight limits on the transmitter output power, as it is controlled to meet conditions such as variations in signal strength. This ensures that the handset only transmits sufficient power to maintain a reliable connection under the prevailing conditions. As a result, the overall level of noise in the handset bands is reduced to the minimum level. A protocol tester is often used to control the handset and set up the relevant scenarios, along with RF measurement and generation equipment that may include signal generators, power meters, analysers and noise generators. Special fading simulators are additionally used to check operation of the handset with multipath and fading.

The Test System illustrated in Figure 3.1 consists of a number of equipments connected together into a system for the purpose of making one or more measurements on a UE in accordance with the test case requirements. A test system may include one or more System Simulators if additional signalling is required for the test case. The System Simulator is system that simulates Node B signalling and analysing UE signalling responses on one or more RF channels, in order to create the required test environment for the UE under test. It will also include the capability of controlling the UE power, measurement of Rx BLER and BER, measurement of signalling timing and delays and the ability to simulate UTRAN.

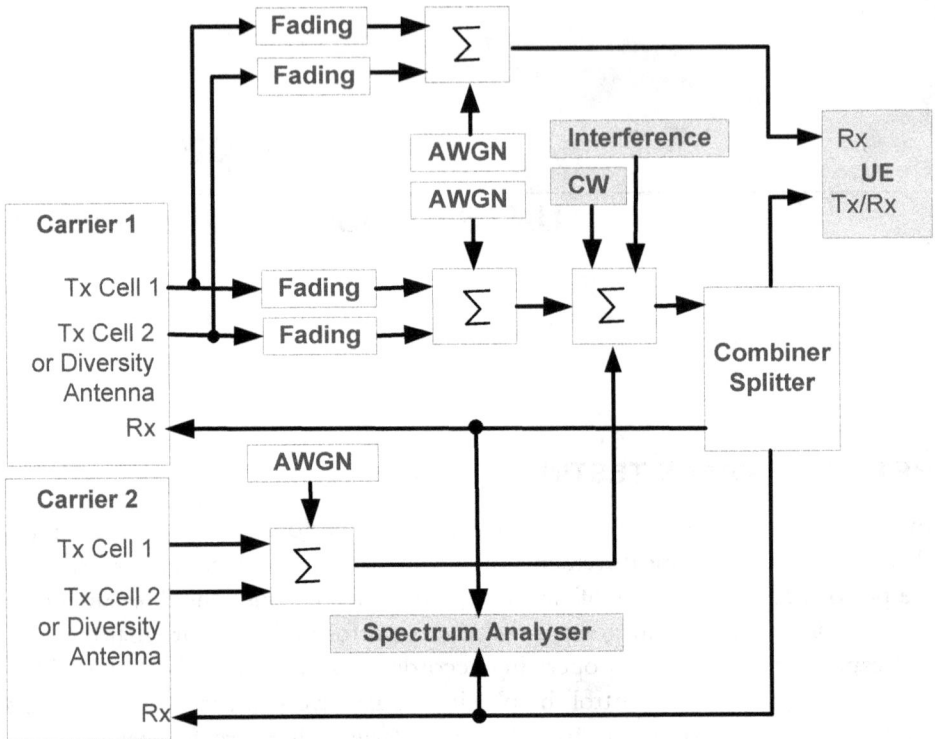

Figure 3.1   Test System setup

Test Covered are:

Single/Multiple single/multiple
carrier
Tx Intermodulation
Rx with Interference
Rx with additional CW signal
Static propagation
Multiple path Fading propagation
Tx and Rx Diversity

UE RF Testing
Transmitter Characteristics
Output Power and Frequency
The following tests are set to check the transmitter output power and frequency:

Maximum Output Power. This test is to verify that the error of the UE maximum output power does not exceed the range prescribed by the nominal maximum output power and tolerance. An excess maximum output power has the possibility to interfere to other channels or other systems. A small maximum output power decreases the coverage area.

Maximum Output Power with HS-DPCCH. This test is to verify that the error of the UE maximum output power with HS-DPCCH does not exceed the range prescribed by the maximum output power and tolerance.

Maximum Output Power with HS-DPCCH and E-DCH. This test is to verify that the error of the UE maximum output power with HS-DPCCH and E-DCH does not exceed the range prescribed by the maximum output power and tolerance.

UE relative code domain power accuracy. This test is to verify that the UE relative code domain power accuracy meets the requirements.

Frequency Error. This test is to verify that the UE carrier frequency error does not exceed the requirement. This test verifies the ability of the receiver to derive correct frequency information for the transmitter, when locked to the DL carrier frequency.

Change of TFC. This test is to verify that the DTX ON/OFF power levels versus time meets the specified mask.

Power setting in uplink compressed mode. This test is to verify that the changes in uplink transmit power in compressed mode are within the prescribed tolerances.

HS-DPCCH power control. This test is to verify that the changes in uplink transmit power when transmitting the HS-DPCCH (Ack/Nack and CQI) and the power between HS-DPDCH transmissions are within the allowed power step tolerances.

Occupied Bandwidth (OBW). This test is to verify that the UE occupied channel bandwidth is less than 5 MHz based on a chip rate of 3,84 Mcps. Excess occupied channel bandwidth increases the interference to other channels or to other systems.

Spectrum emission mask. This test is to verify that the power of UE emission does not exceed the prescribed limits.

Spectrum Emission Mask with HS-DPCCH. This test is to verify that the power of UE emission does not exceed the prescribed limits even in the presence of the HS-DPCCH.

Spectrum Emission Mask with E-DCH. This test is to verify that the power of UE emission does not exceed the prescribed limits even in the presence of the E-DCH.

Adjacent Channel Leakage Power Ratio (ACLR). This test is to verify that the UE ACLR does not exceed prescribed limit. Excess ACLR increases the interference to other channels or to other systems.

Adjacent Channel Leakage Power Ratio (ACLR) with HS-DPCCH. This test is to verify that the UE ACLR does not exceed prescribed limit.

Adjacent Channel Leakage Power Ratio (ACLR) with E-DCH. This test is to verify that the UE ACLR does not exceed prescribed limit

Spurious Emissions. This test is to verify that the UE spurious emissions do not exceed described value. Excess spurious emissions increase the interference to other systems.

Transmit Intermodulation. This test is to verify that the UE transmit intermodulation does not exceed the described value. An excess transmit intermodulation increases transmission errors in the up link own channel when other transmitter exists nearby.

Output Power Dynamics in the Uplink

Power control is a key feature of UMTS and is used to limit the interference level. A number of tests are necessary to make sure the functionality is behaving correctly and the parameters are set according to the specifications. Following are the key tests:

Open Loop Control Power in the Uplink. The power measured by the UE of the received signal and the signalled BCCH information are used by the UE to control the power of the UE transmitted signal with the target to transmit at the lowest power acceptable for proper communication. The test stresses the ability of the receiver to measure the received power correctly over the receiver dynamic range.

Inner Loop Power Control in the Uplink. This test is to verify that TPC_cmd is correctly derived from received TPC commands.

Minimum Output Power. This test is to verify that the UE minimum transmit power is less than the specified value.

Out-of-synchronisation handling of output power. This test is to verify that the UE monitors the DPCCH quality and turns its transmitter on or off according to DPCCH level profile.

Transmit ON/OFF Power

Transmit OFF Power. This test is to verify that the UE transmit OFF power is less than a specified value.

Transmit ON/OFF Time mask. This test is to verify that the power ON/OFF ratio of the PRACH meets the requirements.

Transmit Modulation.

Transmit modulation defines the modulation quality for expected in-channel RF transmissions from the UE.

Error Vector Magnitude (EVM), with HS-DPCCH and phase discontinuity with HS-DPCCH. This test is to verify that the EVM does not exceed a specified value. An excess EVM increases transmission errors in the up link own channel.

Peak code domain error. This test is to verify that the UE peak code domain error does not exceed a specified value. An excess peak code domain error increases transmission errors in the up link own channel.

UE phase discontinuity. This test is to verify that the UE phase discontinuity is within the limits. This test is to verify that any timeslot used in the calculation of a phase discontinuity result also passes the frequency error and EVM requirements.

PRACH preamble quality. The test purpose is to verify that the transmission quality of the first PRACH preamble meets the minimum requirements for modulation quality, carrier frequency, access slot and signature as specified.

Receiver Characteristics

Reference Sensitivity Level. This test is to verify that the UE BER should not exceed the specified limit. The lack of the reception sensitivity decreases the coverage area at the far side from Node B.

Maximum Input Level. This test is to verify that the UE BER should not exceed a specified limit. An inadequate maximum input level causes loss of coverage near the Node B

Maximum Input Level for HS-PDSCH Reception (16QAM). This test is to verify that the UE HSDPA throughput meets the minimum requirements specified for the DL channel. An inadequate maximum input level causes loss of coverage near the Node B.

Adjacent Channel Selectivity (ACS). This test is to verify that the UE BER does not exceed a specified limit. The lack of the ACS decreases the coverage area when other transmitter exists in the adjacent channel.

Blocking Characteristics. This test is to verify that the UE BER does not exceed a specified limit. The lack of the blocking ability decreases the coverage area when other transmitter exists (except in the adjacent channels and spurious response).

Spurious Response. This test is to verify that the UE BER does not exceed a specified limit. The lack of the spurious response ability decreases the coverage area when other unwanted interfering signal exists at any other frequency.

Intermodulation Characteristics. This test is to verify that the UE BER does not exceed a specified limit. The lack of the intermodulation response rejection ability decreases the coverage area when two or more interfering signals, which have a specific frequency relationship to the wanted signal, exist.

Spurious Emissions. This test is to verify that the UE spurious emission meets the specifications. Excess spurious emissions increase the interference to other systems.

Performance Tests

The purpose of the following tests is to check the demodulation of DCH in different types of links in different propagation conditions as follows:

Demodulation in Static Propagation conditions. Demodulation of Dedicated Channel (DCH). This test is to verify the ability of the receiver to receive a predefined test signal, representing a static propagation channel for the wanted and for the co-channel signals from serving and adjacent cells, with a BLER not exceeding a specified value.

Demodulation of DCH in Multi-path Fading Propagation conditions. Single Link Performance. This test is to verify the ability of the receiver to receive a predefined test signal, representing a multi-path fading propagation channel for the wanted and for the co-channel signals from serving and adjacent cells, with a BLER not exceeding a specified value.

Demodulation of DCH in Moving Propagation conditions. Single Link Performance. This test is to verify the ability of the receiver to receive a predefined test signal, representing a moving propagation channel for the wanted and for the co-channel signals from serving and adjacent cells, with a BLER not exceeding a specified value.

Demodulation of DCH in Birth-Death Propagation conditions. Single Link Performance. This test is to verify the ability of the receiver to receive a predefined test signal, representing a birth-death propagation channel for the wanted and for the co-channel signals from serving and adjacent cells, with a BLER not exceeding a specified value.

Demodulation of DCH in downlink Transmit diversity modes

Demodulation of DCH in open-loop or closed loop transmit diversity mode. This test is to verify that UE reliably demodulates the DPCH of the Node B while open loop or closed loop transmit diversity is enabled during the connection.

Demodulation in Handover conditions

Demodulation of DCH in Inter-Cell Soft Handover. This test is to verify that the BLER does not exceed a specified value at the DPCH_Ec/Ior.

Combining of TPC commands from radio links of different radio link sets. This test is to verify that the combining of TPC commands received in soft handover results in TPC_cmd being derived so as to meet the requirements.

Combining of reliable TPC commands from radio links of different radio link sets. This test is to verify that the combining of reliable TPC commands received in soft handover results in TPC_cmd being derived so as to meet the requirements.

Power control in downlink

Power control in the downlink is the ability of the UE receiver to converge to required link quality set by the network while using as low power as possible in downlink. If a BLER target has been assigned to a DCCH, then it has to be such that outer loop is based on DTCH and not on DCCH. Tests of power control in the DL are as follows:

Power control in the downlink, constant BLER target. This test is to verify that the UE receiver is capable of converging to required link quality set by network while using as low power as possible.

Power control in the downlink, initial convergence. This test is to verify that DL power control works properly during the first seconds after DPCH connection is established.

Power control in the downlink, wind up effects. This test is to verify that the UE downlink power control does not require too high downlink power during a period after the downlink power is limited by the UTRAN.

Power control in the downlink, different transport formats. This test is to verify that the UE outer loop power control works properly with different transport formats.

Downlink compressed mode

Downlink compressed mode is used to create gaps in the downlink transmission, to allow the UE to make measurements on other frequencies.

Single link performance. The purpose of this test is to verify the reception of DPCH in a UE while downlink is in a compressed mode. The UE needs to preserve the BLER using sufficient low DL power. It is also verified that UE applies the Delta SIR values, which are signalled from network, in its outer loop power control algorithm.

Blind transport format detection

This test is to verify the ability of the blind transport format detection to receive a predefined test signal, i.e. ability of receiver to decode without TFCI information. Representing a static propagation channel for the wanted and for the co-channel signals from serving and adjacent cells, with a block error ratio (BLER) and false transport format detection ratio (FDR) not exceeding a specified value

Demodulation of Paging Channel (PCH)

This test is to verify that average probability of missed paging (Pm-p) does not exceed a specified value.

Detection of Acquisition Indicator (AI)

This test is to verify that average probability of false detection of AI and average probability of missed AI do not exceed specified values.

RRM Testing
Idle Mode

The cell re-selection delay is defined as the time from a change of cell levels to the moment when this change makes the UE camp on a new cell, and starts to send preambles on the PRACH for the RRC CONNECTION REQUEST message to perform a Location Updating procedure (MM) or Routing Area Updating procedure (GMM) on the new cell. The following scenarios are tested to verify that the delay is within a specified limit.

1. Cell Re-Selection, Single and Multi Carrier
2. UTRAN to GSM Cell Re-Selection, both or UTRA only level changed
3. FDD/TDD Cell Re-selection

UTRAN Connected Mode Mobility

The active set update delay of the UE is defined as the time from the end of the last TTI containing an RRC message implying soft handover to the switch off of the old downlink DPCH. The hard handover delay of the UE is defined as the time from the end of the last TTI containing an RRC message implying hard handover to the transmission of the new uplink DPCCH. The UTRAN to GSM cell handover delay is defined as the time from the end of the last TTI containing an RRC message implying hard handover to the transmission on the channel of the new RAT. The following scenarios are tested to verify that the delay is within a specified limit.

a) FDD/FDD Soft Handover
b) FDD/FDD Hard Handover to intra-frequency cell
c) FDD/FDD Hard Handover to inter-frequency cell
d) FDD/TDD Handover
e) Inter-system Handover from UTRAN FDD to GSM

Cell Re-selection in CELL_FACH. The cell re-selection delay is defined as the time between the occurrence of an event which will trigger Cell Reselection process and the moment in time when the UE starts sending the preambles on the PRACH for sending RRC CELL UPDATE message to the UTRAN. The following scenarios are tested to verify that the delay is within a specified limit.

a) One frequency present in neighbour list
b) Two frequencies present in the neighbour list
c) Cell Reselection to GSM

Cell Re-selection in CELL_PCH. One or Two frequencies present in the neighbour list. The cell re-selection delay is defined as the time from a change of cell levels to the moment when this change makes the UE camp on a new cell, and starts to send preambles on the PRACH for the CELL UPDATE message with cause value "cell reselection" in the new cell. This test is to verify that the UE meets the minimum requirements and is capable of camping on to a new cell, within the required time, when the preferred cell conditions change.

Cell Re-selection in URA_PCH. One or Two frequencies present in the neighbour list. The cell re-selection delay is defined as the time from a change of cell levels to the moment when this change makes the UE camp on a new cell, and starts to send preambles on the PRACH for the URA UPDATE message with cause value "URA reselection" in the new cell. This test is to verify that the UE meets the minimum requirement and is capable of camping on to a new cell, within the required time, when the preferred cell conditions change.

Serving HS-DSCH cell change. When the UE receives a RRC message implying HS-DSCH cell change with the activation time "now" or earlier than RRC procedure delay seconds from the end of the last TTI containing the RRC command, the UE should be ready to receive the HS-SCCH channel from the new cell within Dcell_change seconds from the end of the last TTI containing the RRC command. The purpose of this test is to verify the requirement for the delay when performing the serving HS-DSCH cell change in CELL_DCH state specified.

RRC Connection Control

RRC Re-establishment delay

The UE Re-establishment delay requirement (TUE-RE-ESTABLISH-REQ) is defined as the time between the moment when radio link failure is considered by the UE, to when the UE starts to send preambles on the PRACH. TUE-RE-ESTABLISH-REQ is depending on whether the target cell is known by the UE or not. A cell is known if either or both of the following conditions are true: the UE has had radio links connected to the cell in the previous (old) active set, the cell has been measured by the UE during the last 5 seconds. The phase reference is the primary CPICH.

The UE Re-establishment delay requirement (TUE-E-ESTABLISH-REQ) is defined as the time between the moment when radio link failure is considered by the UE, to when the UE starts to send preambles on the PRACH. TUE-RE-ESTABLISH-REQ is depending on whether the target cell is known by the UE or not. A cell is known if either or both of the following conditions are true: the UE has had radio links connected to the cell in the previous (old) active set, the cell has been measured by the UE during the last 5 seconds. The phase reference is the primary CPICH.

Random Access

The random access procedure is used when establishing the layer 1 communication between the UE and UTRAN. The random access provides a fast access but without disturbing ongoing connections. The purpose of this test is to verify that the behaviour of the random access procedure is according to the requirements. The following is to be tested:

a) Correct behaviour when receiving an ACK
b) Correct behaviour when receiving an NACK
c) Correct behaviour at Time-out
d) Correct behaviour of the PRACH power when reaching maximum UL transmit power

Transport format combination selection in UE

Interactive or Background, PS, UL: 64 kbps When the UE estimates that a certain TFC would require more power than the maximum transmit power, it shall limit the usage of transport format combinations for the assigned transport format set. The purpose is to verify the UE blocks (stops using) a currently used TFC when the UE output power is not sufficient to support that TFC.

E-TFC restriction in UE

Restriction of E-DCH E-TFC in 10ms TTI. When the UE estimates that a certain TFC and E-TFC would require more power than the maximum transmit power, it shall limit the usage of transport format combinations for the assigned transport format set. The purpose is to verify the UE stops using a currently used E-TFC when its remaining power margin is not sufficient to support that E-TFC.

2ms TTI E-DCH E-TFC restriction. When the UE estimates that a certain TFC and E-TFC would require more power than the maximum transmit power, it shall limit the usage of transport format combinations for the assigned transport format set. The purpose is to verify the UE stops using a currently used E-TFC when its remaining power margin is not sufficient to support that E-TFC.

Timing and Signalling Characteristics

UE Transmit Timing. The UE transmit timing is defined as the timing of the uplink DPCCH/DPDCH frame relative to the first detected path (in time) of the corresponding downlink DPCCH/DPDCH frame from the reference cell. The reference point is the antenna connector of the UE. The purpose of this test is to: verify that the UE initial transmit timing accuracy is within the specified limits, verify that the UE transmit timing accuracy remains within the specified limits, after receipt of the ACTIVESET UPDATE message, verify that the maximum amount of timing change in one adjustment, and the minimum and maximum adjustment rate are within the specified, verify that after convergence on the new reference cell the UE is within the specified.

UE Measurements Procedures

FDD intra frequency measurements. In the event triggered reporting period the measurement reporting delay is defined as the time between any event that will trigger a measurement report until the UE starts to transmit over the Uu interface. This requirement assumes that the measurement report is not delayed by other RRC signalling on the DCCH. This measurement reporting delay excludes a delay uncertainty resulted when inserting the measurement report to the TTI of the uplink DCCH. The delay uncertainty is twice the TTI of the uplink DCCH. The following cases are tested:

a. Event triggered reporting in AWGN propagation conditions
b. Event triggered reporting of multiple neighbours in AWGN propagation condition
c. Event triggered reporting of two detectable neighbours in AWGN propagation condition

FDD inter frequency measurements. In the event triggered reporting period the measurement reporting delay is defined as the time between any event that will trigger a measurement report until the UE starts to transmit the measurement report over the Uu interface. This requirement assumes that the measurement report is not delayed by other RRC signalling on the DCCH. This measurement reporting delay excludes a delay uncertainty resulted when inserting the measurement report to the TTI of the uplink DCCH . The delay uncertainty is twice the TTI of the uplink DCCH. The purpose of the following tests is to verify that the UE meets the minimum requirements. The following cases are tested:

a. Correct reporting of neighbours in AWGN propagation condition
b. Correct reporting of neighbours in fading propagation condition
c. Correct reporting of GSM neighbours in AWGN propagation condition
d. Correct reporting of neighbours in AWGN propagation condition. This is to verify that the UE makes correct reporting of an event when doing combined inter frequency and GSM measurements.

Measurements Performance Requirements

CPICH RSCP (received signal code power)

Intra frequency measurements absolute accuracy. The absolute accuracy of CPICH RSCP is defined as the CPICH RSCP measured from one cell compared to the actual CPICH RSCP power from same cell. The purpose of this test is to verify that the CPICH RSCP absolute measurement accuracy is within the specified limits. This measurement is for handover evaluation, DL open loop power control, UL open loop control and for the calculation of path loss.

Intra frequency measurements relative accuracy. The relative accuracy of CPICH RSCP is defined as the CPICH RSCP measured from one cell compared to the CPICH RSCP measured from another cell on the same frequency. The purpose of this test is to verify that the CPICH RSCP relative measurement accuracy is within the specified limits. This measurement is for handover evaluation, DL open loop power control, UL open loop control and for the calculation of path loss.

Inter frequency measurement relative accuracy. The relative accuracy of CPICH RSCP in inter frequency case is defined as the CPICH RSCP measured from one cell compared to the CPICH RSCP measured from another cell on a different frequency. The purpose of this test is to verify that the CPICH RSCP relative measurement accuracy is within the specified limits. This measurement is for handover evaluation, DL open loop power control, UL open loop control and for the calculation of path loss.

CPICH Ec/Io

Intra frequency measurements absolute accuracy. The absolute accuracy of CPICH Ec/Io is defined as the CPICH Ec/Io measured from one cell compared to the actual CPICH_Ec/Io power ratio from same cell. The purpose of this test is to verify that the CPICH Ec/Io absolute measurement accuracy is within the specified limits. This measurement is for Cell selection/re-selection and for handover evaluation. Where Io is the total received power spectral density, including signal and interference, as measured at the UE antenna connector, and Ec is the average energy per PN chip.

Intra frequency measurements relative accuracy. The relative accuracy of CPICH Ec/Io is defined as the CPICH Ec/Io measured from one cell compared to the CPICH Ec/Io measured from another cell on the same frequency. The purpose of this test is to verify that the CPICH Ec/Io relative measurement accuracy is within the specified limits. This measurement is for Cell selection/re-selection and for handover evaluation.

Inter frequency measurement relative accuracy. The relative accuracy of CPICH Ec/Io in the inter frequency case is defined as the CPICH Ec/Io measured from one cell compared to the CPICH Ec/Io measured from another cell on a different frequency. The purpose of this test is to verify that the CPICH Ec/Io relative measurement accuracy is within the specified limits. This measurement is for Cell selection/re-selection and for handover evaluation.

UTRA Carrier RSSI

Absolute measurement accuracy. The absolute accuracy of UTRA Carrier RSSI is defined as the UTRA Carrier RSSI measured from one frequency compared to the actual UTRA Carrier RSSI power of that same frequency. The purpose of this test is to verify that the UTRA Carrier RSSI measurement is within the specified limits. This measurement is for inter-frequency handover evaluation.

Relative measurement accuracy. The relative accuracy requirement is defined as the UTRA Carrier RSSI measured from one frequency compared to the UTRA Carrier RSSI measured from another frequency. The purpose of this test is to verify that the UTRA Carrier RSSI measurement is within the specified limits. This measurement is for inter-frequency handover evaluation.

GSM Carrier RSSI.

The GSM carrier RSSI measurement is used for handover between UTRAN and GSM. The purpose of this test is to verify that the GSM Carrier RSSI measurement accuracy in CELL_DCH state, for UE that needs compressed mode to perform GSM measurements, is within the specified limits. This measurement is for UTRAN to GSM handover evaluation.

UE transmitted power.

The UE transmitted power absolute accuracy is defined as difference between the UE reported value and the UE transmitted power measured by test system. The reference point for the UE transmitted power shall be the antenna connector of the UE. The purpose of this test is to verify that for any reported value of UE Transmitted Power in the range PUEMAX to PUEMAX-10 that the actual UE mean power lies within the specified range.

SFN-CFN observed time difference

Intra frequency measurement. The intra frequency SFN-CFN observed time difference is defined as the SFN-CFN observed time difference from the active cell to a neighbour cell that is in the same frequency. The reference point for the SFN-CFN observed time difference shall be the antenna connector of the UE.

The purpose of this test is to verify that the SFN-CFN observed time difference measurement accuracy is within the specified. This measurement is for handover timing purposes to identify active cell and neighbour cell time difference.

Inter frequency measurement. The inter frequency SFN-CFN observed time difference is defined as the SFN-CFN time difference from the active cell to a neighbour cell that is in a different frequency. The reference point for the SFN-CFN observed time difference shall be the antenna connector of the UE. The purpose of this test is to verify that the SFN-CFN observed time difference measurement accuracy is within the specified limits. This measurement is for handover timing purposes to identify active cell and neighbour cell time difference.

SFN-SFN observed time difference

SFN-SFN observed time difference type 1. The reference point for the SFN-SFN observed time difference type 1 shall be the antenna connector of the UE. The purpose of this test is to verify that the measurement accuracy of SFN-SFN observed time difference type 1 is within the limit specified. This measurement is for identifying time difference between two cells.SFN-SFN observed time difference type 2 without IPDL period active. The reference point for the SFN-SFN observed time difference type 2 shall be the antenna connector of the UE. The purpose of this test is to verify that the SFN-SFN observed time difference type 2 measurement accuracy without IPDL period active is within the specified.

SFN-SFN observed time difference type 2 with IPDL period active. This measurement is specified. The reference point for the SFN-SFN observed time difference type 2 shall be the antenna connector of the UE. The purpose of this test is to verify that the SFN-SFN observed time difference type 2 measurement accuracy without IPDL period active is within the specified limits.

UE Rx-Tx time difference

UE Rx-Tx time difference type 1. The UE Rx-Tx time difference is defined as the time difference between the UE uplink DPCCH/DPDCH frame transmission and the first detected path (in time) of the downlink DPCH frame from the measured radio link. The reference point of the UE Rx-Tx time difference shall be the antenna connector of the UE. The purpose of this test is to verify that the measurement accuracy of Rx-Tx time difference is within the specified limit. This measurement is used for call setup purposes to compensate propagation delay of DL and UL.

UE Rx-Tx time difference type 2. The UE Rx-Tx time difference is defined as the time difference between the UE uplink DPCCH/DPDCH frame transmission and the first detected path (in time) of the downlink DPCH frame from the measured radio link. The reference point of the UE Rx-Tx time difference shall be the antenna connector of the UE. The purpose of this test is to verify that the measurement accuracy of Rx-Tx time difference type 2 is within the specified.

UE Transmission Power Headroom.

The accuracy requirements for the UE transmission power headroom depends on the total power transmitted by the UE. The requirements and this test apply to Release 6 and later releases for all types of UTRA for the FDD UE that support E-DCH and HSDPA. The purpose of this test case is to verify that the UE transmission power headroom measurement report accuracy is within the specified limits.

Performance requirements for HSDPA

Demodulation of HS-DSCH (Fixed Reference Channel)
Single Link Performance

The following tests define the Single Link Performance tests for the different H-Sets for the different HS-DSCH Categories.

Single Link Performance - QPSK/16QAM, Fixed Reference Channel (FRC) H-Set 1/2/3. The receiver single link performance of the High Speed Physical Downlink Shared Channel (HS-PDSCH) in different multi-path fading environments are determined by the information bit throughput R. The requirements and this test apply to Release 5 and later releases for all types of UTRA for the FDD UE that support HSDPA UE capability categories 1 to 6. This test is to verify the ability of the receiver to receive a predefined test signal, in a multi-path fading channel with information bit throughput R not falling below a specified value. The test stresses the multicode reception and channel decoding with incremental redundancy.

Single Link Performance - QPSK, Fixed Reference Channel (FRC) H-Set 4/5. The receiver single link performance of the High Speed Physical Downlink Shared Channel (HS-DSCH) in different multi-path fading environments are determined by the information bit throughput R. The requirements and this test apply to Release 5 and later releases for all types of UTRA for the FDD UE that support HSDPA UE capability categories 11 and 12. This test is to verify the ability of the receiver to receive a predefined test signal, representing a multi-path fading channel with information bit throughput R not falling below a specified value. The test stresses the multicode reception and channel decoding with incremental redundancy.

Single Link Performance - QPSK/16QAM, Fixed Reference Channel (FRC) H-Set 6/3. The receiver single link performance of the High Speed Physical Downlink Shared Channel (HS-DSCH) in different multi-path fading environments are determined by the information bit throughput R. The requirements and this test apply to Release 6 and later releases for all types of UTRA for the FDD UE that support HSDPA UE capability categories 7 and 8. This test is to verify the ability of the receiver to receive a predefined test signal, representing a multi-path fading channel with information bit throughput R not falling below a specified value. The test stresses the multicode reception and channel decoding with incremental redundancy.

Single Link Performance - Enhanced Performance Requirements Type 1 - QPSK/16QAM, Fixed Reference Channel (FRC) H-Set 1/2/3. The receiver single link performance of the High Speed Physical Downlink Shared Channel (HS-DSCH) in different multi-path fading environments are determined by the information bit throughput R. The requirements and this test apply for Release 6 and later releases to all types of UTRA for the FDD UE that support: the HSDPA UE capability categories 1 to 6 and the optional enhanced performance requirements type 1. This test is to verify the ability of the receiver to receive a predefined test signal, representing a multi-path fading channel with information bit throughput R not falling below a specified value. The test stresses the multicode reception and channel decoding with incremental redundancy.

Single Link Performance - Enhanced Performance Requirements Type 1- QPSK/16QAM, Fixed Reference Channel (FRC) H-Set 6/3. The receiver single link performance of the High Speed Physical Downlink Shared Channel (HS-DSCH) in different multi-path fading environments are determined by the information bit throughput R. The requirements and this test apply for Release 6 and later releases to all types of UTRA for the FDD UE that support: HSDPA UE capability categories 7 and 8; and the optional enhanced performance requirements type 1. This test is to verify the ability of the receiver to receive a predefined test signal, representing a multi-path fading channel with information bit throughput R not falling below a specified value. The test stresses the multicode reception and channel decoding with incremental redundancy.

Single Link Performance - Enhanced Performance Requirements Type 2 - QPSK/16QAM, Fixed Reference Channel (FRC) H-Set 6/3. The receiver single link performance of the High Speed Physical Downlink Shared Channel (HS-DSCH) in different multi-path fading environments are determined by the information bit throughput R. The requirements and this test apply for Release 6 and later releases to all types of UTRA for the FDD UE that support: HSDPA UE capability categories 7 and 8 and the optional enhanced performance requirements type 2. This test is to verify the ability of the receiver to receive a predefined test signal, representing a multi-path fading channel with information bit throughput R not falling below a specified value. The test stresses the multicode reception and channel decoding with incremental redundancy.

Single Link Performance - Enhanced Performance Requirements Type 3 - QPSK/16QAM, Fixed Reference Channel (FRC) H-Set 6/3. The receiver single link performance of the High Speed Physical Downlink Shared Channel (HS-DSCH) in different multi-path fading environments are determined by the information bit throughput R. The requirements and this test apply for Release 7 and later releases to all types of UTRA for the FDD UE that support: HSDPA UE capability categories 7 and 8 and the optional enhanced performance requirements type 3. This test is to verify the ability of the receiver to receive a predefined test signal, representing a multi-path fading channel with information bit throughput

R not falling below a specified value. The test stresses the multicode reception and channel decoding with incremental redundancy.

Open Loop Diversity Performance

The following tests define the Open Loop Diversity Performance tests for the different H-Sets for the different HS-DSCH Categories.

Open Loop Diversity Performance - QPSK/16QAM, Fixed Reference Channel (FRC) H-Set 1/2/3. The receiver open loop transmit diversity performance of the High Speed Physical Downlink Shared Channel (HS-DSCH) in multi-path fading environments are determined by the information bit throughput R. The requirements and this test apply to Release 5 and later releases for all types of UTRA for FDD UE that support HSDPA UE capability categories 1 to 6. The requirements and this test apply for Release 6 and later releases to all types of UTRA for the FDD UE that support HSDPA UE capability categories 7 and 8. This test is to verify the ability of the receiver to receive a predefined test signal, representing a multi-path fading channel with information bit throughput R not falling below a specified value. The test stresses the multi-code reception and channel decoding with incremental redundancy.

Open Loop Diversity Performance - QPSK, Fixed Reference Channel (FRC) H-Set 4/5. The receiver open loop transmit diversity performance of the High Speed Physical Downlink Shared Channel (HS-DSCH) in multi-path fading environments are determined by the information bit throughput R. The requirements and this test apply to Release 5 and later releases for all types of UTRA for FDD UE that support HSDPA UE capability categories 11 and 12. This test is to verify the ability of the receiver to receive a predefined test signal, representing a multi-path fading channel with information bit throughput R not falling below a specified value. The test stresses the multi-code reception and channel decoding with incremental redundancy.

Open Loop Diversity Performance - Enhanced Performance Requirements Type 1 - QPSK/16QAM, Fixed Reference Channel (FRC) H-Set 1/2/3. The receiver open loop transmit diversity performance of the High Speed Physical Downlink Shared Channel (HS-DSCH) in multi-path fading environments are determined by the information bit throughput R. The requirements and this test apply for Release 6 and later releases to all types of UTRA for the FDD UE that support HSDPA UE capability categories 1 to 8 and the optional enhanced performance requirements type 1. This test is to verify the ability of the receiver to receive a predefined test signal, representing a multi-path fading channel with information bit throughput R not falling below a specified value. The test stresses the multi-code reception and channel decoding with incremental redundancy.

Open Loop Diversity Performance - Enhanced Performance Requirements Type 2 - QPSK/16QAM, Fixed Reference Channel (FRC) H-Set 3. The receiver open loop transmit diversity performance of the High Speed

Physical Downlink Shared Channel (HS-DSCH) in multi-path fading environments are determined by the information bit throughput R. The requirements and this test apply to Release 6 and later releases for all types of UTRA for FDD UE that support HSDPA UE capability categories 7 and 8 and the optional enhanced performance requirements type 2. This test is to verify the ability of the receiver to receive a predefined test signal, representing a multi-path fading channel with information bit throughput R not falling below a specified value. The test stresses the multi-code reception and channel decoding with incremental redundancy.

Closed Loop Diversity Performance

The following tests define the Closed Loop Diversity Performance tests for the different H-Sets for the different HS-DSCH Categories.

Closed Loop Diversity Performance - QPSK/16QAM, Fixed Reference Channel (FRC) H-Set 1/2/3. The receiver closed loop transmit diversity performance of the High Speed Physical Downlink Shared Channel (HS-DSCH) in multi-path fading environments are determined by the information bit throughput R. The requirements and this test apply to Release 5 and later releases for all types of UTRA for FDD UE that support HSDPA UE capability categories 1 to 6. The requirements and this test apply for Release 6 and later releases to all types of UTRA for the FDD UE that support HSDPA UE capability categories 7 and 8. This test is to verify the ability of the receiver to receive a predefined test signal, representing a multi-path fading channel with information bit throughput R not falling below a specified value. The test stresses the multi-code reception and channel decoding with incremental redundancy.

Closed Loop Diversity Performance - QPSK, Fixed Reference Channel (FRC) H-Set 4/5. The receiver closed loop transmit diversity performance of the High Speed Physical Downlink Shared Channel (HS-DSCH) in multi-path fading environments are determined by the information bit throughput R. The requirements and this test apply to Release 5 and later releases for all types of UTRA for FDD UE that support HSDPA UE capability categories 11 and 12. This test is to verify the ability of the receiver to receive a predefined test signal, representing a multi-path fading channel with information bit throughput R not falling below a specified value. The test stresses the multi-code reception and channel decoding with incremental redundancy.

Closed Loop Diversity Performance Enhanced Performance Requirements Type 1, QPSK/16QAM, Fixed Reference Channel (FRC) H-Set 1/2/3. The receiver closed loop transmit diversity performance of the High Speed Physical Downlink Shared Channel (HS-DSCH) in multi-path fading environments are determined by the information bit throughput R. The requirements and this test apply to Release 6 and later releases for all types of UTRA for FDD UE that support HSDPA UE capability categories 1 to 8 and the optional enhanced performance requirements type 1. This test is to verify the ability of the receiver to receive a predefined test signal, representing a multi-path fading channel with information bit throughput R not falling below a specified value. The test stresses the multi-code reception and channel decoding with incremental redundancy.

Closed Loop Diversity Performance - Enhanced Performance Requirements Type 2 - QPSK/16QAM, Fixed Reference Channel (FRC) H-Set 6/3. The receiver closed loop transmit diversity performance of the High Speed Physical Downlink Shared Channel (HS-DSCH) in multi-path fading environments are determined by the information bit throughput R. The requirements and this test apply for Release 6 and later releases to all types of UTRA for the FDD UE that support HSDPA UE capability categories 7 and 8 and the optional enhanced performance requirements type 2. This test is to verify the ability of the receiver to receive a predefined test signal, representing a multi-path fading channel with information bit throughput R not falling below a specified value. The test stresses the multi-code reception and channel decoding with incremental redundancy.

Reporting of Channel Quality Indicator

Single Link Performance - AWGN Propagation Conditions. The reporting accuracy of channel quality indicator (CQI) under AWGN environments is determined by the reporting variance and the BLER performance using the transport format indicated by the reported CQI median. This test is to verify that the variance of the CQI reports when using TF based on CQI 16 is within the limits defined and that a BLER of 10% falls between the TF based on Median CQI-1 and the TF based on Median CQI TF or between the TF based on Median CQI and the TF based on Median CQI+2.

Single Link Performance - Fading Propagation Conditions. The reporting accuracy of the channel quality indicator (CQI) under fading environments is determined by the BLER performance using the transport format indicated by the reported CQI median. This test is to verify that when using the TF based on the Median CQI that the BLER for blocks associated with CQI reports of Median CQI is less than or equal to 60% and that the BLER for blocks associated with CQI reports of Median CQI+3 is less than or equal to 15%.

Open Loop Diversity Performance - AWGN Propagation Conditions. The reporting accuracy of channel quality indicator (CQI) under AWGN environments is determined by the reporting variance and the BLER performance using the transport format indicated by the reported CQI median. This test is to verify that the variance of the CQI reports when using TF based on CQI 16 is within the limits defined and that a BLER of 10% falls between the TF based on Median CQI-1 and the TF based on Median CQI TF or between the TF based on Median CQI and the TF based on Median CQI+2.

Open Loop Diversity Performance - Fading Propagation Conditions. The reporting accuracy of the channel quality indicator (CQI) under fading environments is determined by the BLER performance using the transport format indicated by the reported CQI median. This test is to verify that when using the TF based on the Median CQI that the BLER for blocks associated with CQI reports of Median CQI is less than or equal to 60% and that the BLER for blocks associated with CQI reports of Median CQI+3 is less than or equal to 15%.

Closed Loop Diversity Performance - AWGN Propagation Conditions. The reporting accuracy of channel quality indicator (CQI) under AWGN environments is determined by the reporting variance and the BLER performance using the transport format indicated by the reported CQI median. This test is to verify that the variance of the CQI reports when using TF based on CQI 16 is within the limits defined and that a BLER of 10% falls between the TF based on Median CQI-1 and the TF based on Median CQI TF or between the TF based on Median CQI and the TF based on Median CQI+2.

Closed Loop Diversity Performance - Fading Propagation Conditions. The reporting accuracy of the channel quality indicator (CQI) under fading environments is determined by the BLER performance using the transport format indicated by the reported CQI median. This test is to verify that when using the TF based on the Median CQI that the BLER for blocks associated with CQI reports of Median CQI is less than or equal to 60% and that the BLER for blocks associated with CQI reports of Median CQI+3 is less than or equal to 15%.

HS-SCCH Detection Performance

Single Link Performance. The detection performance of the HS-SCCH is determined by the probability of event Em, which is declared when the UE is signalled on HS-SCCH-1, but DTX is observed in the corresponding HS-DPCCH ACK/NACK field. The probability of event Em is denoted P(Em). This test is to verify that P(Em) does not exceed the specified limit.

Single Link Performance – Enhanced Performance Requirements Type 1. The detection performance of the HS-SCCH is determined by the probability of event Em, which is declared when the UE is signalled on HS-SCCH-1, but DTX is observed in the corresponding HS-DPCCH ACK/NACK field. The probability of event Em is denoted P(Em). This test is to verify that P(Em) does not exceed the specified limit.

Open Loop Diversity Performance. The detection performance of the HS-SCCH is determined by the probability of event Em, which is declared when the UE is signalled on HS-SCCH-1, but DTX is observed in the corresponding HS-DPCCH ACK/NACK field. The probability of event Em is denoted P(Em). This test is to verify that P(Em) does not exceed the specified limit.

Open Loop Diversity Performance - Enhanced Performance Requirements Type 1. The detection performance of the HS-SCCH is determined by the probability of event Em, which is declared when the UE is signalled on HS-SCCH-1, but DTX is observed in the corresponding HS-DPCCH ACK/NACK field. The probability of event Em is denoted P(Em). This test is to verify that P(Em) does not exceed the specified limit.

Performance requirement (E-DCH)

The MAC header transmission on HS-DSCH for all E-DCH test cases should use a correct MAC-hs header consistent with the actual HSDPA transmission.

Detection of E-DCH HARQ ACK Indicator Channel (E-HICH)

Single link performance. The receive characteristics of the E-DCH HARQ ACK Indicator Channel (E-HICH) in different multi-path fading environments are determined by the missed ACK and false ACK values. The requirements and this test apply to Release 6 and later releases for all types of UTRA for the FDD UE that support E-DCH and HSDPA. This test is to verify that the average probability for missed ACK and false ACK do not exceed the specified values.

Detection in Inter-Cell Handover conditions

RLS not containing the Serving E-DCH cell. The receive characteristics of the E-DCH HARQ ACK Indicator Channel (E-HICH) is determined during an inter-cell soft handover by the missed ACK and false ACK error probabilities. During the soft handover a UE receives signals from different cells. A UE has to be able to detect E-HICH signalling from different cells belonging to different RLS, not containing the Serving E-DCH cell. This test is to verify that during an inter-cell soft handover the average probability for missed ACK and the average probability for false ACK does not exceed specified values.

    RLS containing the Serving E-DCH cell. The receive characteristics of the E-DCH HARQ ACK Indicator Channel (E-HICH) is determined during an inter-cell soft handover by the missed ACK and false ACK error probabilities. During the soft handover a UE receives signals from different cells. A UE has to be able to detect E-HICH signalling from different cells belonging to different RLS, containing the Serving E-DCH cell. This test is to verify that during an inter-cell soft handover for cell 1 the average probability for missed ACK and the average probability for false ACK does not exceed specified values.

Detection of E-DCH Relative Grant Channel (E-RGCH)

Single link performance. The receive characteristics of the E-DCH Relative Grant Channel (E-RGCH) in multi-path fading environment is determined by the missed UP/DOWN and missed HOLD. The requirements and this test apply to Release 6 and later releases for all types of UTRA for the FDD UE that support E-DCH and HSDPA. This test is to verify that average probability for missed up down and average probability for missed hold do not exceed specified values.

    Detection in Inter-Cell Handover conditions. The receive characteristics of the E-DCH Relative Grant Channel (E-RGCH) is determined during an inter-cell soft handover by the missed UP/DOWN and missed HOLD error probabilities. During the soft handover a UE receives signals from different cells. A UE has to be able to detect E-RGCH signalling from different cells, Serving E-DCH cell and Non-serving E-DCH RL. This test is to verify that during an inter-cell soft handover the average probability for missed HOLD and the average probability for missed DOWN do not exceed specified values.

Demodulation of E-DCH Absolute Grant Channel (E-AGCH)

Single link performance. The receive characteristics of the E-DCH Absolute Grant Channel (E-AGCH) in multi-path fading environment is determined by the missed detection probability. The requirements and this test apply to Release 6 and later releases for all types of UTRA for the FDD UE that support HSDPA and E-DCH.

NODE-B RF TESTING

Node-B output power

Output power of the Node-B is the mean power of one carrier delivered to a load with resistance equal to the nominal load impedance of the transmitter. Rated output power of the Node-B is the mean power level per carrier that the manufacturer has declared to be available at the antenna connector. The following tests are to make sure that the power is as specified.

Node-B maximum output power. Maximum output power of the Node-B is the mean power level per carrier measured at the antenna connector in specified reference condition. The test purpose is to verify the accuracy of the maximum output power across the frequency range and under normal and extreme conditions for all transmitters in the NODE-B.

The CPICH power accuracy. Primary CPICH power is the code domain power of the Common Pilot Channel. Primary CPICH power is indicated on the BCH. CPICH power accuracy is defined as the maximum deviation between the Primary CPICH code domain power indicated on the BCH and the Primary CPICH code domain power measured at the TX antenna interface. The requirement is applicable for all NODE-B types. The purpose of the test is to verify, that the NODE-B under test delivers Primary CPICH code domain power within margins, thereby allowing reliable cell planning and operation.

Frequency error

Frequency error is the measure of the difference between the actual NODE-B transmits frequency and the assigned frequency. The same source should be used for RF frequency and data clock generation. It is not possible to verify by testing that the data clock is derived from the same frequency source as used for RF generation. This may be confirmed by a manufacturer's declaration. This test is to verify that the Frequency Error is within the limit of the minimum requirement.

Output power dynamics

Power control is used to limit the interference level. The NODE-B transmitter uses a quality-based power control on the downlink.

Inner loop power control. Inner loop power control in the downlink is the ability of the NODE-B transmitter to adjust the code domain power of a code channel in accordance with the corresponding TPC symbols received in the uplink.

Power control steps. The power control step is the required step change in the code domain power of a code channel in response to the corresponding power control command. The combined output power change is the required total change in the DL transmitter output power of a code channel in response to multiple consecutive power control commands corresponding to that code channel. Inner loop power control in the downlink is the ability of the NODE-B transmitter to adjust the transmitter output power of a code channel in accordance with the corresponding TPC symbols received in the uplink. The power control step is the required step change in the DL transmitter output power of a code channel in response to the corresponding power control command. The combined output power change is the required total change in the DL transmitter output power of a code channel in response to multiple consecutive power control commands corresponding to that code channel. This test is to verify those requirements for the power control step size and response are met as specified.

Power control dynamic range. The power control dynamic range is the difference between the maximum and the minimum code domain power of a code channel for a specified reference condition. Transmit modulation quality should be maintained within the whole dynamic range as specified. The test is to verify that the minimum power control dynamic range is met as specified by the minimum requirement.

Total power dynamic range. The total power dynamic range is the difference between the maximum and the minimum output power for a specified reference condition. The test is to verify that the total power dynamic range is met as specified by the minimum requirement. The test is to ensure that the total output power can be reduced while still transmitting a single code. This is to ensure that the interference to neighbouring cells is reduced.

IPDL time mask. To support IPDL location method, the Node B should interrupt all transmitted signals in the downlink (i.e. common and dedicated channels). The IPDL time mask specifies the limits of the NODE-B output power during these idle periods. The test purpose is to verify the ability of the NODE-B to temporarily reduce its output power below a specified value to improve time difference measurements made by UE for location services.

Output RF spectrum emissions

The occupied bandwidth is the width of a frequency band such that, below the lower and above the upper frequency limits, the mean powers emitted are each equal to a specified percentage Beta/2 of the total mean transmitted power.

Out of band emissions are unwanted emissions immediately outside the channel bandwidth resulting from the modulation process and non-linearity in the transmitter but excluding spurious emissions. This out of band emission limit is specified in terms of a spectrum emission mask and adjacent channel leakage power ratio for the transmitter.

Spectrum emission mask. The mask defined in the 3GPP specifications may be mandatory in certain regions. In other regions this mask may not be applied. This test measures the emissions of the NODE-B, close to the assigned channel bandwidth of the wanted signal, while the transmitter is in operation.

Adjacent Channel Leakage power Ratio (ACLR). Adjacent Channel Leakage power Ratio (ACLR) is the ratio of the RRC filtered mean power centred on the assigned channel frequency to the RRC filtered mean power centred on an adjacent channel frequency. This test is to verify that the adjacent channel leakage power ratio requirement should be met as specified by the minimum requirement.

Spurious emissions. Spurious emissions are emissions which are caused by unwanted transmitter effects such as harmonics emission, parasitic emission, intermodulation products and frequency conversion products, but exclude out of band emissions. This is measured at the Node-B RF output port. The test apply at frequencies within the specified frequency ranges, which are more than 12.5 MHz under the first carrier frequency used or more than 12.5 MHz above the last carrier frequency used. The requirements should apply whatever the type of transmitter considered (single carrier or multi-carrier). It applies for all transmission modes foreseen by the manufacturer's specification. Unless otherwise stated, all requirements are measured as mean power (RMS). This test measures conducted spurious emission from the NODE-B transmitter antenna connector, while the transmitter is in operation.

Transmit intermodulation

The transmit intermodulation performance is a measure of the capability of the transmitter to inhibit the generation of signals in its non linear elements caused by presence of the wanted signal and an interfering signal reaching the transmitter via the antenna. The transmit intermodulation level is the power of the intermodulation products when a WCDMA modulated interference signal is injected into an antenna connector at a mean power level of 30dB lower than that of the mean power of the wanted signal. The frequency of the interference signal are +5MHz, −5MHz, +10MHz, −10MHz, +15MHz and −15MHz offset from the subject signal carrier frequency, but exclude interference frequencies that are outside of the allocated frequency band for UTRA-FDD. The test purpose is to verify the ability of the NODE-B transmitter to restrict the generation of intermodulation products in its non linear elements caused by presence of the wanted signal and an interfering signal reaching the transmitter via the antenna to below specified levels.

Transmit modulation

Error Vector Magnitude. The Error Vector Magnitude is a measure of the difference between the reference waveform and the measured waveform. This difference is called the error vector. Both waveforms pass through a matched Root Raised Cosine filter with bandwidth 3.84 MHz and roll-off 0.22. Both waveforms are then further modified by selecting the frequency, absolute phase, absolute amplitude and chip clock timing so as to minimise the error vector. The EVM result is defined as the square root of the ratio of the mean error vector power to the mean reference power expressed as a percentage. The measurement interval is one timeslot as defined by the C-PICH (when present) otherwise the measurement interval is one timeslot starting with the beginning of the SCH. The test is to verify that the Error Vector Magnitude is within the limit specified by the minimum requirement.

Peak Code Domain Error. The Peak Code Domain Error is computed by projecting the error vector onto the code domain at a specific spreading factor. The Code Domain Error for every code in the domain is defined as the ratio of the mean power of the projection onto that code, to the mean power of the composite reference waveform. This ratio is expressed in dB. The Peak Code Domain Error is defined as the maximum value for the Code Domain Error for all codes. The measurement interval is one timeslot as defined by the C-PICH (when present), otherwise the measurement interval is one timeslot starting with the beginning of the SCH. It is the purpose of this test to discover and limit inter-code cross-talk.

Time alignment error in Tx Diversity. In Tx Diversity, signals are transmitted from two antennas. These signals should be aligned. The time alignment error in Tx Diversity is specified as the delay between the signals from the two diversity antennas at the antenna ports. This test is only applicable for Node B supporting TX diversity transmission. The test is to ensure that the timing alignment error in TX diversity is within a specified limit.

Receiver Reference sensitivity level

The reference sensitivity level is the minimum mean power received at the antenna connector at which the BER should not exceed the specific value indicated by the minimum requirement. The test is set up according to Figure 3.1 and performed without interfering signal power applied to the NODE-B antenna connector. The test is to verify that at the NODE-B Reference sensitivity level the BER should not exceed the specified limit.

Receiver Dynamic range

Receiver dynamic range is the receiver ability to handle a rise of interference in the reception frequency channel. The receiver should fulfil a specified BER requirement for a specified sensitivity degradation of the wanted signal in the presence of an interfering AWGN signal in the same reception frequency channel. The test purpose is to verify the ability of the NODE-B to receive a single-code test signal of maximum with a BER not exceeding a specified limit.

Adjacent Channel Selectivity (ACS) of the Receiver

Adjacent channel selectivity (ACS) is a measure of the receiver ability to receive a wanted signal at assigned channel frequency in the presence of an adjacent channel signal at a given frequency offset from the centre frequency of the assigned channel. ACS is the ratio of the receiver filter attenuation on the assigned channel frequency to the receive filter attenuation on the adjacent channel(s). The interference signal is offset from the wanted signal by the frequency offset. The interference signal should be a W-CDMA signal. The test purpose is to verify the ability of the NODE-B receiver filter to suppress interfering signals in the channels adjacent to the wanted channel.

Receiver Blocking characteristics

The blocking characteristics are a measure of the receiver ability to receive a wanted signal at its assigned channel frequency in the presence of an unwanted interferer on frequencies other than those of the adjacent channels. The test stresses the ability of the NODE-B receiver to withstand high-level interference from unwanted signals at frequency offsets of 10 MHz or more, without undue degradation of its sensitivity.

Receiver Intermodulation Characteristics

Third and higher order mixing of the two interfering RF signals can produce an interfering signal in the band of the desired channel. Intermodulation response rejection is a measure of the capability of the receiver to receiver a wanted signal on its assigned channel frequency in the presence of two or more interfering signals which have a specific frequency relationship to the wanted signal. The test is to verify the ability of the NODE-B receiver to inhibit the generation of intermodulation products in its non-linear elements caused by the presence of two high-level interfering signals at frequencies with a specific relationship to the frequency of the wanted signal.

Receiver Spurious Emissions

The spurious emission power is the power of the emissions generated or amplified in a receiver that appears at the NODE-B antenna connector. The requirements apply to all NODE-B with separate receiver and transmitter antenna port. The test should be performed when both transmitter and receiver are on with the transmitter port terminated. The test purpose is to verify the ability of the NODE-B to limit the interference caused by receiver spurious emissions to other systems.

Receiver Verification of the internal BER calculation

Node-B System with internal BER calculation can synchronise it's receiver to known pseudo-random data sequence and calculates bit error ratio from the received data. This test is performed only if Node-B System has this kind of feature. This test is performed by feeding measurement signal with known BER to the input of the receiver. Locations of the erroneous bits should be randomly distributed within a frame. The test is to verify that the internal BER calculation accuracy should meet requirements for conformance testing.

Performance requirement
Demodulation in static propagation conditions
Demodulation of DCH. The performance requirement of DCH in static propagation conditions is determined by the maximum Block Error Ratio (BLER) allowed when the receiver input signal is at a specified Eb/N0 limit. The BLER is calculated for each of the measurement channels supported by the Node-B. The test is to verify the receiver's ability to receive the test signal under static propagation conditions with a BLER not exceeding a specified limit.

Demodulation of DCH in multipath fading conditions
Multipath fading Case 1. The performance requirement of DCH in multipath fading Case 1 is determined by the maximum Block Error Ratio (BLER) allowed when the receiver input signal is at a specified Eb/N0 limit. The BLER is calculated for each of the measurement channels supported by the Node-B. The test is to verify the receiver's ability to receive the test signal under slow multipath fading propagation conditions with a BLER not exceeding a specified limit.

    Multipath fading Case 2. The performance requirement of DCH in multipath fading Case 2 is determined by the maximum Block Error Rate (BLER) allowed when the receiver input signal is at a specified Eb/N0 limit. The BLER is calculated for each of the measurement channels supported by the Node-B. The test is to verify the receiver's ability to receive the test signal that has large time dispersion with a BLER not exceeding a specified limit.

    Multipath fading Case 3. The performance requirement of DCH in multipath fading Case 3 is determined by the maximum Block Error Ratio (BLER) allowed when the receiver input signal is at a specified Eb/N0 limit. The BLER is calculated for each of the measurement channels supported by the Node-B. The test is to verify the receivers ability to receive the test signal under fast fading propagation conditions with a BLER not exceeding a specified limit.

    Multipath fading Case 4. The performance requirement of DCH in multipath fading Case 4 for Wide Area NODE-B is determined by the maximum Block Error Ratio (BLER) allowed when the receiver input signal is at a specified Eb/N0 limit. The BLER is calculated for each of the measurement channels supported by the Node-B. The test is to verify the receivers ability to receive the test signal under fast fading propagation conditions with a BLER not exceeding a specified limit.

Demodulation of DCH in moving propagation conditions

The performance requirement of DCH in moving propagation conditions is determined by the maximum Block Error Ratio (BLER) allowed when the receiver input signal is at a specified Eb/N0 limit. The BLER is calculated for each of the measurement channels supported by the Node-B. The test is to verify the receiver's ability to receive and track the test signal with a BLER not exceeding the specified limit.

Demodulation of DCH in birth/death propagation conditions

The performance requirement of DCH in birth/death propagation conditions is determined by the maximum Block Error Ratio (BLER) allowed when the receiver input signal is at a specified Eb/N0 limit. The BLER is calculated for each of the measurement channels supported by the Node-B. The test is to verify the receiver's ability to receive the test signal to find new multi path components with a BLER not exceeding the specified limit.

Verification of the internal BLER calculation

Node-B System with internal BLER calculates block error rate from the received blocks. This test is performed only if Node-B System has this kind of feature. This test is performed by feeding measurement signal with known BLER to the input of the receiver. Locations of the erroneous blocks should be randomly distributed within a frame. Erroneous blocks should be inserted into the UL signal. The test is to verify that the internal BLER calculation accuracy should met requirements for conformance testing.

RACH preamble detection in static propagation conditions. The performance requirement of RACH for preamble detection in static propagation conditions is determined by the two parameters probability of false detection of the preamble (Pfa) and the probability of detection of preamble (Pd). The performance is measured by the required Ec/N0 at probability of detection, Pd of 0.99 and 0.999. Pfa is defined as a conditional probability of erroneous detection of the preamble when input is only noise (+interference). Pd is defined as conditional probability of detection of the preamble when the signal is present. Pfa should be 10-3 or less. Only one signature is used and it is known by the receiver. The test is to verify the receiver's ability to detect RACH preambles under static propagation conditions.

RACH preamble detection in multipath fading case 3. The performance requirement of RACH for preamble detection in multipath fading case 3 is determined by the two parameters probability of false detection of the preamble (Pfa) and the probability of detection of preamble (Pd). The performance is measured by the required Ec/N0 at probability of detection, Pd of 0.99 and 0.999. Pfa is defined as a conditional probability of erroneous detection of the preamble when input is only noise (+interference). Pd is defined as conditional probability of detection of the preamble when the signal is present. The Pfa should be 10^{-3} or less. Only one signature is used and it is known by the receiver. The test is to verify the receiver's ability to detect RACH preambles under multipath fading case 3 propagation conditions.

Demodulation of RACH message in static propagation conditions. The performance requirement of RACH in static propagation conditions is determined by the maximum Block Error Ratio (BLER) allowed when the receiver input signal is at a specified Eb/N0 limit. The BLER is calculated for each of the measurement channels supported by the Node-B. The test is to verify the receiver's ability to receive the test signal under static propagation conditions with a BLER not exceeding a specified limit.

Demodulation of RACH message in multipath fading case 3. The performance requirement of RACH in multipath fading case 3 is determined by the maximum Block Error Ratio (BLER) allowed when the receiver input signal is at a specified Eb/N0 limit. The BLER is calculated for each of the measurement channels supported by the Node-B. The test is to verify the receiver's ability to receive the test signal under multipath fading case 3 propagation conditions with a BLER not exceeding a specified limit.

Performance of signalling detection for HS-DPCCH

The performance requirement of HS-DPCCH signalling detection is determined by the two parameters: the probability of false detection of ACK; P(DTX->ACK) and the probability of mis-detection of ACK; P(ACK->DTX or NACK).

ACK false alarm in static propagation conditions. ACK false alarm is defined as a conditional probability of erroneous detection of ACK when input is only DPCCH and DPDCH (+interference). The performance requirement of ACK false alarm in static propagation conditions is determined by the maximum error ratio allowed when the receiver input signal is at a specified Ec/N0 limit. ACK false alarm: P(DTX->ACK) should be 10-2 or less. The test is to verify the receiver's ability to detect HS-DPCCH signalling (ACK/NACK) under static propagation conditions.

ACK false alarm in multipath fading conditions. ACK false alarm is defined as a conditional probability of erroneous detection of ACK when input is only DPCCH and DPDCH (+interference). The performance requirement of ACK false alarm in multipath fading conditions is determined by the maximum error ratio allowed when the receiver input signal is at a specified Ec/N0 limit. ACK false alarm: P(DTX->ACK) should be 10-2 or less. The test is to verify the receiver's ability to detect HS-DPCCH signalling (ACK/NACK) under multipath fading case 3 propagation conditions.

ACK mis-detection in static propagation conditions. The probability of ACK mis-detection is defined a probability of ACK mis-detected when ACK is transmitted. The performance requirement of ACK mis-detection in static propagation conditions is determined by the maximum error ratio allowed when the receiver input signal is at a specified Ec/N0 limit. The test is to verify the receiver's ability to receive the test signal under static propagation conditions with an error ratio not exceeding a specified limit.

ACK mis-detection in multipath fading conditions. The probability of ACK mis-detection is defined a probability of ACK mis-detected when ACK is transmitted. The performance requirement of ACK mis-detection in multipath fading conditions is determined by the maximum error ratio allowed when the receiver input signal is at a specified Ec/N0 limit. The test is to verify the receiver's ability to receive the test signal under multipath fading propagation conditions with an error ratio not exceeding a specified limit.

Demodulation of E-DPDCH in multipath fading conditions

The performance requirement of the E-DPDCH in multi path fading condition is determined by the minimum throughput. The test is to verify the receiver's ability to receive the test signal under slow multipath fading propagation conditions with a throughput not below a specified limit.

Performance of signalling detection for E-DPCCH in multipath fading conditions

The performance requirement of E-DPCCH signalling detection is determined by the two parameters: the probability of false detection of codeword; P(DTX -> codeword) and the probability of missed detection of codeword; P(codeword -> DTX).

E-DPCCH false alarm in multipath fading conditions. E-DPCCH false alarm is defined as a conditional probability of detection of codeword when input is only DPCCH (+interference). The E-DPDCH and E-DPCCH is turned off. The performance requirement of E-DPCCH false alarm in multipath fading conditions is determined by the maximum detection probability allowed when the receiver input signal is at a specified Ec/N0 limit. The test is to verify the receiver's ability to detect E-DPCCH signalling under multipath fading propagation conditions.

E-DPCCH missed detection in multipath fading conditions. The probability of E-DPCCH missed detection is defined a probability of E-DPCCH missed detected when E-DPCCH is transmitted. The performance requirement of E-DPCCH missed detection in multipath fading conditions is determined by the maximum missed detection probability allowed when the receiver input signal is at a specified Ec/N0 limit. The test is to verify the receiver's ability to receive the test signal under multipath fading propagation conditions with a missed detection probability not exceeding a specified limit.

REFERENCES

[1] [1] 3GPP TS 34.121-1 V7.3.0, Technical Specification, 3rd Generation Partnership Project; Technical Specification Group Radio Access Network; User Equipment (UE) conformance specification; Radio Transmission and Reception (FDD); Part 1: **Conformance specification**

[2] 3GPP TS 34.121-2 V7.3.0, Technical Specification, 3rd Generation Partnership Project; Technical Specification Group Radio Access Network; User Equipment (UE) conformance specification; Radio Transmission and Reception (FDD); Part 2: **Implementation Conformance Statement (ICS)**

[3] 3GPP TS 25.141 V7.6.0 , Technical Specification, 3rd Generation Partnership Project; Technical Specification Group Radio Access Network; **Base Station (BS) conformance testing (FDD)**

Chapter Four

Testing Types and Stages

TESTING TYPES OF MOBILES AND BASE STATIONS

Testing has become one of the most important parts of mobile and base station development process. It is necessary to test software and systems in order to develop and deliver a good, reliable and bug/fault free mobile and infrastructure equipments.

Testing can be defined as: Testing is an activity that helps in finding out any non-conformities bugs/defects/errors in a system under development, in order to provide a conformant bug/fault free and reliable system/solution to the customer. The reason for this is that if you fail to deliver a reliable, good and problem free equipment, you fail in your project and probably you may loose your client. So in order to make sure, that you provide your client a proper software/hardware solution, you go for TESTING. You check out if there is any problem, any error in the system, which can make software unusable by the client. You make software testers test the system and help in finding out the bugs in the system to fix them on time. You find out the problems and fix them and again try to find out all the potential problems.

But you must know the fact that, if you make something, you hardly feel that there can be something wrong with what you have developed. It's a common trait of human nature, we feel that there is no problem in our designed system as we have developed it and it is perfectly functional and fully working. So the hidden bugs or errors or problems of the system remain hidden and they raise their head when the system goes into production or operation.

On the other hand, its a fact that, when one person starts checking something which is made by some other person, there are 99% chances that checker/observer will find some problem with the system (even if the problem is with some spelling that by mistake has been written in wrong way.)

Even though it is wrong in terms of human behaviour, this thing has been used for the benefit of software projects (or you may say, any type of project). When you develop something, you give it to get checked (TESTED) and to find out any

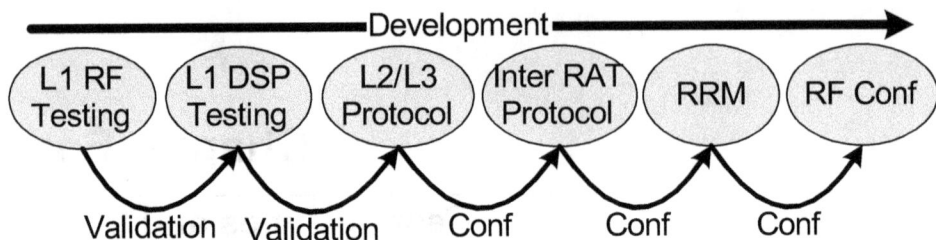

Figure 4.1   Testing stages

problem, which never aroused while development of the system. Because, after all, if you could minimise the problems with the system you developed, it's beneficial for yourself. Your client will be happy if your system works without any problem and will generate more revenues for you.

A tester is a person, software or hardware who tries to find out all possible errors/bugs in the system with the help of various inputs to it. A tester plays an important part in finding out the problems with system and helps in improving its quality. If you could find all the bugs and fix them all, your system becomes more and more reliable. A tester has to understand the limits, which can make the system break and work abruptly.

There are a number of testing stages the system (whether UE or BS) need to go through throughout the lifecycle of the product as illustrated in Figure 4.1 and explained in the following sections. Development testing is essential throughout the lifecycle of the product. Initial system validation of Layer 1 implementation is required at early stages of system development. Conformance testing has to cover L1/L2 protocol, Inter System Handover protocol, RRM protocol and RF performance and implementation of the system.

The following sections state all familiar types of testing that a product goes through throughout its life. Terminology may differ but the intended purposes of these test still widely used. Some of the tests may overlap in some cases for some products. This section provides good coverage of all tests at all stages.

Pre-acceptance testing

Pre-acceptance tests are intended to provide system owners, project managers, and other information system development and maintenance professionals with guidance in preparation for the acceptance activities. Initiation of acceptance activities begins with successful completion of the system testing. Information for systems acceptance preparation is generally characterised as a process to ensure everything is in place to begin the acceptance process activities. Pre-acceptance test requires a Test Plan, which should be approved by the system owner, user, and other project stakeholders prior to conducting any acceptance tests. Testing is conducted to verify the design features were incorporated into the system and the results were recorded.

Conformance testing

Conformance testing in general terms allows end users to practice preventative maintenance without the expenses normally associated with preventative maintenance programs. In addition, conformance testing eliminates ownership issues that result when an open system experiences technical difficulties. Conformance testing also considered as an independent user's guide to help end users differentiate the vast number of products and vendors that have inundated the marketplace.

Conformance testing procedure isn't flawless in its current state, passing the conformance test does not "guarantee" interoperability among products as far as percentages are concerned. However, this is where specifying conformance tested products will once again pay dividends, delivering secondary, even tertiary benefits for the end user. As companies demand conformant products and testing becomes widely practiced, each successive version of the test will evolve based upon feedback. The accuracy and quality of the test will increase exponentially as a result, and end users will see better, technically advanced products. And as the testing process becomes streamlined, improved products will be available in the market sooner. Moreover, feedback from the testing process could be applied to the UMTS technology overall, improving the foundation as well as the products. The ensuing benefits are endless.

Conformance testing is paramount to the end user and the reason can be distilled to the fact that testing eliminates up to 80% of field errors as statistics and experience suggest. Also end users benefit without making a financial investment and testing will improve the technology available.

Production testing

Once a UE or BS is in production, testing needs to be performed on it to ensure that is leaving the production line performs correctly. The testing undertaken at this point in the process is aimed at checking that the unit is built correctly. As this testing adds no value, but only ensures the unit is operating correctly. At this stage it is necessary to ensure that no development tests are undertaken; only those that check the build of the equipment.

The type of test equipment used for production testing of UMTS phones and base stations is quite different to that needed for, say, conformance testing. Tests need to be performed very quickly, and therefore the test equipment is honed to ensure that it can undertake a limited number of tests very quickly.

Production tests do not repeat or replace type approval testing, which includes many software tests. That is because software does not vary among samples of a production series.

Requirements of a mobile production testing are fast production test by providing quick test results, production throughput with fast and accurate measurements and compliant with the relevant standards. Parameter capture during test is essential using triggering techniques to switch between technologies, e.g. GSM, WCDMA and HSPA parameters from a single mobile-under-test.

Complete built in call processing capability is required for mobile production test. Test control calls using different data rates, and fast reference measurement channels, Call control calling with voice quality assessment (VQA) capabilities is a great advantage for quick functional verification. Therefore production testers are required to perform fast terminal testing and scalable. Production testing could be performed using specific test equipment or a collection of conventional generic test equipments, e.g. Power Meters, Spectrum Analysers, etc.

Production must be optimised through increasing work flow and then design tests to ensure that the manufactured product performs to specification, that hardware components work correctly, and are properly connected and aligned.

Because time is money, particularly in a production line, production tests should identify those components and circuits that are most critical to performance. Tests with the least amount of overlap to ensure that the mobile conforms to quality standards when it leaves the factory are required. In addition, over the lifetime of the mobile, tests that unearthed no failures in isolation are eliminated to maintain or improve throughput.

A typical functional test involves call setup; this can be a decisive test involving virtually all the components of the mobile, including the antenna, power amplifier, receiver, oscillators, mixers and audio parts. Such a functional test, usually applied at the final stage before packaging, typically measures power accuracy, frequency error and receiver sensitivity. If an RF connector is available, these functional tests can be performed through that connector, offering the advantage of higher accuracy and repeatability. A test connector alone, however, is not enough. Omitting an antenna test can be risky because that would overlook an antenna that was not properly soldered.

The most common alignment procedure in the mobile is applied at the power amplifier, which often has a nonlinear gain. The mobile must transmit at defined power-level steps with low tolerances and high accuracy, so a communication test set measures the output power of the mobile at all the power steps. A controlling computer compares each measured value with the nominal output power, calculates a correction value and sends it to the mobile, which in turn stores the correction values. The process is repeated for several frequencies in each band the mobile supports.

The production tests start after components are installed, the pc board is connected to a power supply and the first software is downloaded. First, a comprehensive test is performed to ensure the components and connections are working correctly. This includes measurements of current consumption under different conditions to identify short circuits. In the receiver, the received-signal strength indicator is an important circuit to ensure the correct gain. Finally, circuits designed for electronic correction are aligned. If a test fails, the board leaves the production line to enter the repair loop.

Depending on the target test, tests and measurements can be taken either in call mode or in non-call, or test mode. An overall functional test is performed in call mode. To save time, the other tests may be done in test mode. Alternatively, some test equipment offers a stripped-down call mode, where the tester and mobile use a traffic channel without call setup procedures. The test equipment required in manufacturing includes a high-speed, high-accuracy communication test set that simulates a base-station. This test set should include most functions for RF and audio testing. In addition, there is a trend toward small, versatile test systems comprising just a few instruments, sometimes including a power supply and ammeter.

Functional Testing

To increase the accuracy of development test activities, and to keep the development projects time and costs under control, you need to adopt industry standard functional testing. Availability of protocols, interfaces and complex test scenarios including simulation and emulation of the key network components gives UE and BS vendors a competitive advantage and allows them to bring their products to market earlier and with higher reliability. UTRAN simulation and emulation, in depth decoding of protocols, interface protocols are the applications that are needed to meet the challenges of UTRAN functional testing.

Functional testing in general terms is geared to verify and validate if the software application or product under test conforms to all functional specifications. Functional testing are specifically tailored and designed to suit users needs and are used at any stage of the software development life cycle.

Functional testing can use both white box and black box testing techniques and ensure that there are no bugs after testing. Functional testing includes the following:

Unit testing

Unit testing and code walkthrough are aimed at extracting the obvious coding errors. Basic as these errors may appear, but if missed, could have a negative impact on the quality of the application. Even though compilers, visual debuggers and other such sophisticated widely used tools are used, they have not been as effective as a good review. In particular they often do not cover the subtle logical conflicts. Experienced testers will look for problems with loop termination, simple internal parameter passing, proper assignment statements, simple recovery routine logic and errors in functions and subroutines.

Functionality testing (White and Black box testing)

Functionality testing is intended to ensure that applications function according to its design and functional specifications. It is intended to increase confidence level in the application or product.

Functionality testing are specifically tailored and designed to suit manual testing needs as well as automatic. Checks for any defects or bugs in the application under test are carried out through any or all stages of development. Functionality testing uses both white box and black box testing techniques to ensure that there are no bugs after testing.

Functionality test engineers focus on validating the feature of an entire function or a component of the application. Test plans are prepared to describe areas and the ways of testing.

Regression testing

Regression testing is a time consuming and tedious exercise that involves testing after making a functional addition or improvement or repair to the system. Regression tests are similar in scope to functional tests and aim for consistent, repeatable validation of each new release of an application.

Regression testing ensures that the reported product defects are corrected for each new release and that no new quality problems were introduced in the maintenance process. Regression testing could be manual or automated. Regression testing tends to be automatic especially for conformance test cases as there are hundreds of them and you have to make sure that no single test case is broken due to addition of new feature or functionality to your product.

The following approaches are used for regression testing:

- Rerunning the test cases that have a higher risk of failure or test cases whose results have a greater impact in terms of business, or representative (e.g. first in class) test cases
- Running the entire test suite
- Exploratory or ad-hoc testing
- Automation is widely used to reduce the test cycle time

I&C, maintenance, live and field-trial testing

Comprehensive on-site testing of base stations or mobiles using real-world scenarios, both at initial installation and then during ongoing maintenance, plays a vital role in preventing and solving performance problems before they impact subscribers. Thorough independent testing gives operators greater confidence in the quality of the deployed network. Poorly performing base stations or mobiles have a significant impact on the quality of service experienced by users of the higher data rate services available on 3G networks. Whether this is caused by incorrect installation, a gradual degradation in performance or complete failure of a particular module, the end result is that users will suffer poor performance and will be less satisfied with their service provider.

Test equipments generally are selected on the basis of test coverage, differentiator features versus price. However, today, operators additionally need to consider field technician productivity and skill level – favourable changes in either of these can have a significant impact on the cost of maintaining the network and more than offset increased spending on test equipment. This means that test coverage, test speed and ease of use are absolutely critical and compromising on any of these is likely to be a false economy. Inadequate test coverage can also mean that routine maintenance misses potential problems resulting in repeated visits and additional costs from return visits to the same site.

For I&C testing, there are a number of other requirements that are not covered by the 3GPP conformance tests. For example, each operator will want the Node B to be configured in a specific manner. To ensure that an incorrect configuration (perhaps because two modules are swapped over) does not cause network problems, it is useful to conduct functional tests before the Node B is connected to the network. Also, VSWR or distance-to-fault tests should be carried out on antenna feeders to make sure that they are also working correctly.

The value of live or Field Trials or live testing is that it is real code in real environment. But the problem is it could be very late in the development cycle, also is often difficult or too expensive to create a real test environment of any significant size. Real environment tests also tend not to be reproducible, making it difficult to analyse problems when found. Field Trial tests are required to ensure confidence in the performance of the equipment in the operational network environment. Field Trials have proven to be a valuable test tool, which exercise terminals under live conditions. Field Trials are intended do address the mobile behaviour in a dynamic environment, which cannot be achieved by simulator tests under laboratory conditions. Additionally experience has shown that comparable results are achieved in multiple network(s) infrastructures.

Last, but not least, plug fests do give an insight into customer experience/satisfaction which has been, and will be, the main driver for performing Field Trials. This also implies that certain equipment/configuration requirements can be applicable to perform field trials. In order to support the development of mobiles to the maximum extent possible, some organisations has put considerable effort in applying the experience and knowledge of the operator community in to a set of Field Trials tests.

Interoperability testing and plug-fests

A plug-fest is a live session with technical solution providers to test interoperability of data in real life between several systems. It is another variant of live test, which provide the following benefits to the equipment vendors:

- Test products extensively for interoperability with other vendors' products. This has traditionally been a great way to understand interoperability scenarios and flush out any interoperability-related bugs.
- Talks and informative sessions organised by the host organisation about topics that affect the cellular community.
- Opportunities to meet with system developers and get answers to questions about any technical issues.

The Plug-Fest provides an environment for mobile and base station developers to assess the interoperability of their implementations. Interoperability among equipments is essential to the development of complex systems analysis. Such work often involves the collaboration of many disparate organisations. This work environment usually does not permit organisations to standardise on single equipment, and often not even on the same standard. The Plug-Fest in some cases provides an environment for evaluating the interoperability of tools and equipments within the context of a single exchange standard, and across standards.

Application and end-to-end testing

The number and type of mobile applications envisioned for enterprise and mass market users grows daily. With each of these types of applications, there are a number of factors that must be considered when testing for conformance to functional, user and performance requirements. Testing should pay attention to the basics, which can include:

- Verification of baseline functionality and features;
- Checking the design and proof-of-concept solutions against user requirements early in the development cycle;
- Testing under tightly controlled conditions to validate executable code against design during later stages of the development lifecycle;
- Compatibility testing all known, planned variations in the software and hardware configurations within which the application will run;
- Exposing the entire system or application to unexpected events, faults in dependent databases, networks or applications, or unpredictable user behavior;
- Subjecting the software to volume, load and stress conditions to gauge performance at the boundaries of its designed capacity and measure actual limitations of that performance; and
- Determining if the application or system not only meets formal design requirements, but also whether it will be usable and meet the (perhaps undocumented) needs of its users.

Not all of these factors are within the scope of all testing engagements, but a good test plan will at least address them so as to make all assumptions about test scope visible.

When it comes to mobile enabled applications, however, there are factors that require the testing strategy to be modified significantly in order to ensure that hardware and software elements of the solution under test are suitable for its target environment and users.

A strategic approach to testing cellular solutions takes into account a number of characteristics unique to the mobile paradigm:

- The increased complexity of emerging UEs;
- Increased Complexity in UEs;
- The greater sensitivity to security and load related problems in mobile infrastructure.

More attention is given to usability and form factors than with traditional desktop applications. Smaller screens, slower processors, lack of persistent data store and lower bandwidth datalinks all require different testing methods. Developers use special methods to compensate for these factors and develop strategies to test those methods as well. Testing multiple layers of the mobile application's software model is essential. The diversity of device hardware platforms, operating systems, micro-browsers and applications middleware require test experts to be aware of the compatibility issues impacting functionality and performance of mobile applications.

Mobile applications require testers to increase the focus on end-to-end testing as interaction between UE applications and enterprise data stores increases data processing complexity. The number and type of UE applications indicate the diversity of usage mobile networks will support in the future. In terms of functional verification of these applications, the following should be given more attention:

1. End to end testing will require real and live test facilities. Testing conducted through emulators or simulators in labs cannot accomplish the same level of quality verification of the user experience as field-based testing will. Emulators and simulators are useful for validating functionality and compatibility under controlled conditions, particularly during the development cycle and for whitebox testing. However, field verification is necessary to ensure proper validation of services in a real-world environment.
2. Interoperability will require extra focused testing ensuring service and application continuity across networks is important.

Network optimisation and drive testing

Drive tests are intended to perform network optimisation, benchmarking and service monitoring. This is to enable mobile network operators to ensure quality of service (QoS) and quality of end-user experience (QoE). Cellular core infrastructure performance test systems help network operators and equipment manufacturers to identify performance ceilings, enable vendor selections, and plan network capacity prior to deployment and to facilitate smooth service delivery.

Real-time and automated post-processing test software enables rapid verification of key WCDMA performance indicators. Handset data and WCDMA scanner measurements are used to identify missing neighbours and pilot pollution.

Load and Stress Testing

Load and stress testing include testing the application for normal load, heavy load (stress), sudden increase in load (spike) and sustained load (endurance). Performance Engineering Labs should be able simulate a heavy load of concurrent users.

Test beds are set up and load, stress and endurance tests are executed to check if the application can meet your performance objectives before going live. Monitors are deployed on all components of the application under test. Scripts are written to extract vital data during these tests. This data is used to analyse how key parameters can be optimised and root-cause analysis is done to look for bottlenecks.

Non-functional testing

Non-functional testing is designed to evaluate the readiness of the system under test according to several criteria not covered by functional testing. Non-functional testing addresses all testing needs encompassing test planning, strategy and test execution for:

- Integration testing. Integration testing is designed for ensuring that every component system, from source to destination interacts seamlessly with other components
- Compatibility testing. Compatibility testing ensures that a product works as per specifications with a set of other applications in a specific operating environment. Various hardware and software versions are mapped against which the software is to be tested, categorises combinations based on priority and develops a compatibility test plan. The exact user environment is simulated to test the behaviour of the application.
- Platform testing. Platform testing test application across a wide variety of operating systems, browsers, databases, hardware, etc.
- User acceptance testing. This is to ensure that the system that is implemented meets the needs of its users, i.e. agreed User Requirements.

Verification and Validation testing (VVT)

Once the design of the product has been completed, it will be necessary to ensure that it meets its requirements the requirements set down in the development specification, and that it also operates satisfactorily. These tests need to be performed formally to a signed-off test specification to ensure that the development has been satisfactorily completed to the agreed System Requirements.

General development testing

During the development of any product, test equipment is required to ensure that all the circuitry and SW operate as they should. The test equipment that will typically be required will include generic equipments such as oscilloscopes and RF measuring equipment including spectrum analysers, RF power meters and the like. In addition to this test equipment will be needed to aid debugging the software. This is likely to include emulators of various types. Thorough testing is required at this stage as it becomes progressively more difficult to correct any problems. Specialised and sophisticated test equipment may also be required during development.

Development Testing will primarily support the design development and integration processes by confirming design concepts, evaluating alternative design concepts, and investigating the availability of needed technology. This test phase will also support system verification by demonstrating system requirements that cannot be easily confirmed in a pre-operational site environment. Another objective of development testing is to resolve any outstanding design issues. This may include resolving critical but unverified design parameters, and conducting modelling and analyses. This test phase will also employ proof of concept prototype testing to reduce design and integration risk by investigating new technologies or design solutions.

In any development programme of electronic equipment it is necessary to thoroughly test the product at various stages in the development. The same is true for UMTS mobile phones and base stations. Testing takes various forms at different stages of the development, and often requires different types of test equipment, some of which can be particularly specialised. Following are the main areas that are to be tested during UE and BS development.

- RF Parametric Testing during RF development
- Physical Layer Functional Testing during L1 development
- RF Interface and L1/PHY functional and performance testing during PHY integration
- Protocol Host Testing during L2/MAC development
- RF Interface and protocol and performance testing during PHY/L2 integration
- Protocol Host Testing during higher layer RRC development
- RF, protocol, performance and parametric testing during product integration

Chapter Five

Conformance Testing and TTCN

CONFORMANCE TESTING

Conformance testing in general terms allows end users to practice preventative maintenance without the expenses normally associated with preventative maintenance programs. In addition, conformance testing eliminates ownership issues that result when an open system experiences technical difficulties. Due to the fact that the testing process is governed by 3GPP and GCF (instead of a specific vendor) and the actual test is conducted in an independent lab (Test House), the results are equally unbiased. Simply put, a product that passes the test is conformant and interoperable. As a result, the test does away with any potential fingerpointing between vendors. Likewise, if a problem should occur, interoperability is no longer an issue and any resulting error most likely lies elsewhere (improper implementation, etc). By crossing interoperability off the list of potential problems, diagnosis is more accurate and usually takes much less time.

Conformance testing also considered as an independent user's guide to help end users differentiate the vast number of products and vendors that have inundated the open systems' marketplace. If a vendor has a product that is conformant to 3GPP standards, testing shouldn't pose a problem. It is non-tested products, on the other hand, that end users should be wary of. In fact, conformance testing can only be seen as a good business practice from both the end user and vendor's perspective.

It should be born in mind, nonetheless, that the testing procedure isn't flawless in its current state, passing the conformance test does not "guarantee" interoperability among products as far as percentages are concerned. However, this is where specifying conformance tested products will once again pay dividends, delivering secondary, even tertiary benefits for the end user. As companies demand conformant products and testing becomes widely practiced, each successive version of the test will evolve based upon feedback. The accuracy and quality of the test will increase exponentially as a result, and end users will see better, technically advanced products. And as the testing process becomes streamlined, improved products will be available in the market sooner. Moreover, feedback from the testing process could be applied to the UMTS technology overall, improving the foundation as well as the products. The ensuing benefits are endless.

Conformance testing is paramount to the end user and the reason can be distilled to the fact that testing eliminates up to 80% of field errors. Also end users benefit without making a financial investment and testing will improve the technology available.

PROTOCOL CONFORMANCE TESTING

Definition

Conformance Testing is defined as the process of verifying that an implementation performs in accordance (conforms) to a particular standard/specification, in UMTS case, 3GPP specifications. This type of testing is used throughout the initial stages of development to ensure the accuracy of a protocol implementation. It is also used in regression testing after initial product deployment to further verify any changes in the implementation/enhancement are backward compatible.

Purpose
Black Box testing of external behaviour
It deals with what is going in and what is coming out. This testing does not cover performance, reliability, fault tolerance, efficiency, etc.

Implementations Testing
This intends to test how a protocol is implemented in a system. A product has to go through different kinds of testing in its lifecycle. Conformance Testing is generally carried out if the product is based on an open standard and interacts with other entities. Figure 5.1 briefly depicts the testing sequence for a product based on open standards. Conformance testing can be used as a first step in determining interoperability with other implementations. Product that passes the conformance test is ready for interoperability and field trial tests. As a result, the test eliminates any potential blame between vendors.

Approach
In general conformance testing follows four major steps.

- 3GPP defines the process or procedure for conformance testing so that SS manufacturers and UE manufacturers are aligned.
- Products are designed according to the protocol's specification and the interface is exposed for conformance. This provides a basic framework for product conformance.
- Once test solution is available the UE manufacturers may use these solutions to test their implementation.

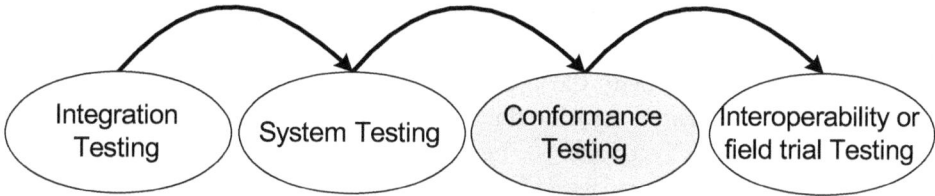

Figure 5.1   Testing stages

- UE manufacturers and standard bodies jointly formed an association (GCF) which independently verifies/certify different implementations for conformance after having sufficient evidence from a third party test house.

A typical conformance test suite is constructed to cover the requirements of a complete specification or significant portion of a specification. These specifications typically come from standard bodies like 3GPP. The same specification that implements the protocol is also used to develop the conformance test suite. A developer will review the protocol specification and construct a test plan for requirements outlined in that specification. Once these requirements are fulfilled the conformance test suites break down a protocol specification into discrete and atomic test cases. The test cases specify information describing initial requirements, inputs, test procedures and the expected response from the implementation under test.

Conformance Test and Tester Architecture

Conformance testing in general as mentioned earlier is a Black Box testing. A protocol conformance tester, which is also referred to as System Simulator is an implementation in which a predefined Conformance Test Suites (CTS) can run to test the desired Implementation Under Test (IUT). The Interface between a conformance tester SS and IUT should be the standard interface defined for the IUT for the real environment as depicted in Figure 5.2.

The architecture of an SS varies with Conformance Test requirements and the SS manufacturer. Implementation of a Conformance Tester should comply with the protocol specification. A system simulator has to mainly provide two interfaces, SS-IUT interface and SS-CTS interface (Figure 5.2). IUT is connected to Conformance

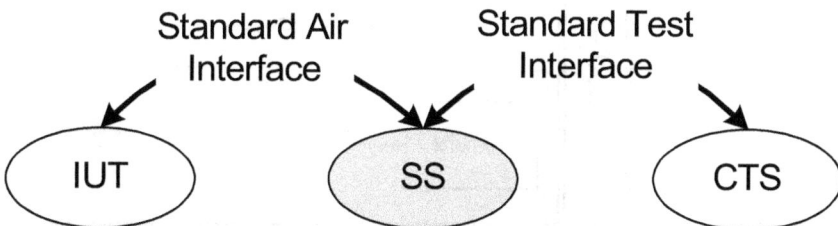

Figure 5.2   Conformance system interfaces

tester via the first interface and its behaviour should be the same as the standard air interface as in real environment. The conformance tester should emulate the behaviour of the real environment. On the other hand, the second interface is defined such that the implementation inside the SS can be used to fulfil the requirement to develop Conformance Test Suite. See Figure 5.3 for typical test system architecture.

Cellular protocol conformance testing can be categorised into two, UE conformance tester and BS Conformance Tester. A UE Conformance Tester aims to test the different layers of the UE protocol implementation. This includes testing of basic lower layer signalling protocols defined for communication between a UE and the network. It basically emulates a real network or part of the real network depending upon test coverage supported by the tester. The interface between a UE and SS is the standard RF Interface defined in core protocol specification as interface between UE and a real network. In addition to testing lower layers system simulators can also support testing of application layer protocols like Mobile IP, MMS, SMS, VoIP, VT etc.

On the other hand Base Station or (network) conformance tester aims to test different part or parts of a real cellular network. The IUT is part of the network. Its interface with the conformance tester is the defined interface between that part or those parts with other network entities. Conformance testing is performed for different parts of the real network such as Base Station (BS), Radio Network Controller (RNC) and other core network components.

Usually conformance test equipment developer provides an Application Programming Interface (API) for control and exchange of information with Conformance Tester. Conformance test suites are either based on the custom interface or plugged in via this interface. As different Conformance test solutions uses different

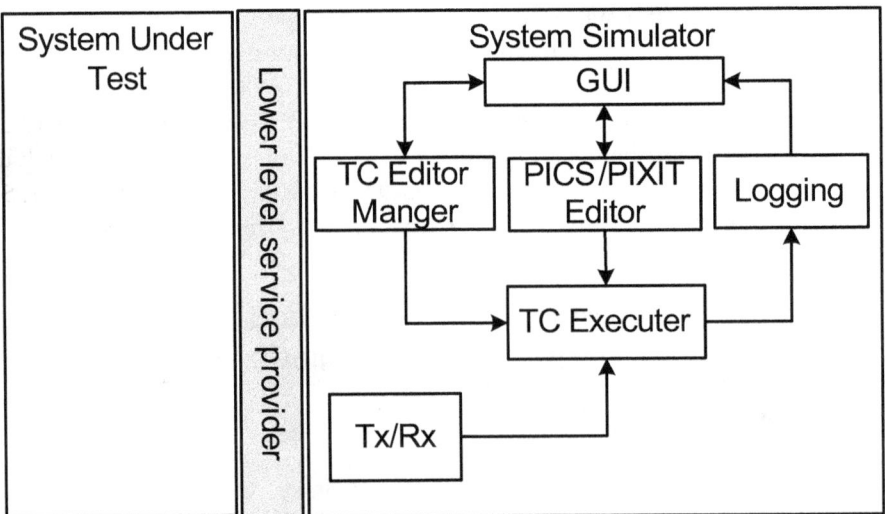

Figure 5.3   Example of high level test system architecture

custom APIs every conformance test equipment manufacturer spends lots of time and effort on development and validation of the tester and the test suites. In addition to Test Cases developed by ETSI, each manufacturer develop their own test cases (e.g. RF or USIM test cases) based on common test specifications but there is a certain degree of variance in implementation of these test cases across different vendors. Because of these issues some standard bodies are using a common interface for defining the test interface and it is taken care that the interface is independent of the platform and languages/tools used for tester development.

Test interfaces are often defined as connection points, signals and expected behaviour instead of defining procedural requirements for the conformance tester. Signals are defined more or less in a similar way as it is defined in protocols for communication between two entities. In other words the test interface is defined as a set of signals (a bit pattern or Octet String), which carry a defined meaning, along with the defined rules about when it can be sent and what are the expected response. This mechanism provides clear abstraction between a Conformance Test Suite and Tester. SS manufacturers and ETSI develop conformance test suite is more or less independent of the tester's platform and tester will have to expose an interface to transfer a bit pattern or octet string on a connection point. A conceptual test setup is illustrated in Figure 5.4.

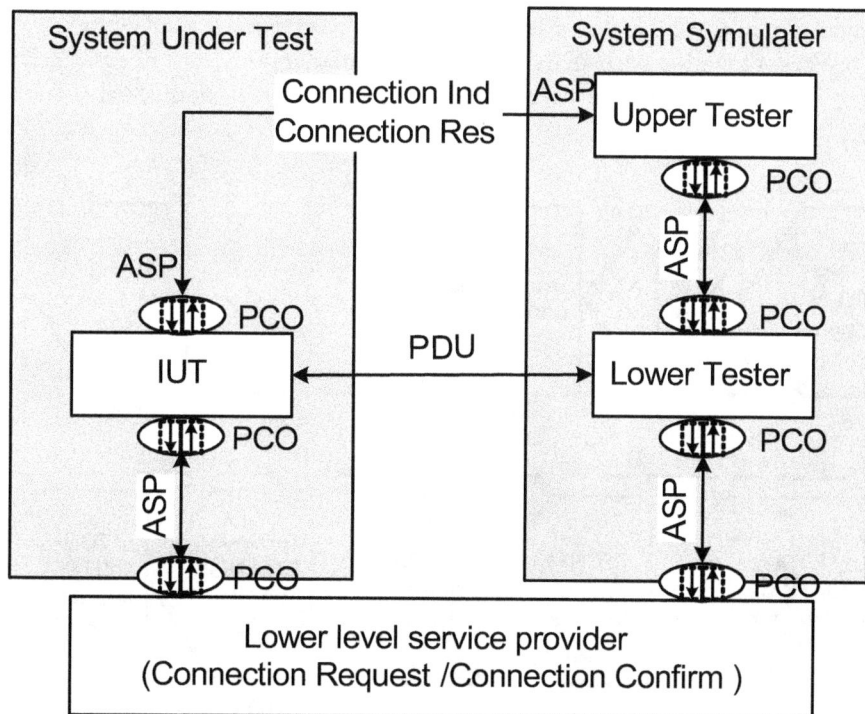

Figure 5.4   Conceptual Test architecture

To define such a test interface there is a requirement of notation which defines Signals as a set of data structure at a level above the programming languages and also able to carry the required semantics. This is also referred to as Abstract Syntax Notation. ASN.1 (Abstract Syntax Notations One) is one of the many such notations but is most commonly used.

Conformance Testing Languages and Tools

This section covers different languages and tools used in the conformance industry to carry out conformance testing. Standard bodies are using special languages like TTCN-2, TTCN-3 to specify the conformance test specifications but conventional languages like Basic/C/C++/Perl/Shell scripts still capture some of the market share for test suite implementation. The following sections briefly covers TTCN-2 and TTCN-3 languages/tools. Figure 5.5 illustrates the history of these two languages.

TTCN-2

TTCN-2 stands for Tree and Tabular Combined Notation version two. As the name implies, the test language combines trees of information with tables of exact details. In its simplest form, TTCN-2 can be considered as being a hierarchy of data definitions, combined with a hierarchy of test cases and supporting test functions (steps). Within each test case or step there is a table describing the actual behaviour of the test (i.e. the order of test events). Test suite is a collection of various test cases together with all the declarations and components it needs. Each test case is described as an event tree and this tree defines the conditional sequential

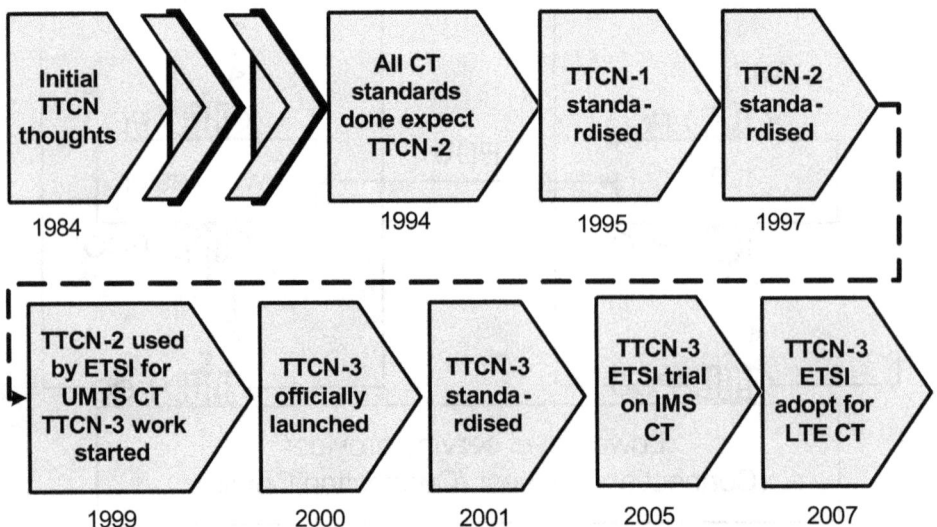

Figure 5.5 TTCN history

behaviour. Figure 5.6 shows an example of test case behaviour. Because TTCN-2 is precise and formal language it is possible to automate the generation of an executable test from a definition constructed in TTCN-2.

TTCN-2 is the standardised test notation recommended by ISO/IEC 9646 for writing abstract tests. 3GPP/ETSI has adopted TTCN-2 for defining conformance tests for UMTS protocols. GCF further mandated the use of TTCN-2 for the certification of UMTS protocol conformance testers.

As we can see from Figure 5.5 in the mid 80s, the standardisation community realised that something needed to be done to "abstract" tests from the system under test (SUT) to facilitate the re- use of tests across multiple implementations of the same specification. TTCN-1 and then a follow- up version, TTCN-2 were defined to standardise the approach for developing conformance tests for Open Systems Interconnection (OSI) based protocols. Tests could finally be written in a language that more closely resembled the requirements against which the tests were being written. Free, off- the-shelf, tests could now be taken from organisations such as ETSI or purchased from other companies. The TTCN-2 language has now been accepted as the de facto test notation within the communications industry. The use of TTCN-2 in other sectors such as the military and aerospace industry has been limited because of the widespread use of telecommunications-specific terminology and concepts within TTCN-2. Thus, although TTCN-2 allows an effective test process to be used, its use is only accepted within a small community.

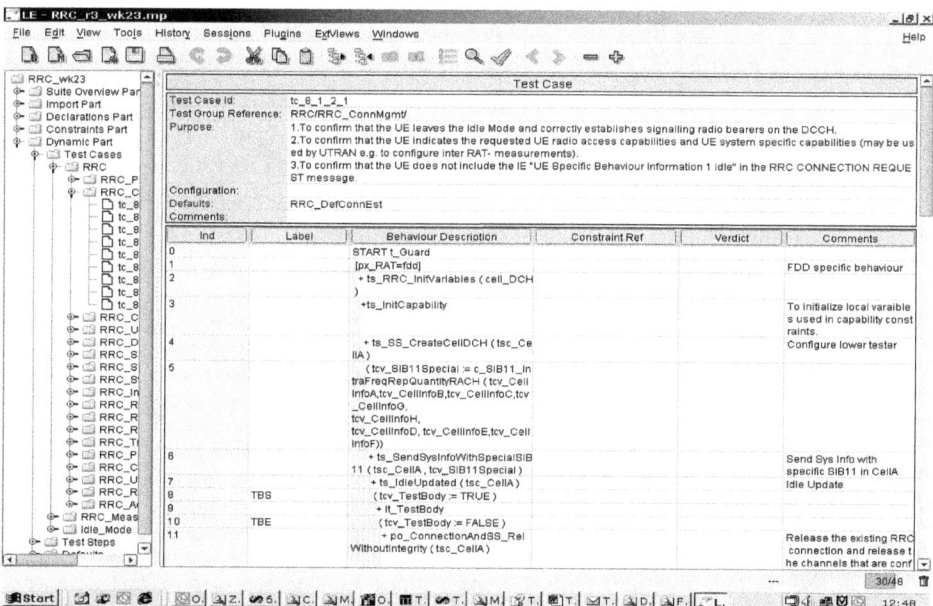

Figure 5.6   Example test case written in TTCN-2 displayed using Leonardo Editor

When the development of the TTCN language was initiated, its founders had in mind the development of a conformance testing service that would guarantee the quality of tested products by a process that could be seen to be fair and independent. This high ideal was brought back to reality by the problems of legal liability, for both passed and failed products, the problems of adequately defining product requirements and the fact that testing cannot guarantee the absence of faults. However, there is no realistic alternative to effective testing. This has led different markets to initiate different testing models.

TTCN-2 Model

TTCN-2 is abstract in the sense of being test system independent. This means that a test suite in TTCN-2 for one application (e.g. protocol, RF, etc.) can be used in any test environment for that application. Use of TTCN-2 has increased tremendously during the last few years. This has been augmented by the significant amount of test suites released by various standardisation bodies, e.g. ETSI MCC-160.

The specifications of the messages being sent and received can be defined using either the native form of TTCN-2 or by using ASN.1 (Abstract Syntax Notation One). A TTCN-2 specification describes an abstract test suite (ATS) that is independent of test system, hardware and software. The ATS defines the test of the implementation under test (IUT). A TTCN-2 ATS can be transformed into an executable test suite (ETS) using the TTCN suite. This ETS is downloaded into the test system (the system performing the test) to execute the test behaviour. The test system performs the test by executing the ETS against the system under test (SUT), which contains the implementation under test. During execution the ETS will report any errors and log events for on-line or post-test evaluation.

The Generic Compiler Interpreter (GCI) interface standardizes the communication between a TTCN-2 component supplied by a vendor (e.g. Telelogic) and other test system components supplied by the SS manufacturer. The GCI interface separates TTCN-2 behaviour from protocol and test equipment specific behaviour. The TTCN-2 Runtime Behaviour and the Test Adaptation must provide services to each other. Some services (e.g. handling values etc) are provided by the TTCN-2 Runtime Behaviour and some services (e.g. send, snapshot, timer functions, logging etc) are provided by the Test Adaptation. To generate the executable from the ATS, the services which are provided by the Test Adaptation need to be implemented. Test Adaptation is the implementation (based on the system) of the services required by TTCN-2 Runtime Behaviour using the GCI interface. This means, that the interface between TTCN-2 Runtime Behaviour and Test Adaptation is GCI (standardised). The above and some of the responsibilities of the two GCI parties (TTCN-2 Runtime Behaviour and the Test Adaptation) are described in Figure 5.7.

As stated earlier, during the implementation of test Adaptation, it might require encoding or decoding the values received from TTCN-2 Runtime Behaviour or IUT. As TTCN-2 supports ASN.1 definitions, it might require encoding or decoding

TTCN Runtime Behaviour	Test Adaptation
SEND: Creates the GCI value and requests the Test Adaptation to send this to the IUT	SEND: Receives the GCI value and sends it to the IUT. Test Adaptation might need to encode the value in some form .
RECEIVE: When a GCI value is put on a PCO queue by the Test Adaptation , it remains there until it matches in TTCN	RECEIVE: Receives the value from the IUT and converts it into a GCI value (Test Adaptation might need to decode the value) and sends the GCI value to the TTCN Runtime Behaviour .

(a)

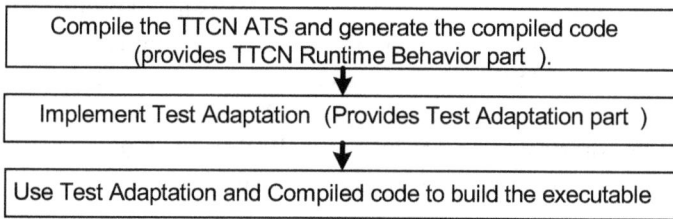

(b)

Figure 5.7   (a) TTCN-2 Runtime behaviour and test adaptation functions, (b) Test case executable generation

the values received using ASN.1 encoding/decoding rules. Test Adaptation should take care of this. As Test Adaptation is taking care of encoding/decoding of ASN.1 definitions, any change in the ASN.1 definitions used by the TTCN-2 will have an impact on the Test Adaptation.

TTCN-2 Specification

TTCN-2 specification is seen to be similar to Pascal or C program but written in a Tree and tabular format. It is Just like Pascal or C, TTCN-2 requires type and data declarations and it uses concepts like functions. TTCN-2 is designed for testing application therefore it contains test specific concepts such as:

- Test case run results are defined via Final and preliminary verdicts
- Handle of alternative outcomes in a test case.
- Preambles and postambles to show how to compose test cases.
- The Module concept supports multi-user test development.
- The Module concept supports the re-use of test components and data structures.

TTCN-2 specification has a standardised layout that produces comprehensive and unambiguous paper printouts. This greatly improves clarity and readability. Test suites are divided into the following four major parts:

- *Overview part*, containing a table of contents and a description of the test suite. Its purpose is mainly to document the test suite to increase clarity and readability.
- *Declarations part*, declaring all messages, variables, timers, data structures and black box interface towards the IUT.
- *Constraints part*, assigning values and creating constraints for inspection of responses from the implementation under test.
- *Dynamic part*, containing all test cases, test steps and default tables with test events and verdicts, i.e. it describes the actual execution behaviour of the test suite

Communication Concept

TTCN-2 uses the concepts of *points of control and observation* (PCOs), *abstract service primitives* (ASPs) and *protocol data units* (PDUs) in order to create an abstract interface towards the implementation under test (IUT) as illustrated in   5.4. A PCO is a point in the abstract interface where the IUT can be stimulated and its responses can be inspected. An ASP or a PDU is either a stimuli or a response that carries information, i.e. parameters and data. Each PCO has two *first in first out* (FIFO) queues for temporary storage of ASPs and PDUs: One queue for send and one queue for receive. These queues are infinite, i.e. they can store any number of ASPs and PDUs.

Tools

There are may Tools available in the market for editing TTCN-2 test suites. Telelogic's Tau, De Vinci's Leonardo, Danet's TTCN-2 Tool Box are few examples of the above. Some vendors also provide compiler and code generators for different languages like C, Java etc. Telelogic, HP, Tektronics, Danet are some of the vendors who claim such features in their TTCN-2 line of products.

Assessment of TTCN-2

Many programmers and system engineers brand TTCN-2 as a complex language. They say it uses too many tables proformas, difficult to learn and has many redundant and unused functionalities. TTCN-2 contains too big and restrictive BNF (Backus Naur Form) and not user friendly as the Browser structure is built into the BNF. Also it is thought to be restrictive due to the fact that it presents the OSI conformance testing view, i.e. limited command set, table proforma format too restrictive, no human understandable text version, and application-specific static semantics.

TTCN-2 was designed to provide an easy and effective way to develop conformance tests for OSI protocols. The broad adoption of TTCN-2 in the communications industry demonstrates how effectively this objective was achieved. However, outside its original intended audience, TTCN-2 has been criticised for its limited command set, which is focused specifically on the testing of communications systems. One of the benefits of TTCN-2 in fact becomes one of its criticisms because other software sectors require different and additional functionality from a test language. One of the strengths of TTCN-2 (Tree and Tabular Combined Notation) is the clear structure it gives to test definitions. For example, the sequences of events that make up a test are specified in one table, which is separated from the signal declarations for the test suite. This very aspect of TTCN-2 limits its applicability because it allows only a single test specification strategy, whereas potential adopters are asking to be able to define their own test specification strategies.

It is obvious that TTCN-2 has not been accepted in many areas outside the telecommunication market. TTCN-2 is, however, still viable in its accepted environment and will live for so many years alongside TTCN-3. This is because a number of existing TTCN-2 tests and test equipment will not be migrated or upgraded. We may even see some environments that support both and having TTCN-2 running alongside TTCN-3.

TTCN-3

More attention is given in this section to TTCN-3 (Testing and Test Control Notation) as is the Conformance Test Language adopted for the next generation of cellular technology LTE (Long Term Evolution). TTCN-3 is a test scripting language that of an international standard. Its main feature is the separation of concern between Abstract Test Suites and the Adaptation Layer which allows full portability of test suites and thus makes them independent of any platform implementation. The Test Adapter handles all platform and implementation languages (java, C, C++) issues for the communication with a System Under Test and also the actual coding and decoding requirements of an application.

TTCN-3 leaves the test development strategy and structure of test specification to both the test engineer and the tool vendor producing test development software. TTCN-3 provides a mechanism for writing tests that more resembles a 3GL (third generation language, which is a programming language designed to be easier for a human to understand, including things like named variables) whilst keeping the language restricted to the test domain. Trees and tables are no longer mandated, although test development systems may still provide them as a mechanism for specifying a test. The following pages examine TTCN-3 in more details and highlight some of the key enhancements it provides over TTCN-2.

TTCN-2 is an OSI-based model and OSI-based terminology like PDUs, ASPs which drives its use in testing OSI-based protocols. The OSI approach adopted by TTCN-2 is easier for engineers to understand and implement the test specifications for protocols. But its specific approach restrains its application only to telecommunication protocol testing and it cannot be used for general protocol testing.

For testing future telecommunication protocols more capabilities were required than TTCN-2 can provide. The telecommunication terminology needed to be

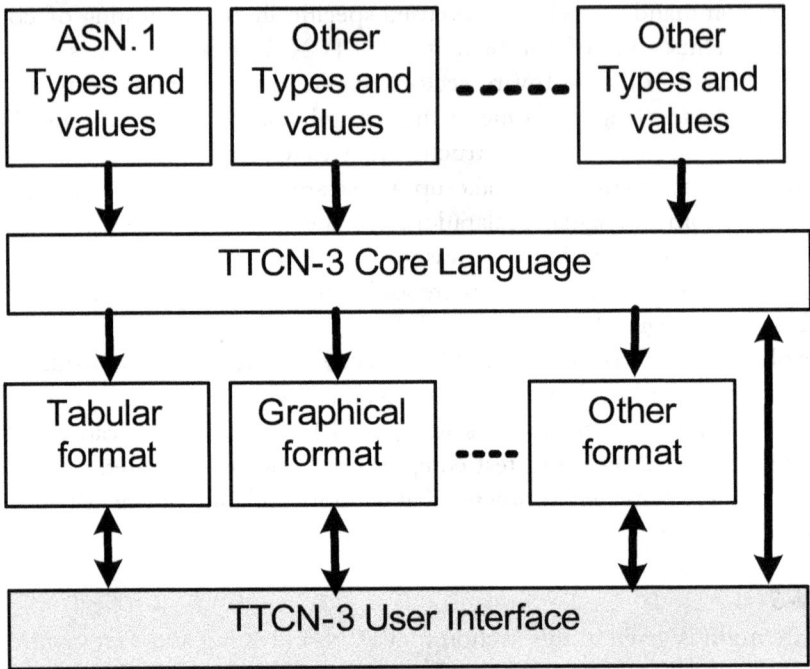

Figure 5.8 Overview of the TTCN-3 language

removed for its application to other protocols testing. ETSI in 1998 started with this concept and implemented TTCN-3, which has all the advantages of TTCN-2, but removing references to all terminologies for OSI based telecommunication protocols. The impact was so high that the meaning of TTCN has been changed to "Testing and Test Control Notation". 3GLs (3rd generation languages) are in use by engineers for most of the testing applications due to the flexibility provided by them. TTCN-3 provides a mechanism to writing test definitions as a 3rd Generation Languages but with its focus on testing domains. TTCN-3 is mainly targeted for testing of wireless applications where test suites need to be integrated with various other software components. Its OOM (Object oriented modelling) and easy integration with programming languages like C/C++, IDL makes it more suitable for other applications and for all kinds of black-box testing for reactive and distributed systems, e.g.

- Telecom systems (ISDN, ATM, DECT)
- Mobile systems (GSM, UMTS, 3G, TETRA, WiMAX)
- Internet (has been applied to IPv6, SIP (Session Initiation Protocol), OSP (Open settlement protocol))
- CORBA based systems

TTCN-3 language was created due to the essential necessity to have universally understood language syntax that is able to describe test behaviour specifications.

Its development was driven by industry and science to obtain a single test notation for all black-box testing needs. In contrast to earlier test technologies, TTCN-3 encourages the use of a common methodology and style which leads to a simpler maintenance of test suites and products. With the help of TTCN-3, the tester specifies the test suites at an abstract level and focuses on the test purpose itself rather than on the test system adaptation and execution. A standardised language provides a lot of advantage to both test suite providers and users. Moreover, the use of a standard language reduces the costs for education and training, as a great amount of documentation and examples is available. It is obviously preferred to always use the same languages for testing than learning different technologies for distinct test classes. Constant use and collaboration between TTCN-3 consumers ensure a uniform maintenance and development of the language.

TTCN-3 enables systematic, specification-based testing for various kinds of tests including e.g. functional, scalability, load, interoperability, robustness, regression, system and integration testing. TTCN-3 is a language to define test procedures to be used for black-box testing of distributed systems. It allows an easy and efficient description of complex distributed test behaviours in terms of sequences, alternatives, and loops of stimuli and responses. The test system can use a number of test components to perform test procedures in parallel. TTCN-3 language is characterised by a well-defined syntax and operational semantics, which allow a precise execution algorithm. The task of description of dynamic and concurrent configuration is easy to perform. The communication can be realised either synchronously or asynchronously. To validate the data transmitted between the entities composing the test system, TTCN-3 supports definition of templates which ensure a powerful matching mechanism. To validate the described behaviours, a verdict handling mechanism is provided. The types and values can be either described directly in TTCN-3 or can be imported from other languages (i.e. ASN.1, XML, IDL). Moreover, in TTCN-3, the parameterisation of types and values is allowed. The selection of the test cases to be executed can be either controlled by the user or can be described within the execution control construct. The external configuration of a test suite through module parameters is also possible.

Figure 5.8 shows an overview of the TTCN-3 language. TTCN-3 has a core language which provides interfaces to reference data defined in other description languages. As the figure shows, one can import types and values specified in ASN.1, but other formats are also supported (IDL, XML etc). The front-end can be either the core language itself or one of the presentation formats (tabular format, graphical format etc). The tabular format and graphical formats are the first in an anticipated set of different presentation formats. These other formats may be standardised presentation formats or they may be proprietary presentation formats defined by TTCN-3 users themselves.

TTCN-3 is based on TTCN-2 and has all the relevant advantages of TTCN-2. In order to increase the usability of tests for multiple systems or different versions of same system, tests are kept portable. Exception handlers are crucial for successful test executions. Exception handlers are defined in the beginning and are available throughout. 3rd generation languages require algorithms for comparing the expected and received data. TTCN-3 provided many in built algorithms for comparing expected and received data.

TTCN-3 provides support for data structures defined in ASN.1, IDL, C etc. A number of new capabilities have been added in TTCN-3 on top of TTCN-2. TTCN-3 provides supports grouping of test cases based of functionality. That means if functionality is not supported by IUT, the whole group of tests can be removed from testing saving the execution time. One of the key drawbacks of TTCN-2 is that it only supports asynchronous communication with SUT. TTCN-3 overcomes this limitation and provides support for synchronous communication. Remote procedure calls (RPC) and callback mechanism along with timers can be used to test many failure scenarios for a module in a automated way.

TTCN 3 Specification

TTCN 3 specification has been divided into two halves. First half describes the actual language and second half defines different way of visualising the test specifications.

Test Format defines the conventional textual way of writing test specification just like C and C++ but the main focus if this is testing. This is mainly designed for normal programmer to reduce the learning curve of the tool.

Tabular Format is designed for existing TTCN 2 user and is very much similar to TTCN-2. It is designed for users that prefer the TTCN style of writing test suites. TFT presents a TTCN-3 module as a collection of tables. The tabular presentation format highlights the structural aspects of a TTCN-3 module and in particular of structures of types and templates.

Message Sequence Chart Format (MSC) or Graphical Format (GFT) aids the visualisation of test behaviour. The graphical format eases the reading, documentation and discussion of test procedures and is also well suited to the representation of test execution and analysing of the test results. For each kind of TTCN-3 behaviour definition, GFT provides a special diagram type, i.e., control diagrams for the control part of a module, function diagrams for functions and test case diagrams for test cases. Presentation Format is designed to graphically define the test in UML Notation. Work is still going on in this direction and is based upon UML 2.0 Notation.

The ETSI standard for TTCN-3 comprises six parts (described below) which are grouped together in the "Methods for Testing and Specification; The Testing and Test Control Notation version 3" document.

1. TTCN-3 Core Language
2. Tabular Presentation Format. TTCN-3 offers optional presentation formats. The tabular format is similar in appearance and functionality to earlier versions of TTCN. It was designed for users that prefer the TTCN-2 style of writing test suites. A TTCN-3 module is presented in the tabular format as a collection of tables.

3. Graphical Presentation Format. It is the second presentation format of TTCN-3 and is based on the MSC format (Message Sequence Charts). The graphical format is used to represent graphically the TTCN-3 behaviour definitions as a sequence of diagrams.
4. Operational semantics.
5. The TTCN-3 Runtime Interface (TRI). A complete test system implementation requires also a platform specific adaptation layer. The TRI specification of a common API interface to adapt TTCN-3 test systems to SUT
6. The TTCN-3 Control Interfaces (TCI). This part provides an implementation interface for the execution environments of TTCN-3. This is the specification of the API of the TTCN-3 execution environments should implement in order to ensure the communication, management, component handling, external data control and logging.

Test Model and Communication Mechanisms

TTCN 3 test configuration is a set of interconnected *test components*. Each test component has a well-defined *communication port* and explicit *test system interface* defining the borders of test systems. There is one *Main Test Component (MTC)* and other test components are called *parallel test components (PTC)*. At the start of each test case, *MTC* is created automatically and the behaviour defined in the test case body is executed in this component. Using *create* and *stop* operations *PTCs* can be created dynamically during execution of a test case.

Communication between test components and test system interface is through ports. A component can have any number of connections but it cannot be connected to itself. Communication partners have to be determined uniquely during test execution as TTCN-3 only supports one-to-one communication. Each port is modelled as an infinite FIFO Queue, which stores the incoming messages or procedure calls until they are processed by the component owning that port.

TTCN-3 ports are either message-based or procedure-based. Message-based ports are used for asynchronous communication by means of message exchange. Procedure-based ports are used for synchronous communication by means of remote procedure calls. Ports are directional and each port may have an **in** list (for the *in* direction), an **out** list (for the *out* direction) or an **inout** list (for both directions) of allowed messages or procedures.

The test adapter and the encoder/decoder illustrated in Figure 5.9 are the glue between the Abstract Test Suite and the System Under Test. It converts the data into the format that the SUT understands. For example, the abstract data is converted into a text string with some command "request x", and pairs of field names and field values. Note that the concrete field names are intentionally different than the abstract types field names. The adapter layer enables us to re-use the same abstract test cases for different test equipment providers. For example while provider A may use text messages, provider B may use an XML representation instead.

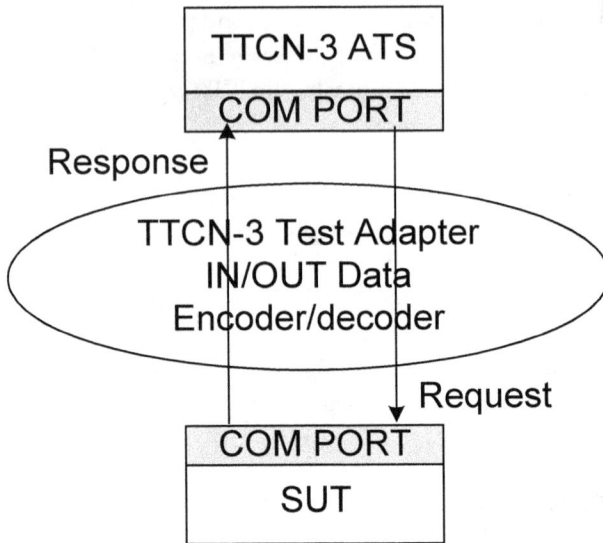

Figure 5.9   TTCN-3 simplified test model and Test Adapter concept

General Structure of a TTCN-3 Test System

A TTCN-3 test system can be conceptually thought of as a set of interacting entities where each entity corresponds to a particular aspect of functionality in a test system implementation. These entities manage test execution, interpreting or executing compiled TTCN-3 code, realise proper communication with the SUT, implement external functions, and handle timer operations.

The structure of a TTCN-3 test system implementation as illustrated in Figure 5.10 above is purely an aid to define TTCN-3 test system interfaces.

The part of the test system that deals with interpretation and execution of TTCN-3 modules, i.e. the Executable Test Suite (ETS), is part of the TTCN-3 Executable (TE). This corresponds either to the executable code produced by a TTCN-3 compiler or a TTCN-3 interpreter in a test system implementation. It is assumed that a test system implementation includes the ETS as derived from a TTCN-3 ATS.

The remaining part of the TTCN-3 test system, which deals with any aspects that cannot be concluded from information being present in the original ATS alone, can be decomposed into Test Management (TM), SUT Adapter (SA), and Platform Adapter (PA) entities. In general, these entities cover a test system user interface, test execution control, test event logging, as well as communication with the SUT and timer implementation.

Test Manager (TM) and Test Adapter (TA)

In the TM part, we can distinguish between functionality related to test execution control and test event logging. This part is responsible for source code generated

```
┌─────────────────────────────────────────┐
│              Test Manager                │
└─────────────────────────────────────────┘
                    ↕
┌─────────────────────────────────────────┐
│              Test Adapter                │
│     (Test Logging, Runtime system)       │
└─────────────────────────────────────────┘
                    ↕
┌─────────────────────────────────────────┐
│           TTCN-3 Executable              │
│     (ETS, Encoder/decoder system)        │
└─────────────────────────────────────────┘
                    ↕
        ┌─────────────────────────┐
        │    SUT and Platform     │
        │        Adapter          │
        └─────────────────────────┘
```

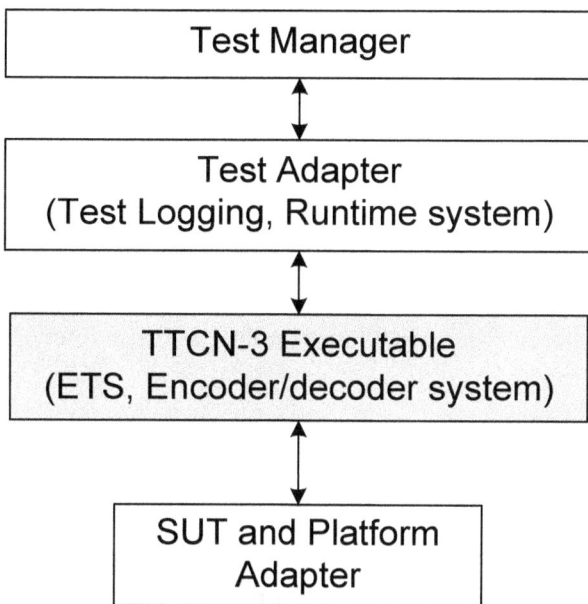

Figure 5.10   Typical TTCN-3 test system

from TTCN-3 files or ATS. Also it is responsible for overall management of the test system. After initialising the test system, test execution starts within this part. Proper invocation of TTCN-3 modules, i.e. propagating module parameters and/ or IXIT information to the TE if necessary are also done in this layer.

Test Logging could be part of the Test Adaptation Layer (Test Adapter) where the test logging is maintained. It is explicitly notified to log test events by the TTCN-3 Executable. The test logging part has a unidirectional interface where any part of the TTCN-3 Executable may post a logging request to the test logging part.

Similarly TTCN-3 Runtime System could be part of the Test Adaptation Layer. TTCN-3 Runtime System part implements the creation and removal of TTCN-3 test components, as well as the TTCN-3 semantics of message and procedure based communication, external function calls, action operations and timers. This includes notifying the SUT Adapter of which message or procedure call is to be sent to the SUT. Also notifying the Platform Adapter of which external function is to be executed or which timers are to be started, stopped, queried, or read. Similarly, the Runtime System notifies the ETS entity of incoming messages or procedure calls from the SUT as well as timeout events.

Prior to sending/receiving messages and procedure calls to/from the SUT Adapter, or prior to handling function calls and action operations in the Platform Adapter for the ETS part, the Runtime system invokes the encoder/decoder system for their encoding or decoding. The Runtime system should implement all message and procedure based communication operations between test components. But only the TTCN-3 semantics of procedure based communication with the SUT, i.e. the possible blocking and unblocking of test component execution, guarding with implicit timers, and handling of timeout exceptions as a result of such communication operations. All procedure based communication operations with the SUT are to be realised and identified (in the case of a receiving operation) in the SUT Adapter as they are most efficiently implemented in a platform specific manner. Note that the timing of any procedure call operation, i.e. implicit timers, is implemented in the Platform Adapter (PA). In short it is responsible for send/receiving messages on ports, timers, calling external functions, encoding and decoding of messages etc.

TTCN-3 Executable

TTCN-3 Executable (TE) is responsible for the interpretation or execution of the TTCN-3 ATS. Conceptually, it can be decomposed into two interacting entities: an ETS and Encoder/Decoder System. The ETS source code generated from ATS or TTCN-3 files handles the execution or interpretation of test cases, the sequencing and matching of test events, as defined in the corresponding TTCN-3 modules. The TE interacts with the Runtime system to send, attempt to receive (or match), and log test events during test case execution, to create and remove TTCN-3 test components, as well as to handle external function calls, action operations, and timers. The encoding/decoding system is responsible for the encoding and decoding of test data, which includes data used in communication operations with the SUT as specified in the executing TTCN-3 module. If no encoding has been specified for a TTCN-3 module the encoding of data values is tool specific. TE is invoked by, and returns to the Runtime system. Note that the encoder/decoder system does not directly interact with the SUT Adapter. The encoder/decoder is provided by the tool vendors and utilised by Target Adaptation Layer.

SUT Adapter and Platform Adapter

The SUT Adapter adapts message and procedure based communication of the TTCN-3 test system with the SUT to a particular execution platform of the test system. It is aware of the mapping of the TTCN-3 test component communication ports to test system interface ports and implements the real test system. It is responsible to propagate send requests and SUT action operations from the TTCN-3 Executable to the SUT, and to notify the TTCN-3 Executable of any received test events by appending them to the port queues of the TTCN-3 Executable.

The Platform Adapter implements TTCN-3 external functions and provides a TTCN-3 test system with a single notion of time. In this part, external functions are to be implemented as well as all timers. Notice that timer instances are created in the TE. The interface with the TTCN-3 Executables enables the invocation of external functions and the starting, reading, and stopping of timers as well as the inquiring of the status of timers using their timer ID. The Platform Adapter notifies the TTCN-3 Executable of expired timers.

TTCN-3 Modules

The TTCN-3 Core Language is a modular language and has a similar look and feel to a typical programming language. In addition to the typical programming constructs, it contains all the important features necessary to specify test procedures and campaigns for functional, conformance, interoperability, load and scalability tests like test verdicts, matching mechanisms to compare the reactions of the SUT with the expected range of values, timer handling, distributed test components, ability to specify encoding information, synchronous and asynchronous communication, and monitoring.

The top-level building-block of TTCN-3 is a module (see example in Figure 5.11). A module contains all other TTCN-3 constructs, but cannot contain sub-modules. It can also import completely or partially the definitions of other modules. The modules are defined with the keyword module. The modules can be parameterised; parameters are sets of values that are supplied by the test environment at runtime. A parameter can be initialised with a default value.

A TTCN-3 module has two parts: the *module definition* part and the *module control* part as depicted in an example script in Figure 5.12. The definition part contains the data defined by that module (functions, test cases, components, types, templates), which can be used everywhere in the module and can be imported from other modules. The control part is the main program of the module, which describes the execution sequence of the test cases or functions. It can access the verdicts delivered by test cases and, according to them, can decide the next steps of execution. The test behaviours in TTCN-3 are defined within functions, *altsteps* and *testcases*. The control part of a module may call any *testcase* or function defined in the module to which it belongs.

```
module ModuleName {
  // DEFINITION PART
  // imports from other modules
  import from OtherModule
  {
    type Type1, Type2;
    template all
  }
  // module parameters
  modulepar { integer par1 := 0, par2; boolean par3 };

  // definitions
  type record myType { charstring field1, integer field2 };
  // CONTROL PART
  control {
    execute(MyTestCase1(0));
  }
}
```

Figure 5.11   Example of TTCN-3 Module

```
// DEFINITION PART
testcase MyTestCase1(integer i) runs on MyTestComponent system
SysComponent {
  // behaviour
}

// CONTROL PART
control {
    execute(MyTestCase1(0));
}
```

Figure 5.12   Example on Test Case Definition and Execution

Test Cases and Test Verdicts

Test cases define test behaviours which have to be executed to check whether the system under test (SUT) passes the test or not. Like a module, a test case is considered to be a self-contained and complete specification that checks a test purpose. The result of a test case execution is a test verdict.

TTCN-3 provides a special test verdict mechanism for the interpretation of test runs. This mechanism is implemented by a set of predefined verdicts, local and global test verdicts and operations for reading and setting local test verdicts. The predefined verdicts are pass, inconc, fail, error and none. They can be used for the judgment of complete and partial test runs.

- pass verdict denotes that the SUT behaves according to the test purpose,
- fail verdict indicates that the SUT violates its specification,
- inconc (inconclusive) verdict describes a situation where neither a pass nor a fail can be assigned
- error verdict indicates an error in the test devices
- none verdict is the initial value for local and global test verdicts, i.e., no other verdict has been assigned yet.

During test execution, each test component maintains its own local test verdict. A local test verdict is an object that is instantiated automatically for each test component at the time of component creation. A test component can retrieve and set its local verdict. The verdict error is not allowed to be set by a test component. It is set automatically by the TTCN-3 run-time environment, if an error in the test equipment occurs. When changing the value of a local test verdict, special overwriting rules are applied. The overwriting rules only allow that a test verdict becomes worse, e.g., a pass may change to inconclusive or fail, but a fail cannot change to a pass or inconclusive.

In addition to the local test verdicts, the TTCN-3 run-time environment maintains a global test verdict for each test case. The global test verdict is not accessible for the test components. It is updated according to the overwriting rules when a test component terminates. The final global test verdict is returned to the module control part when the test case terminates.

TTCN-3 and other Languages

TTCN-3 supports the referencing of objects defined in other languages from within TTCN-3 test suites as illustrated in Figure 5.8. Foreign objects can be used in TTCN-3 only if they have a TTCN-3 view. The term TTCN-3 view can be best explained by considering the case when the definition of a TTCN-3 object is based on another TTCN-3 object. There, the information content of the referenced object shall be available and is used for the new definition. For example, when a template is defined based on a structured type, the identifiers and types of fields of the base type shall be accessible and are used for the template definition. In a similar way, when the base type is a foreign object it shall provide the same information content as would be required from a TTCN-3 type declaration. The foreign object, naturally, may contain more information than required by TTCN-3. The TTCN-3 view of a foreign object means that part of the information carried by the object, which is necessary to use it in TTCN-3. Obviously the TTCN-3 view of a foreign object may be the full set or a subset of the information content of the object but never a superset. There may be foreign objects without a TTCN-3 view, i.e., for some reason no TTCN-3 definition could be based on them.

The use of foreign objects in TTCN-3 modules is supported in two ways. Firstly, the language allows to import and use them (by referencing), secondly special attribute strings are defined which assure that a TTCN-3 module referring foreign objects will be portable to any tool supporting the other language. To make declarations of foreign object visible in TTCN-3 modules, their names shall be imported just like declarations in other TTCN-3 modules. When imported, only the TTCN-3 view of the object will be seen from the TTCN-3 module. There are two main differences between importing TTCN-3 items and objects defined in other languages:

- To import from a non-TTCN-3 module the import statement shall contain an appropriate language identifier string.
- Only foreign objects with a TTCN-3 view are importable into a TTCN-3 module.

Importing can be done automatically using the all directive, in which case all importable objects shall automatically be selected by the testing tool, or done manually by listing names of items to be imported. Naturally, in the second case only importable objects are allowed in the list. Currently, standard language identifier strings are specified for ASN.1, OMG IDL and XML.

More Features Required

By the use of TTCN-3 in a number of test systems, some features have been found missing from the language. These are described in short below.

- Lack of an open type: Although the TTCN-3 type *anytype* is an open type, but its use for generic code in TTCN-3 is limited. By definition, the *anytype* is a union over the known types in a module. Only values of these known types can be stored in a variable of type *anytype*. This is a limitation of this type because it means that it cannot be used in generic definitions. Examples of such generic definitions are for example the definition of the payload in messages of a transport protocol, or the definition of a hash table with arbitrary values. The potential values to be used at these places are usually not known when defining such generic definitions, as a consequence the *anytype* is of no use here. Note the definition of ports that allow the exchange arbitrary messages, i.e. type port message {*inout all*}, has the same limitations as the *anytype*.

- Synchronisation: Almost always, when the behaviour of a test case consists of the execution of several parallel test components, the execution of the behaviour of the parallel test components needs to be synchronised. Although this can be defined in a straightforward way in TTCN-3 and this leads to repetitive code. As synchronisation is a common problem it would ease the development of test suites if synchronisation primitives would be directly available as part of the language.
- Broadcast: Sending the same message to several communication partners, suffers from two problems: The first one is the problem to define a generic solution due to the lack of an open type as explained before. The second one is that sending a message to several other components means that their component reference must be known on the server side. This means that additional code needs to be written for some kind of registering the recipients of broadcast messages before the communication takes place. This is additional code that needs to be written. The second problem could be avoided by using port arrays for communications, but these solutions is not well suited for dynamic test configurations because the size of port arrays must be known at compile time.

Assessment of TTCN 2 and 3

In general TTCN-2/3 provides powerful means for modularising and parameterisation of test cases and test cases that are written in TTCN-2/3 are independent of Test Methods, Layers, Protocols and Test Tools. TTCN-2 .mp file and TTCN-3 core code are textual which allow for electronic storage, portable across different tools, conversion from/to other formats (.ps, .xml, .html) and translation into executable code.

Writing a compiler for TTCN-2/3 is seen to be difficult and TTCN-2/3 tools are expensive. Templates are needed to define each bit in so many test cases. Also the language cannot express hard real-time constraints and must say that long code needs to be written before you can actually do something useful.

Main TTCN-3 capabilities can be summarised as below:

- Test suite parameterisation
- Test case control and selection mechanisms
- Assignment and handling of test verdicts
- Aligned with ASN.1
- More than one presentation formats
- Dynamic concurrent testing configurations
- Various communication mechanisms (sync and async)
- Data and signature templates with powerful matching mechanisms
- Specification of encoding information
- Display and user-defined attributes
- Well-defined syntax, static semantics and operational semantics

VB/VC/Pearl/shell

Conformance test suites are also developed in programming languages (other than TTCN) like VB Scripts, C++, and Java, Pearl and Shell programming. These types of suites are mainly developed for the implementation where test system provides an *Application Programming Interface (API)* and no abstract test interface is defined for the test suite. Standard bodies do generally not drive these types of test suites implementations and are rather vendor specific leading to variation in test case implementation across different vendors. An example of this RF, USIM, USAT and OMA test cases.

Development of test suites is easy and less expansive as specialised tools and skills are not needed to develop the test suites. This also provides user full flexibility to develop the test suite without worrying about the tools limitations. Due to faster debugging capabilities of programming languages, development of test suite is also very fast.

The test suites are developed or customised for a particular implementation of test systems making it difficult to reuse the test suite for different test platforms. As the test suites are developed for a test platform, platform limitations are inherited in the test suite causing possible variance from the technology specifications.

Advantages of TTCN and Conformance Testing

Conformance testing has both cost and performance benefits compared to field testing. User does not require investment on expansive field equipments. It eliminates 80% of field errors in test labs and is performed on test equipment facilitating easy simulation of real scenarios. Conformance test specifications are designed based on the technology not on the implementation of the technology. This leads to improvements and evolution of the technology. Conformance test labs are designed to benefit both the product supplier and end user. It is a faster and economical way of testing enhancement and new functionalities. Conformance test results are comparable and provide mutual recognition.

These benefits have made many end users to only choose products that have passed the Independent lab test. And many more will follow the procedure as they become aware that purchasing conformance-tested products can eliminate most field errors.

TTCN is specifically designed for testing, due to the fact that the syntax and operational semantics of TTCN tests are commonly understood and not related to a particular programming language. Also TTCN tests concentrate on the purpose of the test and are abstracted from particular test system details. Another advantage of using TTCN is that off the shelf tools and TTCN-based test systems are readily available. In addition the language is constantly maintained and developed and maintenance of test suites and products is easier.

Chapter Six

Standardisation and Validation Bodies

TEST CASE DEVELOPMENT

In order to test the UE it is essential to specify the sequences of interactions, or test events, that the test system required to control and observe. A sequence of such events that specify a complete test purpose is called a test case. A set of test cases for a particular protocol is called a test suite.

Summary of Conformance Test Case Lifecycle

Conformance testing performed on a UE must be repeatable and conform to the 3GPP specifications, regardless of the SS equipment being used. To achieve this, formal test cases are written for each test. The tests concentrate on areas that are critical to functionality and interoperability, including testing an implementation's reaction to erroneous behaviour and are broadly referred to as "conformance testing". This chapter is mainly concentrate on the signalling test cases merely for illustration and simplicity.

Many distinct processes apply to the provision of test cases. Test cases are written in prose, describing in detail how the test is carried out and the pass and fail criteria. To ensure that each test is a true representation of the original intent of the test, a validation and approval process has been set in place.

Once the test case is verified by the supplier (3GPP), i.e. is satisfied that the test operates correctly, it is then given to an independent validation organisation to test for conformance with the original test specification and check for proper operation. When it has passed this test, it can be submitted to the relevant industry body for approval. After approval, it can be used in formal mobile terminal testing and certification.

The "outside world" view of the procedure, as well as the equivalent but distinct procedures for the other systems is shown in diagrams 6.1 and 6.2:

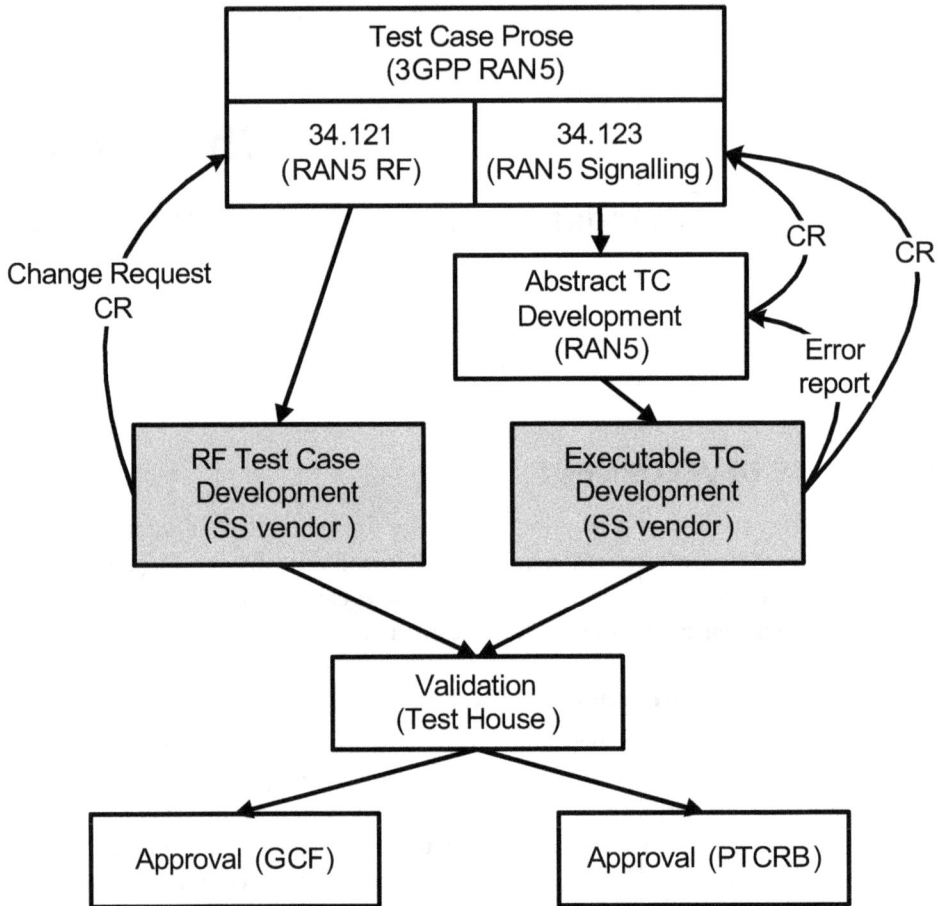

Figure 6.1   Single-RAT test case approval procedures

Test Case production follows the process shown in Figures 6.1, 6.2, 6.3 and 6.4. The diagrams show the major inputs and review activities that lead to delivery of approved test cases. It also makes clear the links between the major internal and external activities, and highlights some of the supporting components (such as version control and change management) that need to be carefully managed.

TTCN-2 Test Cases

All testing performed on a UE must be repeatable and conform to the specifications regardless of the test equipment being used. To achieve this, formal test cases are written for each test. Protocol test cases for UMTS are written in prose and then converted into TTCN-2 code for signalling and any other programming

UMTS to GSM ISHO   GSM to UMTS ISHO

Test Case Prose (3GPP RAN5)	Test Case Prose (GERAN)
34.123 (RAN5 Signalling)	51.010 (RAN5 Signalling)

Change Request
CR

CR

Abstract TC Development
(ETSI)

Error
report

CR

Error
report

CR

Executable TC
Development
(SS vendor)

Validation
(Test House)

Approval (GCF)   Approval (PTCRB)

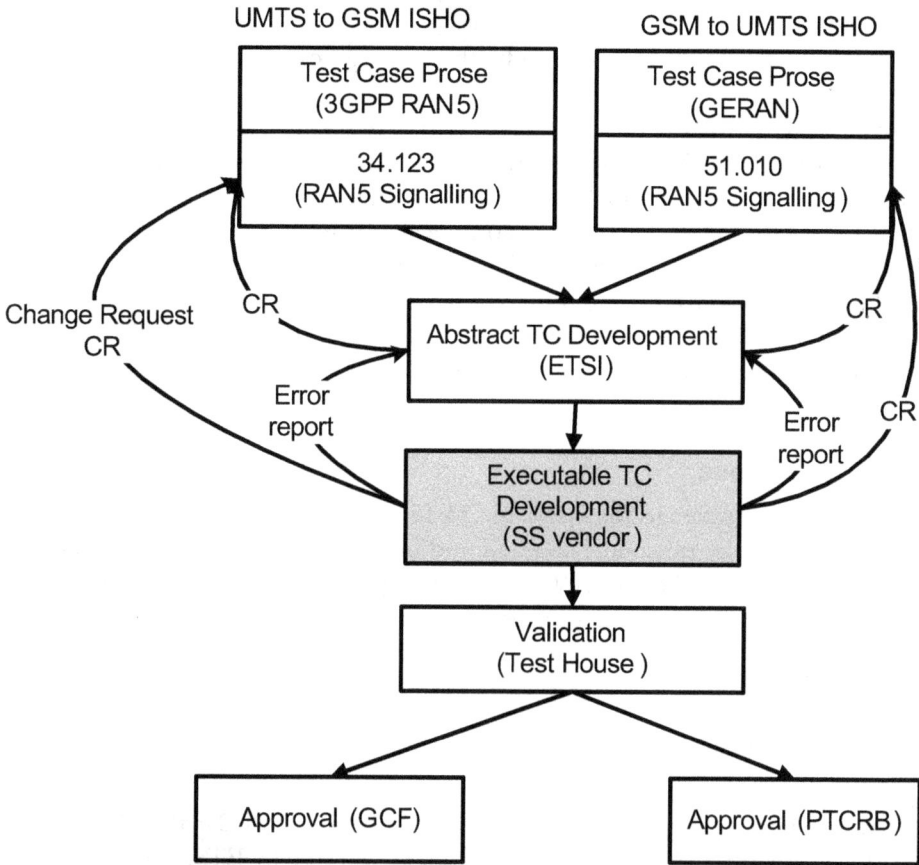

Figure 6.2   Multi-RAT test case approval procedures

language for RF conformance testing. In case of TTCN-2 it enables the test cases
to be compiled by a computer into a format (ETS) that can be run directly on the
target test equipment (SS).

Test cases are implemented in TTCN-2 (Tree and Tabular Combined Notation)
and are developed and approved by RAN5 (UE, SS, Network Operators and ETSI
MCC group). Contributions to the development of the TTCN-2 are made by
different SS and UE manufacturers, but are maintained and controlled by ETSI
including any changes or amendments on behalf of the 3GPP. Basic TTCN-2
code is developed and then reviewed and verified within the 3GPP community
via e-mail. Any subsequent changes to the test case are then handled at RAN5
meetings via the submission of Change Requests (CRs). Test cases for GCF and
PTCRB are the same but working on different frequency bands.

Note that prose CRs are handled at regular RAN5 meetings while the TTCN-2
CRs are handled at any time by submission to ETSI.

TTCN-3 Test Cases

In order for TTCN to continue to be the language of choice for testing future communications protocols, more capabilities were needed than TTCN-2 could provide. ETSI realised this and developed TTCN-3, encompassing all of the benefits of TTCN-2, removing any references to the telecommunications domain and extending the capability of the language to support more technologies than just OSI. TTCN-3 is seen as so radical that even the meaning of TTCN has been changed to "Testing and Test Control Notation". Finally, TTCN-3 facilitates an effective testing process that is applicable to all areas of software development. MCC-160 have decided to deliver IMS test cases in TTCN-3 as an experiment. It is already decided that LTE conformance test cases will be written in TTCN-3

Test Case Prose

The term Prose refers to the standards 34-123 part 1 as referenced in 3GPP. The prose is written by personnel at ETSI and other supporting companies and for 3GPP specification, the MCC160 & 272 are responsible for the conformance and protocol specifications which defines the following:-

- Test purpose
- Method of test
- Test procedure
- Expected sequence of test

Any changes to the prose must be reflected into the TTCN-2 in the 34.123-3, the Abstract Test Suite (ATS). It is the ATS that is delivered by ETSI for verification.

ETSI publish a number of Abstract Test Suites (ATS) which are made available to all interested parties. The System Simulators (SS) manufacturers then compile these ATS's into ETS's and execute these ETS's using handsets connected to the SS platform.

The following table 6.1 summarizes the 3GPP Standards used for Signaling Conformance test cases

34-123-1	User Equipment (UE) conformance specification; Part 1: Protocol conformance specification
34-123-2	User Equipment (UE) conformance specification; Part 2:Implementation Conformance Statement (ICS) proforma specification
34.123-3	User Equipment (UE) conformance specification; Part 3: Abstract Test Suite (ATS)
34.108	Common test environments for User Equipment (UE) - conformance testing

Table 6.1

Verification

This is a RAN5 process. Verification is carried out by the industry and RAN5, working in collaboration with each other.

The purpose of Verification is to check that the implementation of the TTCN-2 code has been done according to the prose in 34.123-1. This is done by executing the relevant test case (sometimes referred to as test script) on the target SS which simulates a real network scenario whereby the UE behaviour can be checked.

The TTCN-2 coding standard must be followed for writing the ETSI TTCN-2 to ensure maintainability in the future. The verification process does not have to use commercially available UE's, unlike the GCF validation process.

The output log which consists of signalling protocol messages is checked to ensure that it is consistent with the prose. In other words, Verification is a check that the test case is consistent with the prose and runs correctly.

TTCN-2 Test Case Life Cycle in Details

Protocol Conformance tests are written in TTCN-2 and are contained in 3GPP specification 34.123-3. Each test case is drafted (prose and TTCN-2), by ETSI or a company within the UE or test industry verification teams, on behalf of 3GPP and is debugged and verified on an SS against a real 3G UE prior to submission to 3GPP RAN5 for approval.

Once approved by 3GPP the test cases are validated in an accredited independent test house (or test lab) in accordance with the GCF rules as set out in GCF-OP. The SS manufacturers are responsible for getting this done. The test houses use the SS/ TP to do the tests. The results are then presented by the test house to the GCF for approval. The validation is performed on Single-RAT System Simulators and ISHO platforms. The process is laid down in the rules rather than being typical.

Below is detailed explanation to the above steps in Figure 6.3:

1 Unapproved 3GPP TTCN-2 Test Case (TC) available from ETSI. This may be developed by ETSI or any SS manufacturer.
2 Visual inspection of TC by ETSI members.
3 TC is made available by ETSI for debug and verification. For example the interim Working Document of week 31 is iWD_wk 31. This is a suite of all the test cases (an ATS), provided at regular intervals and tagged with the current week number.
4 The SS manufacturers are in competition to be the first to submit the TC and CR. Although sometimes the TCs are shared out by mutual agreement. Verified on SS with at least one UE. Change Request (CR) is generated and is posted on 3GPP website for e-mail approval for an agreed consultation period normally 2 or 4 weeks (refer step 6).
5 3GPP members review CR. During this approval time, MCC160 or other SS or UE manufacturers can send their comments and proposed changes, for industry to view and comment on ETSI to incorporate any changes into the

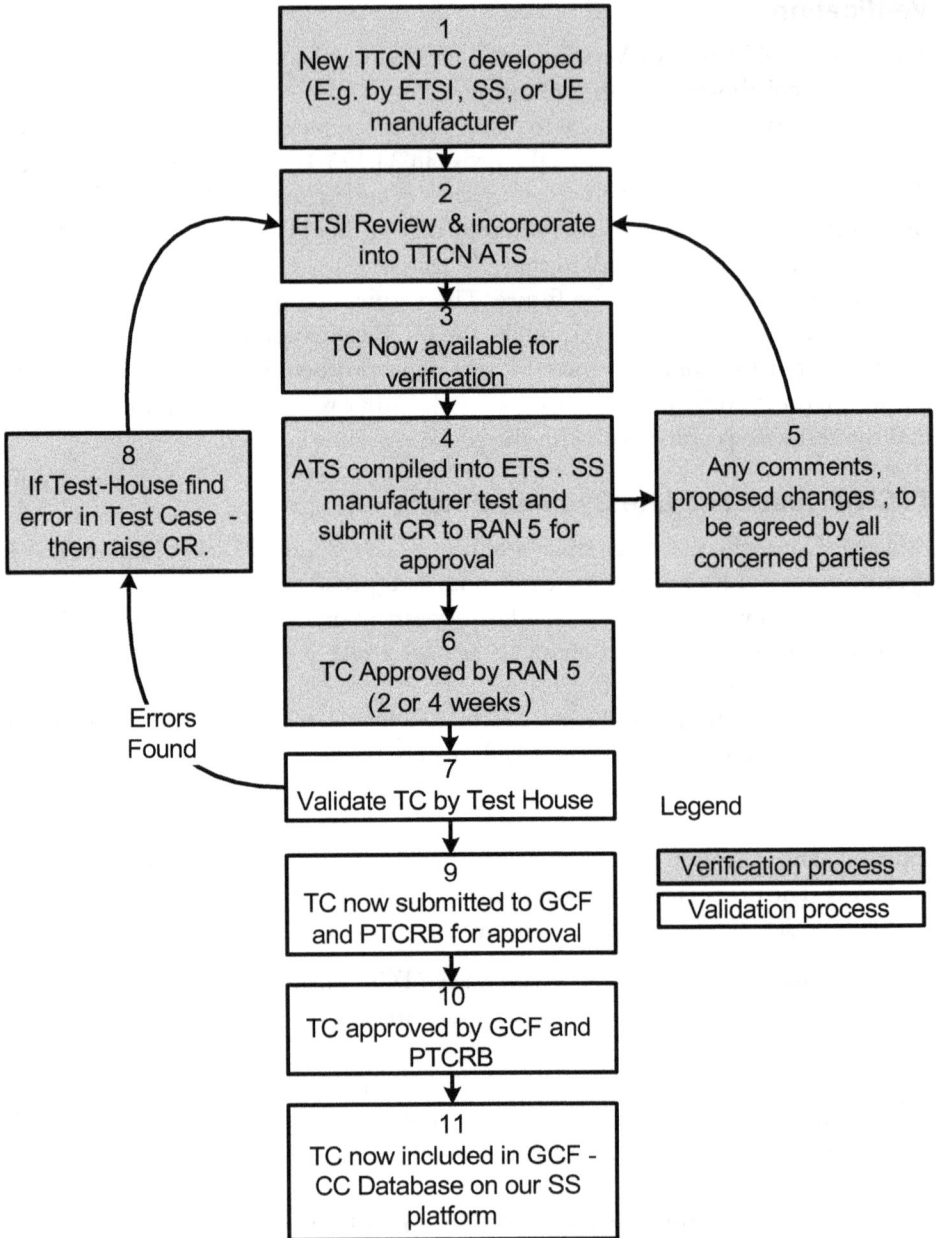

Figure 6.3   Test case lifecycle

ATS and re-issue the ATS. Subsequent amendments and changes to the original CR can be made and submitted again under the discretion of ETSI.

6 TC approved by RAN5. At the end of the consultation period, the test case is deemed approved, if there are no objections.

For RAN5 CR consultation times are:-

- 2 weeks - if TC verified using 2 UEs or using one UE if similar TCs are already approved
- 4 weeks – if verified using one UE only and as long as no other similar TC is approved.

If there are any comments during the consultation period, these are mutually agreed by the parties concerned and the final decision will be made from ETSI MCC160 group. The final status of test case is sent by the RCM from ETSI to the 3GPP reflector to highlight if the test case is approved or still pending approval.

7   TC ready to be validated by Independent Test House
8   If during validation, the Test House finds an error in the test case, then the contributing company will raise either a TTCN-2 or Prose CR to fix the error. The test case is deemed unapproved until the CR is formally approved by RAN5 and therefore, the TC will not be validated. The test house can't restart the validation activity until RAN5 has re-approved the TTCN-2
9   If the test case passes validation, then this is submitted to GCF and PTCRB for approval via a validation report.
10 Test case is approved by GCF or PTCRB
11 If approved (if no challenges from industry on GCF reflector), then this TC is finally included in the GCF-CC database.

The main purpose of this process is for industry to verify and prove the test case by verifying the validity of the test case by running on a System Simulator (SS) with a conformant phone (UE).

Test Case Validation

This is the responsibility of GCF and PTCRB. The purpose of validation is to compare the execution of test cases for different UE's in different SS. Validation is to confirm the correct execution of test cases against the requirements specified in the prose for different UEs (at least 2) with different SS. Each SS manufacturer submits their own test cases to a test house, to confirm that it is OK on their particular SS. The Test Case is checked by external validation bodies to confirm that the test runs correctly.

The already approved test cases are available in TTCN-2 ATS suite form and these can be downloaded from the 3GPP web site. These are then build locally with system simulator software which then can be executed on a System Simulator (SS) using a real phone (UE) and logs generated. The logs detail the test steps and test method followed and should follow the test case specification. The hardware build standard of the SS is defined beforehand and strictly controlled. The TCs are contained within a declared SS software release.

The logs are then checked against the test prose contained in the conformance test specifications for correctness. Any differences found between the logs and test specifications i.e. the test cases are NOT compliant with the Conformance test specifications, a CR is raised by either the Test House or the SS manufacturer.

Validated test cases must have been tested by the test house with at least 2 independently developed handsets (UE) and the test implementation should correctly give Pass verdicts for every single test case path. If a test case fails due to a problem within the TTCN-2 then a CR will be generated to correct the failure.

The following supporting evidence is required by the test house for validation to take place:

- A working .mp file (ETS file)
- An execution (e.g. SS) log
- Any CRs applied to the TTCN-2 of the base version. The base version is the iWD TTCN-2 version used as delivered and downloaded from the 3GPP server. This forms an integral part of the SS software.
- A PIXIT information. PIXIT contains information regarding the physical setup and connection of the test that is not part of the protocol. This information could be the system under test, hardware, socket, etc.

The following table is an extract from the GCF-OP [1] document which details the steps that have to be undertaken by the test house as part of the Validation Process. The test house is required to provide a statement of compliance against each process item. Typical responses have been shown in Table 6.2.

For validation purposes the test house must run the test case using 2 UEs. Recent changes to the validation rules means that both UEs MUST give a 'Pass' verdict, before it was required that only one should pass.

External validation of test cases

For most of the widely adopted mobile radio standards, a standards body grants the formal "approval" to test cases that have been fully reviewed and validated.

To ensure that different suppliers' test solutions produce consistent results, a standardised validation process is defined by GCF and followed by the Validation organisation. Hence, the validation of test implementations is performed by a qualified external Validation Organisation. The validation process used has been set up to be consistent globally across Validation Organisations. A key aim is that Validation shall be reproducible and repeatable.

For approval, test cases are submitted to GCF meetings which occur every three months. Prior to this, test cases are developed and approved within 3GPP RAN5. 3GPP develop TTCN-2 and publish it in a 'verification branch' for industry to prove it. When the TTCN-2 has been proven, it is published within a 'formal branch'. 3GPP define on which version of the core specifications the TTCN-2 should be based for submission to a given meeting.

In order to move things on as rapidly as possible, change requests can be submitted via email and will be accepted unless objections are raised within 5 days, the 'five day rule'.

Validation Process Item (GCF-OP, 8.4.4)	Statement of Compliance
1 Confirm that the Test System is capable to run certain TC (see 8.4.3.3 and 8.4.5.1)	CETECOM confirms that the requirements according to GCF-OP 8.4.3.3 and 8.4.5.1 are fulfilled
2 Confirm that the available UE can be used for validation of certain TC (see 8.4.3.2)	CETECOM confirms that requirements according to GCF-OP 8.4.3.2 are fulfilled
3 Confirm the 3GPP version of the core specification being tested and ensure the correct test specification is being used.	CETECOM confirms that the correct 3GPP core specification baseline and test specification has been used.
4 Confirm the test case source code fully conform to the 3GPP test specifications including thorough verification of initial conditions, the test sequence, message content and timer values Any deviations from the 3GPP test specification shall be documented in the validation report.	CETECOM confirms that the test case source code fully conforms to the 3GPP test specifications including thorough verification of initial conditions, the test sequence, message content and timer values where appropriate.
5 Confirm that message flow in the trace file gives the expected behaviour as described in the relevant 3GPP test specification. (e.g. initial conditions, test procedures, part-verdicts, etc)	CETECOM confirms that the message flow in the trace file gives the expected behaviour as described in the relevant 3GPP test specification.
6 Confirm the test case runs on the test platform against at least 2 independently developed reference terminals. For at least one of these terminals the test implementation should correctly give all pass verdicts. If possible, failure verdicts should also be reached with at least one of these terminals.	CETECOM confirms that each of the listed test cases runs on the test platform described and against at least 2 independently developed reference terminals. For at least one of these terminals the test implementation correctly gave all pass verdicts.
7 Confirm stability of method of test.	CETECOM confirms the stability of the methods of test

Table 6.2   Extract from GCF-OP [1] document details the Validation Process

Fundamentals of the SS Manufacturer Policy

The SS manufacturer's policy is to secure formal external approval for all test cases wherever possible, although it also makes available working test cases that cannot or do not meet the validation requirements for some reason. The deliverables to the validation authority are:

- Adequate User Documentation
- TC manual - to provide all information required for the validation or running of the test case. In general, this document describes all parameters or values used for the test case. TC manual can be part of the User Documentation.
- List of test cases supported from a specific version of the relevant test specification and against a specified version of core specification

- Configuration control information, enabling re-creation of the test system if required
- Readable (by a human) Test Case source code.

More Details in GCF-OP [1]

Validation Workflow

The overall workflow is shown in the diagram below Figure 6.4 [1]. A key issue is the version of the core specification to which the validation applies. Certification can only be performed against that version of core specification.

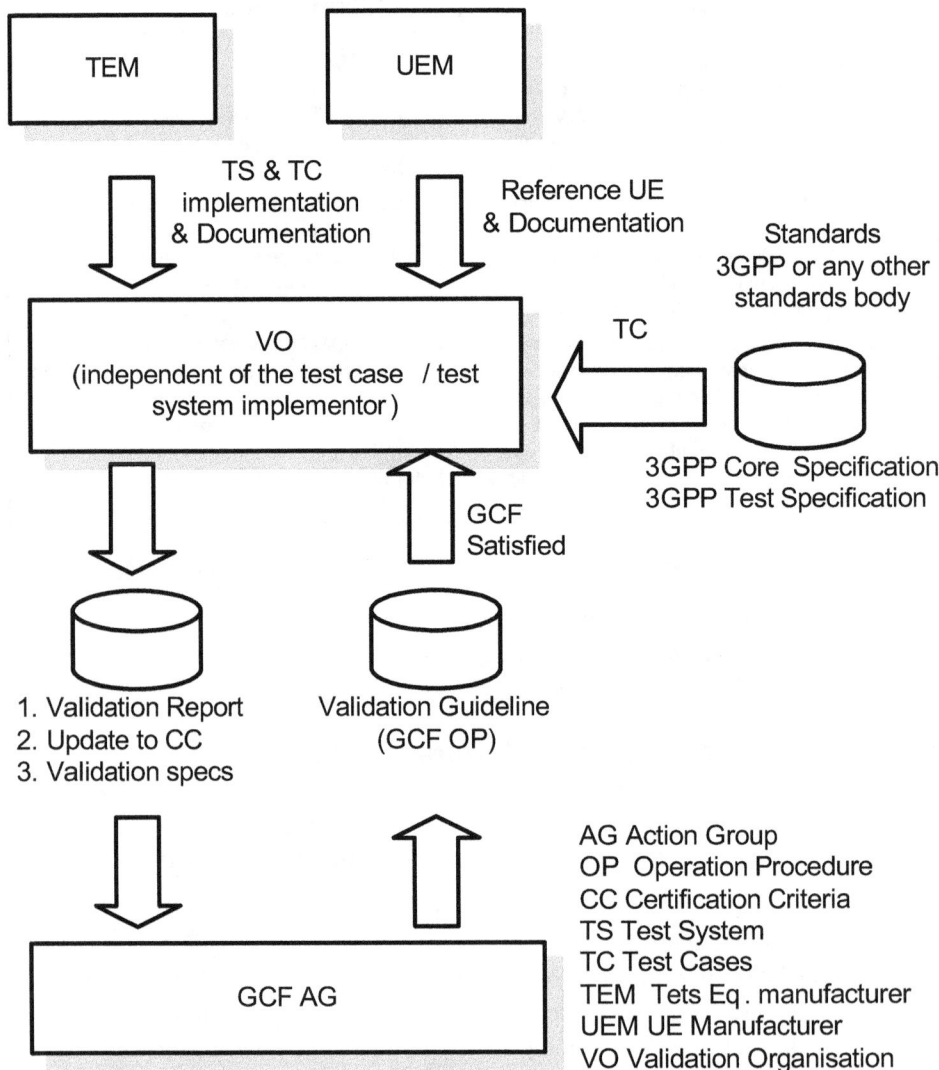

Figure 6.4   Validation process [1]

Validation Reasons

During submission of VRs to GCF by Test Houses, the test cases are categorised within the VR with the following reasons:

1 Test re-validated and GCF-CC category changed back to category "A", "B" or "C"
2 Test re-validated and re-introduced into GCF-CC category "A" since more than 45 days elapsed.
3 Test re-validated and no GCF-CC category is changed.
4 Test is validated at new test platform but no GCF-CC category is involved.
5 Validation is not valid and GCF-CC category and/or TP entry is changed if necessary.
6 Test is not explicitly re-validated, however the previous validation is valid in the new version

Validation Categories

On first validation and first inclusion of a TC into the GCF-CC, it is initially categorised as category "A, B" or "C"

These are the test case categories defined by the GCF.

Submission Rules to GCF

GCF originally introduced these procedures in order to minimise delays in getting new material published and approved.

The process allows documents to be placed publicly on the web-site for review by users and either challenged during this period or, in the absence of any challenges, accepted by default.

In order to notify web site users as soon as new 5-day or 10-day rule documents are available for review the web site automatically distributes e-mails to designated users. (Users wishing to receive 5-day or 10-day rule notifications must ensure that a correct e-mail address is registered on the GCF web site together with the appropriate notification settings)

10-day rule (CAG Meeting Approval)

The 10-day rule applies to new tests and would normally be presented as a Change Request to a meeting.

Validation of new test cases, i.e. promoted from Provisional category (P) to Approved category (A), can only be done at a meeting and must be presented to members 10-days in advance of the meeting.

The 10-day rule is used for submission of documents (also known as Validation Reports - VRs) which require changes to PRD's and other general documents, and these can be submitted at any time to the GCF web-site.

Test cases that have not been validated on another Test Platform (i.e. they are currently at Category P), must be submitted under the cover of a VR using the 10-day rule for approval at the next CAG meeting which are held quarterly.

Category	Status	Notes
Category A (or Class A)	Full Status. Based on tests which have been specified by an appropriate SDO e.g. ETSI and validated on commercially available test equipment.	The test has been validated and is applicable for the purposes of Certification. This class is used for all test purposes covered by validated Means of Test.
Category B (or Class B)	Validation Status. Based on tests which have been specified by an appropriate SDO and validated on commercially available test equipment with exceptions.	This class is used for test purposes which are not completely validated. The parts of the test which have been validated are applicable for the purposes of Certification. Cases where a test may not be wholly validated include: a) Means of Test where test purposes are not fully tested, because parts of the Means of Test are not complete and/or incorrect and therefore not valid for accredited testing. b) Means of Test where a "PASS" verdict is a valid verdict, but a "FAIL" verdict may not be valid.
Category C (or Class C)	Validation Status. Based on tests which have been specified by an appropriate SDO and validated on commercially available test equipment, but do not yet meet the RAN5 Work Item or Test Case certification entry criteria	The test has been validated but is not yet applicable for the purposes of Certification, i.e. the Work Item Entry Criteria or the Test Case Certification Entry Criteria is not yet met. This class is used for all test purposes covered by validated Means of Test.
Category D (or Class D)	Downgrade Status. Based on tests which have been specified by an appropriate SDO from which validation or full status has been removed.	The test has previously been in full or validation status but full or validation status has been removed by "5-day rule" or at an Agreement Group meeting (downgraded), and less than 45 days has elapsed since the change to Downgrade status was approved by the Agreement Group. This category will apply for 45 days, during which time the test case can be re-validated and return to Cat A. If it is not re-validated within 45 days, the test case will automatically go to Cat P, requiring a validation report to be submitted to a CAG meeting.
Category E (or Class E)	Based on tests which have been specified by an appropriate SDO and validated on commercially available test equipment, but do not yet meet the Work Item or Test Case certification entry criteria.	Category E for test cases validated before the 80% criterion is met and how they become Category A. A minimum of 80% rule is applied to calculate the GCF Work Item (WI) for UE certification.

Table 6.3

Category	Status	Notes
Category P (or Class P)	Provisional Status. Based on tests which have been specified by an appropriate SDO and not validated on commercially available test equipment	This class is used for Test purposes which are planned for future implementation within the GCF scheme, or for tests that have been in Downgrade status for more than 45 days.
Category N (or Class N)	Test Not Applicable Status. This class covers the case where a test purpose is not valid for a particular band.	This class covers the case where a test purpose is not valid for a particular band. This class is used for test purposes which are not applicable to a particular GSM frequency band.

Table 6.3 (*Cont.*)

SS manufacturer attend the meeting as observers. 10 working days leading up to the meeting date are required during which the report can be challenged.

NOTE. '5-day rule' means 5 working days, that is 7calendar days. Similarly the '10-day rule' should be interpreted as 10-working days, or 14 calendar days. More details on the definition of these rules is documented in the GCF Organisation Procedures document GCF-OP [1].

5-day Rule

The 5-day rule is usually used in the following situations:

- downgrades to previously approved test cases due to problems
- validation of test cases that aren't new - i.e. a "means of test" has been previously accepted
- upgrade of existing test cases that had previously been approved (within the last 45-days)

The 5-day process can be used at any time, and is not restricted to the period leading up to GCF approval meetings. Often though, there will be a flurry of activity just before, or just after a GCF meeting, so that the latest updates can be made available at the same time as any newly approved test cases.

In summary, for any test case that have already been validated on one or more test platforms the validation report can be submitted for approval at any time, and 5 working days are then allowed during which the report can be challenged. If no challenges are made, the report automatically becomes approved.

110-Day Rule – New Version of GCF-CC Database

Whenever a new version of GCF-CC database is introduced, the previous version will remain valid for a set period referred to as the **Overlapping Period**. Currently this is 110 calendar days. During the Overlapping Period, the UE manufacturer may chose to Test and Certify a phone declare against the previous or the new version of the GCF-CC, but not a combination of versions. After the Overlapping period has elapsed, the previous version of GCF-CC will expire and cannot anymore be used as basis for certification. Note that the GCF 110-day rule is a

9 months period in PTCRB. Any changes to the status of the test cases are maintained in this database.

UE Certification

A UE is deemed certified if it has passed an agreed set of conformance tests and field trials. To test a new handset (UE), the manufacturer generally approaches a qualified test house. The test house will run the required approved test cases and state that the handset is suitable for use on the available networks. Once certified, their phones should operate on any network.

3GPP has overall control of the test cases for GSM and UMTS. The validation and approval of the implemented test cases then are handled by the GCF.

The North American version of GSM running in the 1,900-MHz band (referred to as PCS), is now complemented by another allocation at 850 MHz. The group known as PVG handles the approvals, and its results are ratified by the PTCRB. Phones then are tested against the test cases that, if successful, are certified by the CTIA. To achieve CTIA certification, it is necessary for phones to be tested in CTIA-approved laboratories.

The procedures governing external approval of test cases are established and maintained by each certification body. SS manufacturer's internal activities have to be undertaken to synchronise and comply with these procedures.

Downgrade of Test Cases

Test cases can be downgraded mainly due to the following reasons:-

Prose CR -   At RAN5 meetings a prose CR is approved as a result of test or core specification changes, and has impact on test case

TTCN CR -   These are raised as a result of regression testing on the TTCN-2 delivered by ETSI. Regression testing is performed by SS manufacturers when a new version of TTCN-2 iWD_wkxx is released from ETSI. New versions of this ATS's are released roughly every 2-3 weeks. For example if iWD-wk27 is the current version and the next delivery is iWD-wk31, then any TTCN-2 errors found in wk27 regression will be implemented in wk31.

Multiple Test Case Paths

Some test cases contain a number of different paths. Sometimes a test can be validated for some but not all paths. An example of this Ciphering whether is enabled or not and CS or PS path. Sometimes, single test case can be quite complex containing multiple paths.

Cut-off point for Change Requests

Change requests (CRs) are processed and agreed on an almost continuous basis.

Once a CR is agreed, any related Test Cases needing approval must incorporate or conform to the agreed change. This may involve some last minute amendments to the code, and this situation can be particularly challenging if the deadline for validation and submission is approaching.

In the period leading up to a meeting of the certification body, and in order to allow sufficient time for final validation and submission, there is usually a "final date" for CRs. This is effectively the point where the test specification standard applicable at approval time is confirmed (by the appropriate committee). Any change requests generated after this date will be carried forward to feature in later releases (after the current certification approval meeting). This may be a notional date, or it may be explicitly declared by the relevant authority.

UMTS Test Case Priority

Tests are divided into a number of categories, e.g. RF, protocol, U-SIM, IMS, etc. Test cases, when first defined, are assigned a priority stage (and are classed "provisional"), each category has 4 "priority stages". Industry has to complete at least 80% of a stage before proceeding to next stage. Manufacturers can begin work on the next stage in one category, but validation reports are not accepted until the previous stage is 80% complete for that category.

TC DEVELOPMENT AND RELEASE PROCESS

Test Case SW Requirements

TC SW requirements are the requirements that are key features for a suite of test cases that can be successfully marketed, but that are not covered by the published standards. That is, they are outside the scope of the essential tests specified for verifying inter-working of the mobile terminals. The requirements are usually fairly general, and would normally be applicable across a wide range of application test software. Many of them are inherently covered by the operation and behaviour of the Test System, together with its API and an adaptation SW. Even so, developers must be clear which specific requirements apply to the work in hand since it may have a bearing on the way scripts are written or how they are evaluated. These requirements must cover aspects such as:

- GUI features and intuitiveness
- Campaign management
- Correct interfacing to the Test System API
- Performance / speed of execution
- Reliability
- Exception handling / recovery from errors
- Logging / Presentation of results
- Special modes of execution (e.g. diagnostic)

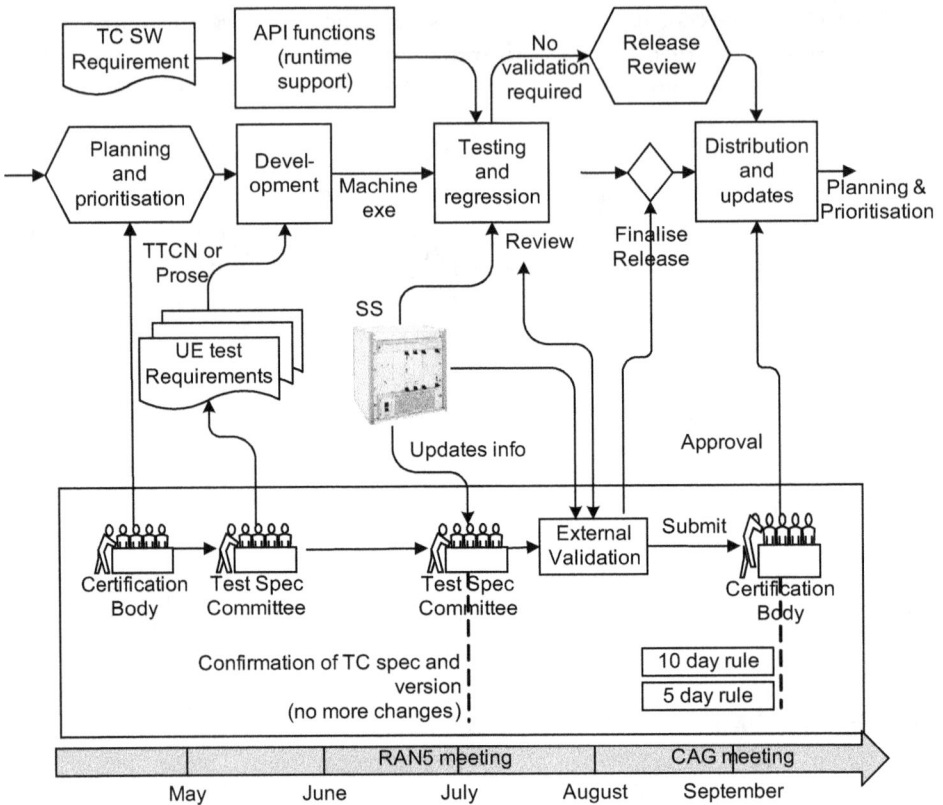

Figure 6.5 Internal and external interaction in TC development cycle

These requirements are not specifically assessed by the validation tests, but nevertheless should form part of the review and release considerations, see Figure 6.5. (In other words, even though test cases have successfully passed the external validation exercise, which is not on its own the only consideration when deciding that something is ready for release.)

API Functions and Runtime support for Test Cases

Test requirements are prepared so that they are independent of any specific test system solution. They are oriented mainly around the expected performance of the mobile terminal, and are often referred to as "abstract tests".

The test system's API (Application programming interface) and related "Adaptation SW" functions map these more generalised test operations into practical machine executable steps, and where possible they provide optimised functions to aid performance and maintainability.

The adapter and library functions also serve the needs of the SS vendor's TC SW Requirements.

Adaptation SW

This component provides the following:

- An adaptation layer between the machine independent C code produced by TTCN-2 compiler and the Test System. It is linked to the machine independent C code to form an executable test suite, or ETS
- A standardised interface both to the machine independent C code and from the executable test suite to the outside world
- The interface between the machine independent C code and the System Simulator via the User API. It also allows information on the executable test suite to be read and test cases to be selected either individually or in groups for execution.

Test Cases - Testing and Regression checks

This stage of the process is where executable test script code comes together with the test system, mobile phone(s), and system API and Library functions.

The test scripts themselves may have been developed internally or by an external partner, or they may have been generated automatically from a compiler tool. In many cases it is dealt with updates to existing test cases, and often there is high confidence in the software being assessed, and typical testing is done on a sample basis. That is, selected tests are performed on selected mobile phones.

Testing and Regression checks performed within an SS vendor are the primary means of proving that the software works and performs to the SS vendor standards. It is important to note that External Validation cannot be regarded as the primary means of proving that the software works. That is an external and independent check, using mobile phones that may possibly not be available within the SS manufacturer premises. But it remains the SS manufacturer's responsibility to develop and assess the software to a good standard.

Prior to external validation, a Final Design Review checkpoint shall be convened. The aim of this is to clarify the overall position and to minimise the risk of inefficient use of the limited and potentially expensive validation effort.

Sample testing

This can be achieved via test records, in the form of logs, Problem Reports or Change Requests to standards.

Review and Release of Test Cases

For Test Cases, the review process and the release process merge together and are closely tied to the external validation activities. In practice, test cases are often checked and validated a few at a time (drip-fed), and so the merged process is used incrementally to oversee the cycle of checking, validating and releasing the software. This cycle culminates in submission for approval, which is completed by the validation authority.

In some cases, particularly where rigorously reviewed TTCN-2 TC is not used as the basis for the tests, feedback is provided by the test house indicating problem areas, and rework is carried out.

- Early in the cycle of external validation, a Final Design Review checkpoint shall be convened. The aim of this is to clarify the overall position and to minimise the risk of inefficient use of the limited and potentially expensive validation effort.
- A record is to be maintained of the Test Case versions that are handed across for validation.
- The SW must be defined, released and available as a prerequisite for approval by certification bodies.
- Software submitted for approval shall be documented as a normal, full release (not a Preview).

A Final Design Review checkpoint takes place early in this stage of the work, ideally before the first test case is handed over to the validation authority.

This checkpoint is mandatory, as SS manufacturer should not offer anything for formal assessment unless it's been through checks and assessment, even if these are limited. Among other things, the review confirms which Test System (simulator) version is applicable to the validation.

This is a necessary coordination checkpoint, to confirm that the SS manufacturers are making best use of calendar time available; avoiding unnecessary cost or wasted effort; make sure the Company's credibility isn't harmed by releasing poor quality software.

The checklist is as follows:

- Understand the intent, i.e. What is expected or ought to do prior to starting validation.
- What CRs are included?
- What new test set features are to be exploited.
- Strategy for regression tests / sampling (e.g. when system or common code changes).
- What release of System Simulator is applicable? Clarify its status.
- Priorities for validation: What test cases can be included (in the limited time available) and which will omit; what is the confidence level? The expected content (draft list of test cases and versions).
- Check and confirm the plans and conventions for remaining testing, problem reporting and external liaison.

(For SW designed in-house) various additional questions...

- Plan or review the Release Notes. What needs to go in them?
- What are the risks / compromises.
- Plans and priorities for next iteration.

The Release is finalised on completion of validation, at the point where the submission is made to the certification body. At this point the versions are frozen, and the Test Suite and related Options are baselined and perform final checks, i.e.:

- Update the master Test Case status database.
- Build final install packages.
- Establish and document the final baseline for the Package (the whole SW suite).
- Record any deficiencies and outstanding actions.
- Finalise and approve the SW Release Approval Form. The form should contain the identity and versions of all the constituent Test Cases.

Note that the external approval status is not recorded on the software release paperwork. This is due to the fact that the SW often not formally confirmed at the time of release and may change over time. Also note that a package of test cases can contain test cases which each have a different validation status.

Distribution and Updates of Test Cases

Often the main point of reference for distribution is the Company's Download Portal. This should only contain software that has been formally released. Snapshots of complete sets on CD-R medium may be made available for backup purposes at GCF approval milestones or quarterly, whichever is the more frequent. These sets shall be documented on the SW release paperwork, which lists the identity and versions of all constituent Options.

Planning and Prioritisation

The potentially large number of test cases to be developed is tackled in phases, largely determined by external certification bodies. Priority is given to the tests that are critical to successful operation of phones on the network, while some of the optional or more specialised features tend to be left to a later date. Emphasis is placed on maintaining visible progress in making validated test cases available in order to meet the overall goal of sustaining a smooth process and minimising delays.

When planning work is carried out by an SS manufacturer, account is also taken of the needs of key customers and the effect of competition from other providers of test solutions.

Test case implementation priorities are based on the following

- Test system features to be promoted, and that will benefit sales of test solutions
- Availability of TTCN-2
- Priority stages from GCF

Tests are divided into a number of categories, e.g. RF, protocol, U-SIM, IMS, etc. Test cases, when first defined, are assigned a priority stage (and are classed "provisional"), each category has 4 "priority stages". Industry has to complete at least 80% of a stage before proceeding to next stage. Manufacturers can begin work on the next stage in one category, but validation reports are not accepted until the previous stage is 80% complete for that category.

Test Case Development

The final objective of this is to produce machine-executable software that correctly implements the test case on a test system.

Projects use a structured design method and tools that promote software reuse and a good maintainability. Peer design reviews shall be used as the primary means of design validation. The main input to the TC development stage in whole process is the set of test specifications, which have been defined and approved by external technical bodies (3GPP). Depending on the communication standard, they are available to us in either prose or TTCN-2 or TTCN-3 form.

TTCN is used for specifying protocol tests for the UMTS systems, and are provided with specific test details that have achieved an initial level of approval. Subsequent development work consists primarily of compiling the TTCN with the relevant test system adapter libraries and then testing the implementation on the physical test platform. Where problems or errors are found in the original specification, these are reported back using the change management system.

Prose for example is used for 3G USIM, RF Test cases and for the GSM, GPRS and CDMA systems. From this prose, scripted procedures are designed and produced in e.g. C or Basic. A design document captures the key high level and low level features of the proposed implementation, and this is reviewed prior to coding. To aid maintainability and efficiency, common library function are created, and Test system API functions are extended and generalised into higher level API functions.

The development of some Test Cases could be undertaken by external suppliers or partners, who carry the majority of the work described here according to their own methods and procedures. The main deliverables from the supplier are:

- Test Script Design document (not required for TTCN-2 scripts)
- Machine-executable code
- Error reports relating to TTCN-2 specifications
- Change requests relating to prose specifications
- TTCN-2 submissions for selected Tests

UE Test Requirements

The essential requirements for the testing to be performed on UEs are determined by expert technical committees working under the authority of the certification and standardisation bodies. The committees have established appropriate review and change management procedures to help ensure that tests can be implemented expediently and accurately. The tests are published as specifications, either as prose or in TTCN-2/3 notation.

Note that other requirements relating to the ease of use, logging and reliability are a commercial concern and are handled internally by the SS manufacturer. Many of these performance attributes are determined by the Test System, together with its API.

SS manufacturers sometime work on TTCN-2 only on behalf of ETSI although some Companies choose to do this in advance, at their own risk. For protocol tests, the TTCN-2 approved by 3GPP TSG is the definitive description of test cases and shall be used in the first instance by the test equipment industry to develop validated test cases.

Other protocol test implementations are allowed as long as they are based on prose for which there is a corresponding approved TTCN-2 test case contained in 3GPP TS 34.123-3 or TS 51.010-5. This approval will take place at the 3GPP RAN5 or TSG-GERAN meetings as appropriate.

TTCN TC Baseline handling

Figure 6.6 illustrates how baseline versions of TTCN-2 are published (e.g. TTCN-2 Ver 370), and are then the subject of change requests. Intermediate versions with corrections and enhancements are then made available in the form of "Interim Working Documents" or iWD. Much effort is spent on testing with various mobile terminals and test systems to ensure the material in the iWD has not regressed.

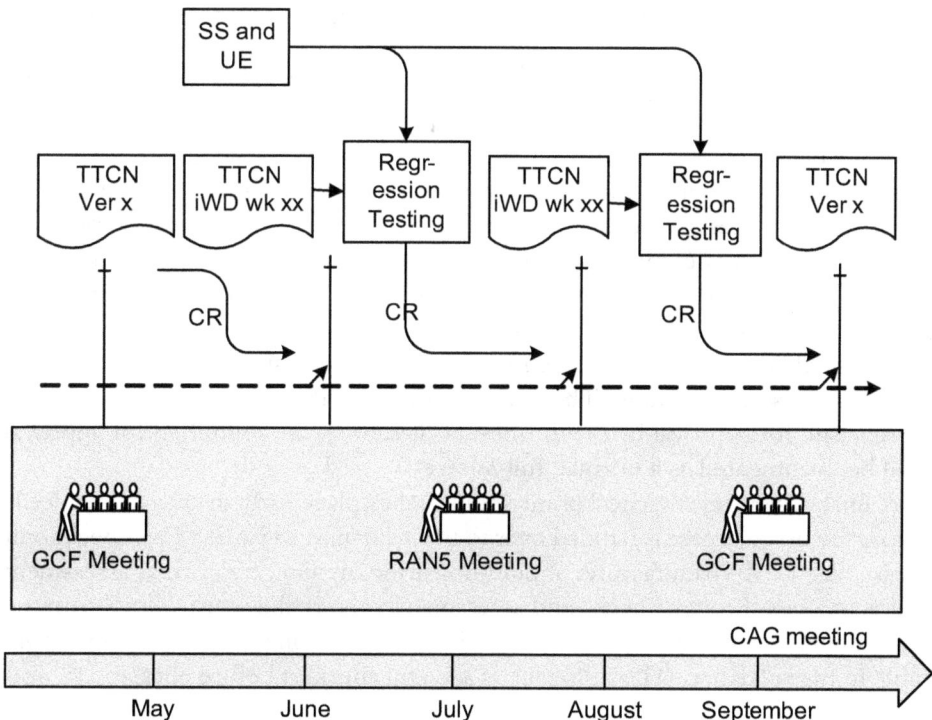

Figure 6.6 Simplified TTCN-2 SW Evolution

Finally, amendments are consolidated into a new release of TTCN-2 ratified by the certification body.

Test System SW/HW Updates

At the time of preparing the Test Case software for release, it is important to be clear which Test System release the software is to operate on.

Test validation covers the complete test environment, that is, both the Test Case software and the Test System on which the tests execute. The test system is also referred to by external parties as the System Simulator (SS) and also as a Test Platform (TP).

Test houses are obliged to audit a test platform to ensure that its configuration is properly controlled etc. The Technical Authority should therefore review the platform to ensure that it is suitable for submission to the test house, and should make sure that it is formally released.

Any changes (hardware or software) could affect validation, and therefore have to be communicated to the validation authority. They judge the significance of the changes and whether it is necessary to re-test previously validated test cases.

Review and Release of Test Cases

For Test Cases, the review process and the release process merge together and are closely tied to the external validation activities. In practice, test cases are often checked and validated a few at a time (drip-fed), and so the merged process is used incrementally to oversee the cycle of checking, validating and releasing the software. This cycle culminates in submission for approval, which is completed by the validation authority.

In some cases, particularly where rigorously reviewed TTCN-2 is not used as the basis for the tests, feedback is provided by the test house indicating problem areas, and rework is carried out. Early in the cycle of external validation, a final design review should be convened. The aim of this is to clarify the overall position and to minimise the risk of inefficient use of the limited and potentially expensive validation effort. A record is to be maintained of the Test Case versions that are handed across for validation. The SW must be defined, released and available as a prerequisite for approval by certification bodies. Software submitted for approval shall be documented as a normal, full release.

A final design review checkpoint should takes place early in this work, ideally before the first test case is handed over to the validation authority. This checkpoint is essential, as SS manufacturer should not offer anything for formal assessment unless its been through checks and assessment, even if these are limited. Among other things, the review confirms which Test System (simulator) version is applicable to the validation. The following is a useful checklist before release:

- Understand the intent, i.e. What SS manufacturer expected or ought to do prior to starting validation

- What CRs are included?
- What new test set features are to be exploited
- Strategy for regression tests / sampling (e.g. when system or common code changes)
- What release of System Simulator is applicable. Clarify its status.
- Priorities for validation: What test cases can SS manufacturer include (in the limited time available) and which will omit; what is our confidence level? The expected content (draft list of test cases and versions)
- Check and confirm the plans and conventions for remaining testing, problem reporting and external liaison.

For SW designed in-house various additional questions...

- Plan or review the SW release rotes. What needs to go in them?
- What are the risks / compromises
- Plans and priorities for next iteration

The Release is finalised on completion of validation, at the point where the submission is made to the certification body. At this point the versions are frozen, and the Test Suite and related Options are baselined. Then update the master Test Case status spreadsheet, build final install packages and establish/document the final baseline for the Package.

APPROVAL OF TEST CASES

For most of the widely adopted mobile radio standards, a standards body grants the formal "approval" to test cases that have been fully reviewed and validated.

At the conclusion of validation testing, the external test house submits the relevant documents for approval by the Certification body.

Assuming it is successful, the approval is signified by means the email reflector, and the GCF CC Database.

An important requirement or condition of approval is that the software must be commercially available, and so the Company's own release processes be concluded at this point.

Note however, that not all standards that are currently followed have a certification body. In these cases the software is published to good commercial standards and is typically reviewed with relevant experts on external committees. It is also worth noting that the Company expects on occasions to deliver Test Cases that have been unsuccessful in gaining formal approval.

The role of the certification bodies

Standards organisations and other related bodies have agreed to co-operate for the production of a complete set of globally applicable Technical Specifications for the 3rd Generation Mobile System based on the evolved GSM core networks and the radio access technologies supported by 3GPP partners. The Project is entitled the "Third Generation Partnership Project" and is known by the acronym "3GPP"

The 3GPP was established in December 1998. The collaboration agreement brings together a number of telecommunications standards bodies which are known as "Organizational Partners" - ARIB, CWTS, ETSI, TTA, TTC, etc.

The original scope of 3GPP was to produce Technical Specifications and Reports for a 3rd Generation Mobile System based on evolved GSM core networks and the radio access technologies that they support (i.e., Universal Terrestrial Radio Access both Frequency Division Duplex (FDD) and Time Division Duplex (TDD) modes). The scope was subsequently amended to include LTE and the maintenance and development of the GSM, GPRS and EDGE.

A permanent project support group called the "Mobile Competence Centre (MCC)" has been established to ensure the efficient day to day running of 3GPP. The MCC is based at the ETSI headquarters in Sophia Antipolis, France.

Currently in Europe, GSM and GPRS standards' testing is the preserve of the GSM Certification Forum (GCF) scheme, and in North America the PCS Type Certification Review Board (PTCRB) scheme is the GCF equivalent. The main difference between them is that the European GCF scheme is conducted on a 'cooperative' voluntary basis and the North American PTCRB scheme is mandatory and driven by the operators.

GCF is a partnership between network operators and terminal manufacturers that has been formed with the objective of establishing an independent programme to ensure global interoperability of GSM terminals.

GCF

The primary function of the GCF is to enable members' phones to get to market as quickly as possible and identifies the most important 3GPP-defined tests for members and expedites delivery of test cases. Members include handset manufacturers and operators and they meet every 3 months

GCF Certification operates through self certification process; certificate supplied by GCF and manufacturer completes and signs. Manufacturer must indicate terminal has been tested against – and is conformant with – any GCF-approved tests. Field trials are compulsory for GCF certification.

Origins of the GCF

The GSM association, which includes many network operators in its membership, was concerned with the risk of reduced regulation and decided to initiate a voluntary certification scheme for mobile phones. Note, however that this scheme is not mandatory and network operators are free to purchase phones that have not been included in this scheme.

Originally named "GSM Certification Forum", it was developed in co-operation between the GSM Association representing the network operator community and terminal manufacturers to ensure the interoperability of GSM terminals worldwide.

The GCF provided a process for GSM terminal manufacturers for the test and verification of terminals against specified technical requirements. This certification is aimed at being globally recognised and aims to avoid multiple testing and create cost efficiencies. This scheme does not cover any commercial or quality aspects between the terminal manufacturer and the customer. New requirements and tests may be incorporated into the Forum once a standards authority has published them and validated test equipment is available.

A further aim of the GCF was to evolve as GSM technology evolves so that it could encompass future technologies such as high-speed data and 3rd generation wireless air interface specifications.

The GSM certification programme included the requirement for a "Means of Test". This had to satisfy the test requirements detailed in the GCF requirement tables and also further requirements which would be published through a relevant standards authority.

PTCRB

The purpose of the PTCRB is to provide the framework within which GSM Mobile Equipment Type Certification can take place for members of the PTCRB.

This includes, but is not limited to, determination of the test specifications and methods to implement the Type Certification process for GSM Mobile Equipment. It is also the group's responsibility to generate input regarding testing of Mobile Stations to standards development organisations.

Several Technical Specification Groups (TSG) have been set up by 3GPP to undertake technical specification development work. Each TSG has the responsibility to develop, approve and maintain the specifications within its terms of reference.

TSGs report to the Project Coordination Group (PCG), and may organise their work in Working Groups and liaise with other groups as appropriate.

The RAN5 group is concerned with mobile Terminals, their interfaces and their conformance to the various standards. RAN5 has sub-working groups "signalling" and "RF" that deal with those specific aspects of conformance testing

The GERAN Technical Specification Group (TSG-GERAN) is responsible for the radio access part for GERAN specifications. The RAN group deals with Radio Access Networks and defines the functions, requirements and interfaces of the UTRA network. This includes layers 1, 2 and 3, access network interfaces (Iu, Iub and Iur); definition of the O&M requirements in UTRAN and conformance testing for Base Stations.

Technical and administrative support to TSGs is provided by the Mobile Competence Centre (MCC) located in Sophia Antipolis, France.

MCC 160

In order to accelerate the TTCN-2 test specification, 3GPP has funded the expert team, ETSI MCC task 160, for the development of the TTCN-2 test cases. The task started in July 2000 and the team consists of the skilled protocol / TTCN-2 experts coming from ten companies of the 3GPP partners. Since then hundreds of TTCN-2 and TTCN-3 test cases have been drafted for the UE conformance testing.

Standards Tracking

Test system manufacturers, test houses, chipset manufacturers and UE suppliers, need to track the activities of all major standards bodies that impact the 3G testing process. In practical terms, this involves participating in working group meetings and monitoring standards and their evolution.

The volatility of test specifications, and rapid processing of change requests, under the 5-day rule, means that projects may face numerous revisions of the standards.

Change requests, versions of TTCN-2 and proposed 3GPP timescales are monitored by reading the appropriate email reflectors and meeting minutes.

Timeliness is a key issue here. Under the '5-day rule' there are 5 days to object to any submitted change request before it becomes automatically accepted.

OVERVIEW OF STANDARDS

The 3rd Generation Mobile System and its capabilities have been developed in a phased approach. 3GPP elaborates, approves and maintains the necessary set of Technical Specifications for the first phase of a 3G system including:

- UTRAN (UMTS Terrestrial Radio Access Network)
- 3GPP Core Network (evolved from GSM)
- Terminals for access to the above, including specifications for a UIM
- System and service aspects of the mobile network
- Test specifications are based on core specifications, and include
- GSM/GPRS 1500+ protocol & RF tests (51.010)
- UMTS 700+ protocol tests (34.123) and 37+ RF tests (34.121)

For approval, test cases are submitted to GCF meetings which occur every three months. Prior to this, test cases are developed and approved within 3GPP RAN5.

3GPP develop TTCN-2 and publish it in a 'verification branch' for industry to prove it. When the TTCN-2 has been proven, it is published within a 'formal branch'. 3GPP define on which version of the core specifications the TTCN-2 should be based for submission to a given meeting.

CR (Change Request) Management

3GPP uses a rigorous change control mechanism for its documents. Once a draft technical specification has reached a reasonably stable state of development within the responsible working group, the parent Technical Specification Group (TSG) will put it under change control. Any technical change whatever that needs to be made to that document thereafter has to be formulated as a change request (CR) explicitly approved by the TSG.

A Change Request can be proposed by any 3GPP member organisation. It is normally submitted for discussion to the TSG Working Group (WG) responsible for the specification: for GSM, changes are handled by the GERAN Working Group for GSM, and UMTS changes are addressed by the RAN5 Group. The validation and approval of the implemented test cases then are handled by the GCF.

Once the TSG's WG has agreed that the Change Request is both valid and required (often it may be revised several times before reaching this stage), it is presented, on behalf of the WG (rather than the originating member organization) as an agreed proposal to the parent TSG plenary (most of which meet four times a year) for final approval. After approval at TSG level, the 3GPP Support Team MCC (the Mobile Competence Centre, based at the ETSI HQ in France) incorporates it and any other CRs into a new version of the specification, normally within 2 weeks of the meeting.

Version Control of Test Cases

The version of a test case or a suite of test cases is defined at the suite level. Individual test cases do not need to have a distinct version of their own. This means that whenever a re-validated suite becomes available as a result of only a few test cases changing, all the constituent test cases effectively assume the version identity of the whole set.

This approach is taken because:

- the creators of approval documentation and the user community finds it convenient to have a single version reference to determine validation status,
- test cases can rarely exist on their own, as they are constructed and maintained in groups, using shared routines for common functions

Interaction with External Bodies

3GPP has overall control of the test cases for GSM and UMTS. Changes, however, are handled by the GERAN Working Group for GSM, and by the RAN5 Group for UMTS. The validation and approval of the implemented test cases then are handled by the GCF.

In North America the group known as PCS validation group (PVG) handles the approvals, and its results are ratified by the PTCRB. Phones are tested against the test cases that, if successful, are certified by the CTIA. To achieve CTIA certification, it is necessary for phones to be tested in CTIA-approved laboratories.

UMTS Test Cases

The experiences and lessons learned through GSM have resulted in a number of changes to the way in which test cases are generated for UMTS. Protocol test cases for UMTS are written in prose and then converted into TTCN-2 code. This language enables the test cases to be compiled into a format that can be run directly on the target test equipment.

This approach saves time for the industry as a whole and reduces costs because generating the test cases is far easier. It also gives far more consistency across the industry because tests no longer are open to the same level of interpretation that they were before. As a result, time is saved during the validation process.

The TTCN-2 test cases have been prepared by a team of industry experts based at ETSI working on behalf of the 3GPP. The team developed the basic TTCN-2 code that was reviewed within the 3GPP community using e-mail reflectors. By reviewing the software at the beginning of the process, the individual test cases do not need to be reviewed each time they are submitted by a test-equipment manufacturer for validation.

To test a new handset, the manufacturer generally approaches a qualified test house. The test house will run the required certified test cases and state that the handset is suitable for use on the available networks.

EXTERNAL ORGANISATIONS

3GPP

The 3GPP was established in December 1998. The collaboration agreement brings together a number of telecommunications standards bodies which are known as "Organizational Partners" - ARIB, CWTS, ETSI, TTA, and TTC.

The original scope of 3GPP was to produce Technical Specifications and Reports for a 3rd Generation Mobile System based on evolved GSM core networks and the radio access technologies that they support (i.e., Universal Terrestrial Radio Access both Frequency Division Duplex (FDD) and Time Division Duplex (TDD) modes). The scope was subsequently amended to include the maintenance and development of the GSM, GPRS and EDGE.

The scope was subsequently amended to include:-

* the maintenance and development of the Global System for Mobile communication (GSM).
* Technical Specifications and Technical Reports, (these were originally developed and maintained by ETSI), including evolved radio access technologies (e.g. General Packet Radio Service (GPRS) and Enhanced Data rates for GSM Evolution (EDGE)).

In order to obtain a consolidated view of market requirements a second category of partnership was created within the project called "Market Representation Partners".

"Observer" status is also possible within 3GPP for those telecommunication standards bodies which have the potential to become Organizational Partners but which, for various reasons, have not yet done so.

3GPP is a global cooperative project in which standardisation bodies in Europe, Japan, South Korea and the United States as founders are coordinating WCDMA issues. It was set up to expedite the development of open, globally-accepted technical specifications for UMTS.

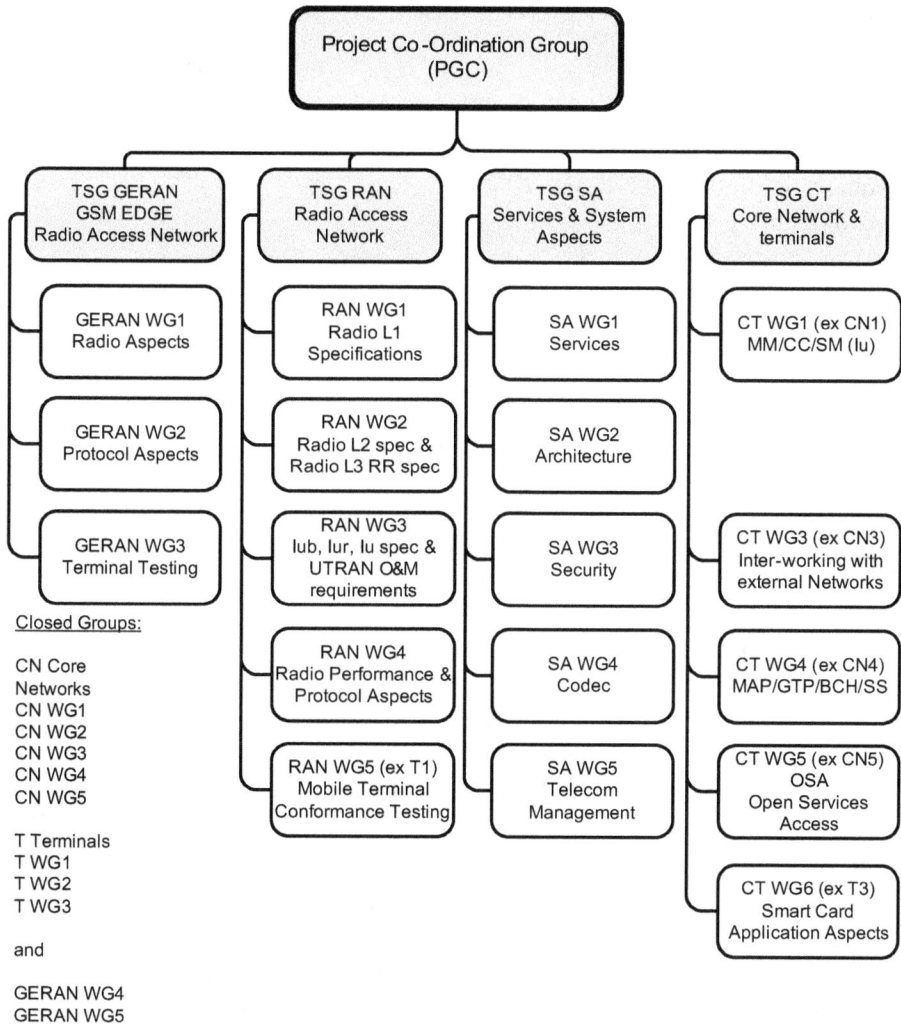

Figure 6.7 PCG Structure [2]

The diagram below shows the relationship and breakdown of TSGs within the 3GPP Project Co-ordination Group (PCG).

RAN5

There are presently 5 Technical Specification Groups (TSGs). Each TSG is made up of several Working Groups (WGs) which deal with specific parts of the TSG's work. The WGs may be further split into Sub Working groups.

RAN5 (RAN WG5 ex T1) works on the specification of conformance testing at the Radio interface (Uu) for the User Equipment (UE). RAN5 is organised in two subgroups, RF subgroup and Signalling subgroup.

Test Houses

Test houses develop and validate test cases. They often offer other services directed at manufacturers, network operators and service providers include:

- driving new technologies
- enhancing product development
- state-of-the-art testing
- world-wide approvals
- research and feasibility studies
- new technology training

CAG

Combined Agreement Group, GCF Combined Agreement Group, formerly U-AG (UTRA) and AG (GSM).

CDG

CDMA Development Group, CDG is an international consortium of companies (service providers and manufacturers) who have joined together to lead the adoption and evolution of CDMA wireless systems around the world.

It is mission is to lead the rapid evolution and deployment of CDMA-based systems, based on open standards and encompassing all core architectures, to meet the needs of markets around the world.

ETSI

European Telecommunications Standards Institute, ETSI is an independent and non-profit organization, based in Sophia Antipolis (France).

It is officially responsible for standardisation of Information and Communication Technologies (ICT) within Europe. These technologies include telecommunications, broadcasting and related areas such as intelligent transportation and medical electronics.

GCF

Global Certification Forum, GCF is a partnership between network operators and terminal manufacturers that has been formed with the objective of establishing an independent programme to ensure global interoperability of GSM terminals. GCF evolved partly in response to the relaxation of the previous regulatory type approval regime in April 2000 within the European Community following the implementation of the Radio & Telecommunications Terminal Equipment (R&TTE) Directive and the repeal of the previous Telecommunications Terminal Equipment (TTE) Directive.

GERAN (TSG-GERAN)

GSM/EDGE Radio Access Network Technical Specifications Group. The GERAN TSG undertakes technical specification development work within 3GPP. TSGs report to the Project coordination Group (PCG), and may organize their work in Working Groups and liaise with other groups as appropriate. Each TSG has the responsibility to develop, approve and maintain the specifications within its terms of reference. The TSG (TSG-GERAN) is responsible for the radio access part for GERAN specifications.

ETSI

European Telecommunications Standards Institute (ETSI) is an independent, non-profit organization, whose mission is to produce telecommunications standards for today and for the future.

Based in Sophia Antipolis (France), ETSI is responsible for standardization of Information and Communication Technologies (ICT) within Europe. These technologies include telecommunications, broadcasting etc

ETSI unites 688 members from 55 countries inside and outside Europe, including manufacturers, network operators, administrations, service providers, research bodies and users - in fact, all the key players in the ICT arena.

ETSI develops a wide range of standards and other technical documentation as Europe's contribution to world-wide ICT standardisation. This activity is supplemented by interoperability testing services and other specialists. ETSI's prime objective is to support global harmonization by providing a forum in which all the key players can contribute actively.

ETSI's Members determine the Institute's work programme, allocate resources and approve its deliverables. As a result, ETSI's activities are closely aligned with market needs and there is wide acceptance of its products.

MCC

Mobile Competence Centre (located in Sophia Antipolis, France)   A permanent project support group called the "Mobile Competence Centre (MCC)" has been established to ensure the efficient day to day running of 3GPP. The MCC is based at the ETSI headquarters in Sophia Antipolis, France.

MCC 160

In order to accelerate the TTCN-2 test specification, 3GPP has funded an expert team, ETSI MCC task 160, for the development of the TTCN-2 test cases. The task started in July 2000 and the team consists of the skilled protocol / TTCN-2 experts coming from ten companies of the 3GPP partners.

A permanent project support group called the "Mobile Competence Centre (MCC)" has been established to ensure the efficient day to day running of 3GPP. The MCC group is based at the ETSI headquarters in Sophia Antipolis, France.

The MCC group is subdivided into two groups, i.e MCC Task Force (TF) 160 and MCC Task Force (TF) 272.

MCC 160 is involved with the following:-

- The MCC160 is responsible of 3GPP RAN5 Group.
- The development of the test specifications found in conformance standards 34.123 parts 1, 2 and 3.
- Stabilise and to maintain all the currently available TTCN-2 Abstract Test Suites (ATS)
- To continue developing new TTCN-2 test cases for R99, Rel-4 and Rel-5 for the specifications.
- Work in accordance with the priorities identified by the Global Certification Forum (GCF) as part of the initiative to achieve certification of 3G handsets

SS manufacturers work in conjunction with MCC160 for the development and maintenance of TTCN-2 test cases. This is done by:

- Submission of corrections to TTCN-2 or prose using CRs
- Submission of new test cases
- Participation in meetings
- Participation in the regression testing of the test suites prior to release

PTCRB

PCS Type Certification Review Board The purpose of the PTCRB is to provide the framework within which GSM Mobile Equipment (ME) Type Certification can take place for members of the PTCRB. This includes, but is not limited to, determination of the test specifications and methods to implement the Type Certification process for GSM Mobile Equipment. This group will also be responsible to generate input regarding testing of Mobile Stations to standards development organizations.

PVG

PCS Validation Group which has similar role to the GCF for the US market.

RAN 5

(TSG-RAN 5) Radio Access Network working group 5. 3GPP's technical specification group dealing with Radio Access Networks. Specifically it has a responsibility for Radio aspects of Terminal Equipment and UTRAN functions (FDD & TDD), requirements and interfaces. RAN5 - a technical specification group set up by 3GPP that defines 3G test cases for mobile terminal conformance tests. RAN5 has sub-working groups "signalling" and "RF" that deal with those specific aspects of conformance testing

UAG

UTRA Agreement Group. The GCF UTRA Agreement Group, being redefined as part of CAG - Combined Agreement Group

WG3

(TSG-GERAN-WG3). Working group 3 of the GERAN technical standards group, dealing with terminal testing.

Testing

Validation must not be regarded as the primary means of proving that the software works. It is an external and independent check, using mobile phones that may possibly not be available to the SS manufacturer. But it remains the Company's responsibility to develop and assess the software to a good standard. Validation, needs to be done against a sound documented baseline. Effort, time and credibility could be lost if a poor standard of code released by the SS manufacturer. Remember also that in some cases software is released that has not successfully gone through an external validation phase.

Validation Control

A record is to be maintained of the Test Case versions that are handed across for validation.

Reviews

Prior to external validation, a Final Design Review checkpoint shall be convened. The aim of this is to clarify the overall position and to minimise the risk of inefficient use of the limited and potentially expensive validation effort.

Release formalities

Owing to the fluid nature of the final stages of validation, and the possible need to make late updates to source code, Release does not occur at a single point in time.

The Release process commences at the Final Design Review checkpoint, which precedes the first Test Case handover to validation authority. This needs to be a mandatory checkpoint.

The Release is finalised on completion of validation, at the point where the submission is made to the certification body. At this point the versions are frozen, and the Test Suite and related Options are baselined.

Note that the SW must be defined, released and available as a prerequisite for approval by certification bodies.

Following formal approval of Test Cases, it only remains for the SS vendor's records to be updated with the latest status of the Tests Cases.

Status

A Release package may contain Test Cases that are either (a) Development (b) Validated or (c) Approved, or a mixture of these.

Development

Test cases that have not been validated by an approved test house or accepted by GCF / PTCRB (where applicable), and may be further modified in a future release of test cases. If they are offered to customers it is only for preliminary assessment.

Validated

Tested and validated by an approved test house, and pending submission to the next available GCF/PTCRB meeting.

Approved

Approved by GCF or PTCRB for 1 or more frequency bands. The SS manufacturers policy is to secure validation and approval for all test cases wherever possible, although it also makes available working test cases that cannot or do not meet the validation requirements for some reason.

For most of the widely adopted mobile radio standards, a standards body grants the formal "approval" to test cases that have been fully reviewed and validated. However, not all standards that are currently followed have a certification body. In these cases the software is published to good commercial standards and is typically reviewed with relevant experts on external committees. It is also worth noting that the SS vendor expects on occasions to deliver Test Cases that have been unsuccessful in gaining formal approval.

External approval of test cases is considered to be an aspect of the software status that is independent of the release process . (and does not affect the release process and decision making.)

In other words, approval (or downgrading does not affect the release status of the software.

Storage

The main point of reference for distribution shall be the SS manufacturer's Download Portal. Snapshots of complete sets on CD-R medium shall be made for backup purposes at Certification milestones or quarterly, whichever is the more frequent. These sets will have been documented on an SRF, which lists the identity and versions of all constituent Options. The composition and definition of Options shall be recorded by the Project, in liaison with the Product Manager.

DEFINITIONS

5-day rule existing test cases offered for review, and if no adverse comment received, accepted by default.

10-day rule 10-day period provided to review new test case material, in advance of agreement group meeting

Approval Approval is the term used for all regulatory processes and requirements (such as Type Approval)

ASN.1 Abstract Syntax Notation One. A notation for describing the structure of protocol messages exchanged between communicating systems. Its use can significantly contribute to the precision and speed of protocol standardisation and product development and maintenance

GCF-CC (Certification Criteria) Database. GCF-CC Database is a central place holding all the information of test cases validated against different platforms.

Certification Certification is used for all voluntary programmes organised by one or different industry groups

Circuit and Packet Domains Two types of networks approaches are specified for UMTS - the R99 core network solution: circuit and packet domains. Iu-cs is an UMTS interface for the circuit domain, equivalent to the GSM "A" interface, Iu-ps is an UMTS interface for packet domain, equivalent to the "Gb" interface in GPRS.

iWD Series of numbered weekly Interim Working Documents, that are incremental instructions from GCF on the latest agreed definitions and changes.

NAPRD.03 North America Program Reference Document

PICS Protocol Implementation Conformance Statement. The manufacturer of a mobile handset to completes a Protocol Implementation Conformance Statement (PICS) stating the features and specification the handset supports. The PICS statement and the relevant Technical Bases for Regulation (TBR's 19, 31) are used to create a test plan, which may also stipulate the test platform to be used for the conformance tests.

Qualification Qualification = Approval + Certification

Verification Verification confirms compliance (of end-user products) with technical test specifications.

Validation Specific test equipment meets the technical test / core specification. It includes the assessment of test case. (This meaning differs from to the way this term is used in regular SW development processes)

Reason 7 (in validation), Test is not explicitly re-validated, however the previous validation is valid in the new version. (e.g. in the case of a minor change to the test system). There are 6 other reasons are listed in Annex F of of GCF-OP

Spectrum for UMTS WRC'92 identified the frequency bands 1885-2025 MHz and 2110-2200 MHz for future IMT-2000 systems, with the bands 1980-2010 MHz and 2170-2200 MHz intended for the satellite part of these future systems.

System Simulator The simulator or test facility that a mobile interacts with during testing and assessment.

Test Case In order to test something, need to specify the sequences of interactions, or test events, that wish the test system to control and observe. A sequence of such events that specify a complete test purpose is called a test case. A set of test cases for a particular protocol is called a test suite.

TTCN-2 Test and Test Control Notation. formerly Tree and Tabular Combined notation

UE, MT User equipment, mobile terminal. (informally, the handset)

UTRA, UTRAN Universal Terrestrial Radio Access Network

UMTS Universal mobile telephony system , refers to the next generation of cellular technology that is also being standardized by European Telecommunications Standard Institute (ETSI).

WCDMA Wideband code division multiple access is the air interface technology selected by ETSI and the major European and Japanese mobile communications operators. This technology is optimized to allow very high-speed multimedia services such as full-motion video, Internet access, and videoconferencing.

REFERENCE DOCUMENTS

[1]   GCF- OP v3.18.0
[2]   3GPP 34.123-1, 2, 3
[3]   3GPP website www.3gpp.org
[4]   GCF website   http://gcf.gsm.org
[5]   GCF-CC Database http://gcftech.org

List of Abbreviations

2G	2nd Generation
3G	3rd Generation
3GPP	Third Generation Partnership Project
ACK	Acknowledgement
ACL	APN Control List
AICH	Acquisition Indicator Channel
AM	Acknowledged Mode
AMC	Adaptive Modulation and Coding
AMR	Adaptive Multi Rate
AMR-WB	Adaptive Multi Rate Wide Band
AN	Access Network
ANSI-41	Cellular Radiotelecommunications Intersystem Operations
AoC	Advice of Charge
AoCC	Advice of Charge Charging
AoCI	Advice of Charge Information
AP	Access preamble
APDU	Application Protocol Data Unit
API	Application Programming Interface
APN	Access Point Name
ARQ	Automatic Repeat ReQuest
AS	Access Stratum
ASC	Access Service Class
ASN.1	Abstract Syntax Notation One
ASP	Abstract Service Primitive
AT command	ATtention Command
ATM	Asynchronous Transfer Mode
ATS	Abstract Test Suite
AUT(H)	Authentication

(*Continued*)

AUTN	Authentication token
AWGN	Additive White Gaussian Noise
BCCH	Broadcast Control Channel
BCD	Binary Coded Decimal
BCH	Broadcast Channel
BDN	Barred Dialling Number
BER	Bit Error Ratio
	Basic Encoding Rules (of ASN.1)
BLER	Block Error Ratio
BMC	Broadcast/Multicast Control
BNF	Backus Naur Form
BPSK	Binary Phase Shift Keying
BS	Base Station
BSC	Base Station Controller
BSS	Base Station Subsystem
BTFD	Blind Transport Format Detection
BTS	Base Transceiver Station
CAN	Control Area Network
CB	Cell Broadcast
CBS	Cell Broadcast Service
CC	Call Control
CCCH	Common Control Channel
CDMA	Code Division Multiple Access
CFN	Connection Frame Number
CIR	Carrier to Interference Ratio
CK	Cipher Key
CKSN	Ciphering Key Sequence Number
CM	Connection Management
	Compressed Mode
CN	Core Network
	TTCN3 Core Notation
CORBA	Common Object Request Broker Architecture
CQI	Channel Quality Indicator
CR	Change Request
CS	Circuit Switched
CSCF	Call Server Control Function
CT	Conformance Testing
CTCH	Common Traffic Channel
CTE	Classification Tree Editor
CTS	Conformance Test Suite
CW	Call Waiting
	Continuous Wave (unmodulated signal)
DCCH	Dedicated Control Channel
DCH	Dedicated Channel

(Continued)

DCS1800	Digital Cellular Network at 1800MHz
DDI	Direct Dial In
DECT	Digital Enhanced Cordless Telecommunications
DHCP	Dynamic Host Configuration Protocol
DL	Downlink (Forward Link)
DPCCH	Dedicated Physical Control Channel
DPCH	Dedicated Physical Channel
DPDCH	Dedicated Physical Data Channel
DSP	Digital Signal Processing
DTCH	Dedicated Traffic Channel
DTD	Dynamic Type Definition
DTM	Dual Transfer Mode
DTX	Discontinuous Transmission
E-AGCH	E-DCH Absolute Grant Channel
Ec/No	Ratio of energy per modulating bit to the noise spectral density
E-DCH	Enhanced-DCH
EDGE	Enhanced Data rates for GSM Evolution
EF	Elementary File (on the UICC)
EGPRS	Enhanced GPRS
EHPLMN	Equivalent Home PLMN
EMF	Eclipse Modelling Framework
EPLMN	Equivalent PLMN
E-RGCH	E-DCH Relative Grant Channel
E-RNTI	E-DCH Radio Network Temporary Identity
ETS	Executable Test Suite
ETSI	European Telecommunications Standards Institute
EVM	Error Vector Magnitude
FACH	Forward Access Channel
FBI	Feedback Information
FDD	Frequency Division Duplex
FDN	Fixed Dialling Number
FPLMN	Forbidden PLMN
FQDN	Fully Qualified Domain Name
GCF	GSM Certification Forum
GCI	Generic Compiler Interpreter
GERAN	GSM EDGE Radio Access Network
GET	Graphical Format for TTCN-3
GFT	Graphical Format
GGSN	Gateway GPRS Support Node
GL	Generation Language
GMM	GPRS Mobility Management
GPRS	General Packet Radio Service
GSM	Global System for Mobile communications

(*Continued*)

H-ARQ	Hybrid Automatic Repeat reQuest
HCS	Hierarchical Cell Structure
HHO	Hard Handover
HLR	Home Location Register
HO	Handover
HOLD	Call hold
HPLMN	Home Public Land Mobile Network
HSDPA	High Speed Downlink Packet Access
HSPA	High Speed Packet Access (HSDPA + HSUPA)
HSUPA	High Speed Uplink Packet Access
HTTP	Hyper Text Transfer Protocol
I&C	Installation and Commissioning
IDL	Interface Definition Language
IE	Information Element
IFD	Interface Device
IMEI	International Mobile Equipment Identity
IMS	IP Multimedia Subsystem
IMSI	International Mobile Subscriber Identity
IP	Internet Protocol
IPv4	Internet Protocol Version 4
IPv6	Internet Protocol Version 6
ISDN	Integrated Services Digital Network
ISO	International Organisation for Standardisation
ITU	International Telecommunication Union
IUT	Implementation Under Test
iWD	Interim Working Document (TTCN-2 or TTCN-3 Delivery from ETSI)
L1	Layer 1 (physical layer)
L2	Layer 2 (data link layer)
L3	Layer 3 (network layer)
LA	Location Area
LCS	Location Services
LLC	Logical Link Control
LPLMN	Local PLMN
LTE	Long Term Evolution
MAC	Medium Access Control (protocol layering context)
MAC-c/sh	MAC handling data transported on common and shared transport channels
MAC-hs	MAC-high speed
MBMS	Multimedia Broadcast Multicast Service
MCCH	MBMS point-to-multipoint Control Channel
MDA	Model Driven Architecture
MIMO	Multi Input Multi Output
MMI	Man Machine Interface

(Continued)

MM	Mobility Management
MMS	Multimedia Services
MO-LR	Mobile Originating Location Request
MSC	Mobile Switching Centre
	Message Sequence Chart Format
MSCH	MBMS Scheduling Channel
MTCH	MBMS traffic channel
NACK	Non-ACK
NAS	Non-Access Stratum
O&M	Operations & Maintenance
OCNS	Orthogonal Channel Noise Simulator
OMG	Object Management Group
OOM	Object oriented modelling
OPLMN	Operator Controlled PLMN (Selector List)
OSP	Open Settlement Protocol
OVSF	Orthogonal Variable Spreading Factor
P-CCPCH	Primary Common Control Physical Channel
PCH	Paging Channel
PCO	Protocol Configuration Options
	Point of Control and Observation
PCPCH	Physical Common Packet Channel
P-CPICH	Primary Common Pilot Channel
PCS	Personal Communication System
P-CSCF	Proxy Call Session (Server) Control Function
PDCH	Packet Data Channel
PDCP	Packet Data Convergence Protocol
PDP	Packet Data Protocol
PDU	Protocol Data Unit
PICH	Page Indicator Channel
PICS	Protocol Implementation Conformance Statement
PID	Packet Identification
PIN	Personal Identification Number
PIXT	Protocol Implementation eXtra information for Testing
PLMN	Public Land Mobile Network
PRD	Permanent Reference Documents
PS	Packet Switched
PSCH	Physical Shared Channel
P-SCH	Primary Synchronisation Channel
PSTN	Public Switched Telephone Network
PTCRB	PCS Type Certification Review Board
P-TMSI	Packet TMSI
PVG	PCS Validation Group
QAM	Quadrature Amplitude Modulation
QoS	Quality of Service

(Continued)

QPSK	Quadrature (Quaternary) Phase Shift Keying
R99	Release 1999
RA	Routing Area
RAB	Radio Access Bearer
RACH	Random Access Channel
RAN	Radio Access Network
RAN5	3GPP Radio Access Network group 5 (or RAN WG5)
RAT	Radio Access Technology
RB	Radio Bearer
RF	Radio Frequency
RL	Radio Link
RLC	Radio Link Control
RNC	Radio Network Controller
RNTI	Radio Network Temporary Identity
RoHC	Robust Header Compression
RPLMN	Registered Public Land Mobile Network
RRC	Radio Resource Control
RRM	Radio Resource Management
RSCP	Received Signal Code Power
RSSI	Received Signal Strength Indicator
RTP	Real Time Protocol
S	Cell Selection value, (dB)
SAP	Service Access Point
SAPI	Service Access Point Identifier
S-CCPCH	Secondary Common Control Physical Channel
SCH	Synchronisation Channel
S-CPICH	Secondary Common Pilot Channel
S-CSCF	Serving CSCF
SDN	Service Dialling Number
SDU	Service Data Unit
SF	Spreading Factor
SFN	System Frame Number
SI	System Information
SIB	System Information Block
SIP	Session Initiated Protocol
SM	Session Management
	Short Message
SMS	Short Message Service
SQN	Sequence number
SRB	Signalling Radio Bearer
SRLS	serving radio link set
SRNS	Serving RNS
SS	Supplementary Service
	System Simulator

(Continued)

S-SCH	Secondary Synchronisation Channel
SUT	System Under Test
TC	Test Case
TCI	Control Interface
TDMA	Time Division Multiple Access
TE	Terminal Equipment
TETRA	Terrestrial Trunked Radio
TFC	Transport Format Combination
TFS	Transport Format Set
TLV	Tag Length Value
TM	Transparent Mode
TMSI	Temporary Mobile Subscriber Identity
TPC	Transmit Power Control
TRI	Runtime Interface
TSG	Technical Specification Group
TSN	Transmission Sequence Number
TTCN-2	Tree and Tabular Combined Notation 2
TTCN-3	Testing and Test Control Notation 3
TTI	Transmission Timing Interval
UCS2	Universal Character Set 2
UDP	User Datagram Protocol
UE	User Equipment
UICC	USIM Integrated Circuit Card
UID	User ID
UL	Uplink (Reverse Link)
UM	Unacknowledged Mode
UMD	Unacknowledged Mode Data
UML	Unified Modelling Language
UMTS	Universal Mobile Telecommunications System
UPLMN	User PLMN
URI	Uniform Resource Identifier
USAT	USIM Application Toolkit
USIM	Universal Subscriber Identity Module
USS	User Security Setting
UTRA	Universal Terrestrial Radio Access
UTRAN	Universal Terrestrial Radio Access Network
VoIP	Voice Over IP
VPLMN	Visited Public Land Mobile Network
VQA	Voice Quality Assessment
VT	Video Telephony
WAP	Wireless Application Protocol
WCDMA	Wideband Code Division Multiple Access
WG	Working Group
WiMAX	Worldwide Interoperability for Microwave Access
XML	eXtensible Markup Language

THE AUTHOR

Dr Faris MUHAMMAD is Senior Technical Consultant and Technical Authority of UMTS and WiMAX technology at Aeroflex Wireless, UK. He is a chartedred engineer and fellow of the Institute of Engineering and Technology. He received his BEng from University of Technology, Baghdad, MSc and PhD from University of Strathclyde, Glasgow. He worked as Senior Lecturer for Manchester Metropolitan University and University of Hertfordshire. His main topics of interest are Wireless and Optical Fibre Communications. He has numerious publications in conferences and technical journals.

Index

www.ingramcontent.com/pod-product-compliance
Lightning Source LLC
Chambersburg PA
CBHW060809220326
41598CB00022B/2575